Milan Burša · Karel Pěč

Gravity Field and Dynamics of the Earth

With 89 Figures

Springer-Verlag Berlin Heidelberg GmbH

Prof. Dr. Milan Burša
Astronomical Institute
Academy of Sciences of the Czech Republic
Bočni II, 1401
141 31 Praha 4
Czech Republic

Prof. Dr. Karel Pěč†
Charles University
Povltavská 2
180 00 Praha 8
Czech Republic

Translated from the Czech by
Dr. Jaroslav Tauer
Sleská 86
130 00 Praha 3
Vinohrady
Czech Republic

Title of the Original Czech Edition
Tihové pole a dynamika Země
© Academia 1988

ISBN 978-3-642-52063-1

Library of Congress Cataloging-in-Publication Data. Burša, Milan. [Tihove pole a dynamika
Země. Czech] Gravity field and dynamics of the Earth/Milan Burša, Karel Pěč; [translated by
J. Tauer] p. cm. Includes bibliographical references and index.
ISBN 978-3-642-52063-1   ISBN 978-3-642-52061-7 (eBook)
DOI 10.1007/978-3-642-52061-7
1. Gravity. 2. Geodesy. I. Pěč,
Karel. II. Title. QB331.B8313 1993 526'.7–dc20 93-4997

© Springer-Verlag Berlin Heidelberg 1993
Originally published by Springer-Verlag Berlin Heidelberg New York in 1993

Typesetting: Macmillan India Ltd., Bangalore-25

32/3145/SPS-543210–Printed on acid-free paper

Preface

Since the Czech edition was published four years ago, the authors have revised the original text to bring it up to date. During these four years, thanks to satellite altimetry the accuracy of the global description of the gravity field (model GEM-T2), of the fundamental astrogeodetic constants, of the principal moments of inertia of the Earth and, in particular, of their differences, of the precession constant, and of a number of other dynamical parameters of the Earth have been improved.

The authors have included most of these improvements in the revised English edition. They have, of course, also made factual, formal and other corrections and have modified some of the figures. Additions to the index and references have also been made.

Praha, Czech Republic
August 1993

M. BURŠA and K. PĚČ

Contents

Introduction

Earth and space sciences are developing very rapidly thanks to contemporary methods of satellite research and to modern measuring and computer facilities. The origin of modern satellite methods dates back to the launching of the first satellite, Sputnik I, in 1957; since then satellite methods have developed very quickly and have had a profound effect on Earth and space sciences. New independent scientific disciplines have originated, e.g. space geodesy and space meteorology, and satellite methods have, especially in geophysics, yielded new and fundamental information about the structure and dynamics of the Earth's magnetosphere. In celestial mechanics this has led to the development of improved theories of satellite motion, and in geodesy it has contributed significantly to accurate determination of the parameters of the figure of the Earth and of the external gravitational field. The data obtained from satellite observations are used in studying inhomogeneities in the Earth's interior and related dynamic processes.

Thanks to satellite methods, the Earth's external gravitational field is known with remarkable accuracy and high resolution, second to no other geophysical quantity subject to lateral variations. The fundamental global parameters of the Earth have been determined with high accuracy: the equatorial radius, the flattening of the Earth's reference ellipsoid, and the product of the Earth's mass and gravitational constant, referred to as the geocentric gravitational constant, GM. Satellite methods also provide data on the position of the axis of rotation with an accuracy better than 10 cm, and enable accurate monitoring of the motion of the Earth's pole. The external gravitational field has been described in terms of geopotential coefficients (spherical harmonic coefficients) up to degree $n = 180$ and this enables the equipotential surfaces of the geopotential, i.e. also the geoid, to be mapped in considerable detail. The current accuracy of global long-wave components of the geoid, corresponding to geopotential coefficients up to degree $n = 4$, is 8 cm and the overall accuracy, inclusive of the other known terms of higher degrees in the determination of the geoid, is about 1m. Although the accuracy of the higher short-wave components of the geoid is lower, it is nevertheless sufficient to locate geoid slopes with large gradients, remarkable in that they are nearly identical with tectonically active areas on the Earth's surface in which stress is accumulated and released; this is reflected in the occurrence of tectonic earthquake foci. The accuracy with which the shape of the gravimetric geoid has been determined can be checked by another, practically independent method, namely satellite altimetry, which consists in measuring the distance between the satellite and ocean surface directly. Very good agreement exists between the geoid determined by satellite altimetry and by gravimetric methods. However, in 1988–1990 satellite altimetry attained such a high degree of accuracy that the gravity field is best described in the regions of oceans and seas, i.e. in regions studied but sporadically in the presatellite era (Marsh et al. 1989, 1990).

The detailed knowledge of the external gravitational field has yielded some new and quite fundamental information which is sure to affect the future development of Earth sciences. It has been proved beyond doubt that the Earth is not in hydrostatic equilibrium, and that it is neither rotationally nor equatorially symmetrical. The substantial deviations from hydrostatic equilibrium led to dramatic revisions of opinions of the configuration of the Earth's interior, opinions which had prevailed for a very long time, from Clairaut's times to the 1960s. The fact that surfaces of equal density are not surfaces of equal pressure and equipotential surfaces of the geoid, reflected in the existence of non-zero tesseral and sectorial terms in the geopotential expansion, can only be explained by lateral variations of density in the Earth's interior. The proved existence of density deviations from hydrostatic equilibrium is responsible for the generation, in a viscous-elastic body such as the Earth, of a dynamic process, convection, which tends to balance these density inhomogeneities by flow. As regards the existence of convective flows, the cause of the density deviations, whether disbalance of chemical substances or the existence of heat sources, is quite irrelevant. The existence of the dynamic processes within the Earth introduces a new parameter into the description of the Earth's gravitational field, i.e. time. Strictly speaking, this means that the parameters of the gravitational field must be considered to be quantities variable in time with a characteristic period, corresponding to the duration of geological cycles. The system of convection in the Earth's interior cannot be observed directly. It is reflected indirectly in the lithosphere, in so-called subduction regions and, of course, in the structure of the system of lateral density deviations. Subduction and other dynamic phenomena in the lithosphere are explained by the theory of tectonics of lithospheric plates which, at this time, is the only unifying theory of observed geophysical and geological phenomena. This theory was founded by Wilson in 1965 and its fundamental concept is that the lithosphere is the product of a closed dynamic cycle. It originates in the region of ocean-bottom spreading from Earth-crust rocks; the specific region in which the oceanic lithosphere originates is the North Atlantic oceanic ridge. The oceanic lithosphere moves relatively freely across the asthenosphere, a region at a depth of 80 to 200 km with relatively low viscosity, and in subduction regions the lithosphere submerges into the mantle again. The Earth's surface is divided into several lithospheric plates whose boundaries are defined by interactions between plates, e.g. by the occurrence of earthquake foci concentrated in narrow belts. Relative motions between plates have been measured by means of distant sources of radio waves and by satellite methods, and their rate of motion varies from 0 to 7 cm/yr. The driving mechanism of this cycle is the convection in the Earth's upper mantle. Although the theory of lithospheric plates has proved to be quite successful, i.e. as regards lithospheric-plate kinematics (Smith et al. 1990) and in explaining the phenomena taking place at their boundaries, it has become evident that plate tectonics would not be able to explain the main features of the geoid, e.g. the existence and structure of large-scale anomalies. It is probable that the sources of global gravity anomalies are deeper in the mantle than the region in which processes related to plate tectonics take place. Nevertheless, the coincidence of the seismically active regions at the boundary of the Pacific plate with the high geoid gradients mentioned is evidence of the coupling of the deep-mantle convection with convection related to plate tectonics.

The current findings of Earth and space sciences prove that the Earth, Moon and other bodies of the Solar System have complicated internal structures reflected in the complicated interactions of their fields. In this book use will be made of detailed knowledge of the coefficients of the external gravitational fields of the Earth and Moon, and primary attention will be given to describing the interrelation of gravitational fields and the dynamics of the Earth–Moon system. The resultant field depends on the mutual positions of these bodies and therefore varies relatively quickly with time. Secular time variations, related to the internal dynamics of the Earth, will be considered only as exceptions. To be able to describe at least the rough features of the Earth's response to the time-variable external field, the detailed internal structure will be disregarded and the Earth will be considered as a homogeneous body capable of elastic linear deformation. This applies, for example, in the chapters on Earth tides (Chap. 4) and on rotation dynamics (Sect. 3.5). This means that all force fields on non-mechanical origin will be disregarded, and that the response of the real Earth will be replaced by the response of an equivalent homogeneous, elastic body. This procedure provides the possibility of deriving the characteristic terrestrial parameters and provides the boundary conditions required for studies in which the detailed density and rheological structure is considered.

The subject was chosen to avoid duplicating text-books of geophysics. The authors have attempted to deal with the part related to the results of satellite determination of the parameters of the Earth's gravitational field and dynamics of the Earth-Moon system. The foundations of the theory of satellite motion (Chap. 1) are only presented to the extent necessary to understand Chapter 2, devoted to the Earth's gravity field and its sources. In orbital rotational dynamics, space geodynamics requires that real and unidealized bodies be considered with regard to their shape and gravitational field. Such a consideration has been attempted in Chapter 3 in which the principal emphasis is placed on deriving the force function of the Earth–Moon–Sun system, the basis for solving all dynamic problems. Chapter 4 presents a detailed account of the theory of tidal disturbances of the Earth's gravitational field; ocean tides have not been included in view of the extent of the subject and its specifics. Chapter 5 is an attempt at interpreting geodynamic phenomena requiring a synthetic approach. It deals with the dynamics of rotational orbital motion of the Earth–Moon system, the theory of the perturbations of the Earth's body and of its gravitational field, connected with the variations of the vector of instantaneous rotation of the Earth. The theory of the effect of the secular variation of the second zonal geopotential coefficient, detected by the LAGEOS satellite, on Earth-pole wandering, is also presented. A critical assessment is given of the hypothesis of the Earth's expansion in the light of contemporary astronomical and satellite data and lunar laser ranging. The final chapter, 6, gives an outline of the dynamic system of the Solar System to the extent required with regard to broader connections among geodynamic phenomena.

It is assumed that readers have mastered the basic course of mechanics of the mass point and rigid body as well as elements of mechanics of the continuum, and that they have a knowledge of mathematics corresponding to the standard of basic university courses of the technical or natural-science type. The only exception is Section 2.5.2 which requires a basic knowledge of differential geometry. Nevertheless, should readers be satisfied with the conclusions of Section 2.5.2 without proof, knowledge of the foundations of differential geometry is not required even in this case.

Naturally, the selection of the subject matter has by no means exhausted the whole realm of applications of satellite methods to the theory of the gravity field and dynamics of the Earth. For example, a description of the experimental parts of the problems has been omitted completely, as has also a detailed account of orbital satellite dynamics. In the theory of Earth dynamics considerable emphasis has been placed on the description of external force fields and their effect on rotation dynamics, the Earth being represented by a model of a rigid or homogeneous elastic body. However, the real Earth is 3-D inhomogeneous, the radial dependence being strongly predominant over the lateral (angular), at least with regard to density and elastic parameters. The study of lateral inhomogeneities has recently come into the limelight of scientific interest, because the knowledge of the lateral variations of elastic parameters and of density is the key to solving the most important problem of internal dynamics of the Earth, i.e. understanding convection within the Earth's mantle. As regards the group of problems related to the lateral variations of Earth parameters, this book only deals with the method of determining lateral variations of density, based on gravimetric and seismological data. The group of problems connected with the theory of response on a non-elastic, laterally and radially inhomogeneous Earth to internal and external force fields has been omitted. These problems are very complicated and their solutions are incipient.

The References contain those referred to directly in the separate chapters. It is not a complete list of papers related to the topic and the reader requiring a complete list of references should turn to fundamental works, e.g. P. Melchior *The Earth's Tides*.

The authors wish to express their gratitude for valuable comments to the reviewer of the original Czech version, Dr. O. Praus DrSc., to the scientific editor, Prof. Dr. V. Červený DrSc., and to Prof. J. Kostelecký CSc., Prof. Z. Martinec DrSc., Prof. O. Novotný CSc., Dr. M. Šidlichovský DrSc., Dipl.-Ing. Z. Šimon DrSc., Dipl.-Ing. J. Vondrák DrSc., Prof. M. Pick DrSC., and

Dr. M.I. Yurkina. All the comments were considered and helped to improve the book.

The English version has been up-dated, especially with regard to satellite and geodynamic data and the results related to the 1986–1991 period. By comparing the Czech and English editions of the book one could determine the progress achieved in the relevant fields.

We would like to thank RNDr. J. Tauer, CSc. who translated the book with an understanding of the subject, and in fact rendered the text more comprehensible in many places. Our thanks are also due to Mrs. Dana Hanšpachová who prepared the English version for press from the technical point of view with exceptional care and patience.

Last but not least, thanks are due to Dr. M.I. Yurkina for a number of additions and for checking the historical priorities of the classical fundamental relations.

1 Fundamentals of Determining the Parameters Defining the Earth's Gravitational Field by Satellite Methods

1.1 Introduction

The satellite epoch, begun but a quarter of a century ago, has impressed research into the shape and gravitational fields of bodies of the Solar System with completely new features, and has brought revolutionary results not even dreamt of in the presatellite epoch.

The progress achieved with regard to uniform global description of the Earth's gravitational field and in deriving the parameters defining its form, is such that the new results multiply exceed, in extent and accuracy, all that was achieved in this field in the presatellite era (Marsh et al. 1989).

Indeed, the presatellite determination of the figure of the Earth and of its external gravity field was based on astro-geodetic and gravity data gleaned from observations made only on the continents, and not even on all the continents, and from measurements that were by no means homogeneously distributed. This unfavourable fact caused the derived shape parameters to be representative but only regionally and not globally. Equatorial flattening and equatorial asymmetry of the Earth was always very inaccurate for this very reason.

Currently, over 1000 parameters describing the Earth's gravitational field and figure (Marsh et al. 1989, 1990) are known from satellite orbit analyses based on millions of accurate (mostly laser and Doppler) observations of satellite positions. Moreover, research is being continued with unabated intensity, and the possibilities provided by satellite dynamic methods have by no means been exhausted.

In this chapter an attempt will be made to present the principles of these methods together with the main results achieved. The explication of satellite orbital dynamics will be presented only to the necessary extent because the subject matter exceeds the scope of this book and would require a separate study.

1.2 Satellite Equations of Motion

We shall assume that the satellite is moving in an exclusively gravitational force field generated by natural celestial bodies; we shall not consider the effect of forces of non-gravitational origin. Consequently, conservative forces with potentials will always be involved.

We shall also assume that the mass of the artificial satellite, as compared with the masses of natural celestial bodies, is always so small that (1) it will practically have no effect on the motion of the natural bodies, and (2) it may be considered a point mass.

We shall simplify the problem from the very beginning by assuming that the artificial satellite is so close to the Earth's body that the effect of the other celestial bodies on the satellite's motion will be substantially smaller than the effect of the Earth, and its nature will be that of perturbations. In particular, we shall only consider the effects of the Moon and Sun and will neglect the effects of planets completely.

The symbols $M_\oplus$, M_D, $M_\odot$ will be used to designate the Earth, Moon and Sun, respectively, (Fig. 1.1) and to represent their total masses; the mass differentials of the Earth, Moon and Sun will accordingly be designated $dm_\oplus$, dm_D, $dm_\odot$, with M_S being the mass of the satellite S. In any inertial coordinate system (in Newton's sense) $\mathbf{R}\{X_j\}$, $j = 1, 2, 3$, the orbits of these four bodies can be described by Newton's equations of motion (Subbotin 1937, 1941; Brouwer and

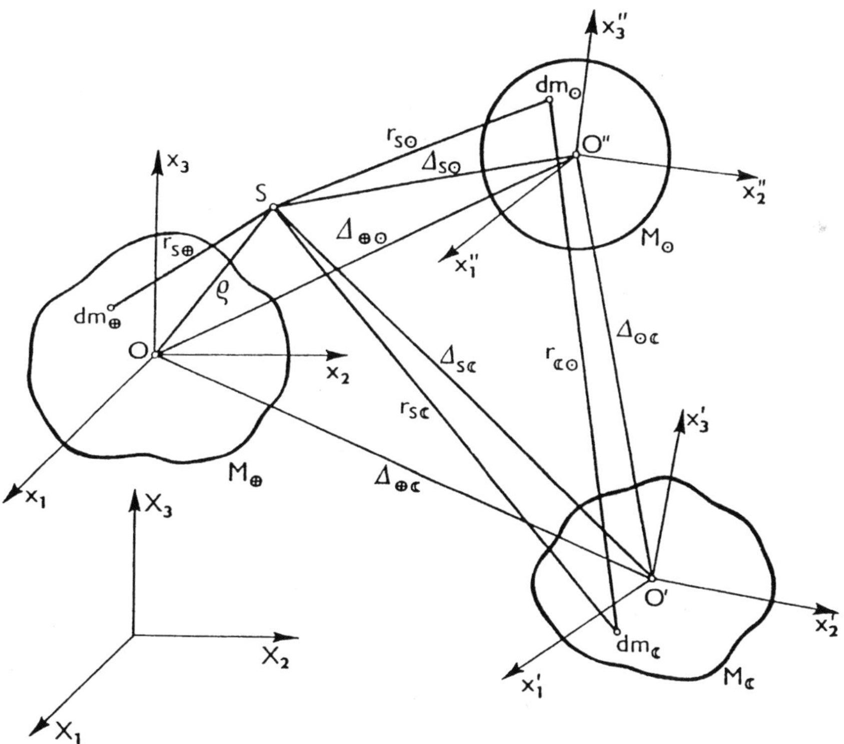

Fig. 1.1. The Earth–Moon–Sun satellite system

Clemence 1961):

$$M_i \ddot{\mathbf{R}}_i = \mathrm{grad}_{\mathbf{R}_i} V, \quad i = S, \oplus, \leftmoon, \odot .$$

(1.1)

The force function V of the system of these four bodies reads

$$V = GM_S \left(\int\limits_{M_\oplus} \frac{dm_\oplus}{r_{S\oplus}} + \int\limits_{M_\leftmoon} \frac{dm_\leftmoon}{r_{S\leftmoon}} + \int\limits_{M_\odot} \frac{dm_\odot}{r_{S\odot}} \right) + V_{\oplus\leftmoon} + V_{\oplus\odot} + V_{\leftmoon\odot} ,$$

(1.2)

where

$$V_{\oplus\leftmoon} = G \int\limits_{M_\oplus} \int\limits_{M_\leftmoon} \frac{dm_\oplus \, dm_\leftmoon}{r_{\oplus\leftmoon}}, \quad V_{\oplus\odot} = G \int\limits_{M_\oplus} \int\limits_{M_\odot} \frac{dm_\oplus \, dm_\odot}{r_{\oplus\odot}} ,$$

$$V_{\leftmoon\odot} = G \int\limits_{M_\leftmoon} \int\limits_{M_\odot} \frac{dm_\leftmoon \, dm_\odot}{r_{\leftmoon\odot}} ;$$

(1.3)

$\mathbf{R}_i$ is the radius-vector of the mass centre of the i-th body, G is the gravitational constant, and r_{pq} the distance between elements dm_p and dm_q. Note that the introduction of the force function and potential (the force function in the case that the mass of one of the bodies is equal to unity) is frequently ascribed to Lagrange or Laplace; however, Huygens, Maupertuis and Euler had already used them earlier (Yurkina 1983).

We shall now separate the parts corresponding to ideal bodies, spherically symmetrical in density, from the individual terms of force function (1.2), and designate the remaining parts, referred to as perturbations of the force function, by the letter R with the appropriate indices. Thus,

$$V = G \left[M_S \left(\frac{M_\oplus}{\Delta_{S\oplus}} + \frac{M_\leftmoon}{\Delta_{S\leftmoon}} + \frac{M_\odot}{\Delta_{S\odot}} \right) + \frac{M_\oplus M_\leftmoon}{\Delta_{\oplus\leftmoon}} + \frac{M_\oplus M_\odot}{\Delta_{\oplus\odot}} + \frac{M_\leftmoon M_\odot}{\Delta_{\leftmoon\odot}} \right]$$

$$+ R_{S\oplus} + R_{S\leftmoon} + R_{S\odot} + R_{\oplus\leftmoon} + R_{\oplus\odot} + R_{\leftmoon\odot} .$$

(1.4)

We shall keep to the designations usually used in classical celestial mechanics: $\Delta_{Si} = |\mathbf{R}_S - \mathbf{R}_i|$ is the distance between satellite S and the mass centre of the i-th body; and $\Delta_{ik} = |\mathbf{R}_k - \mathbf{R}_i|$ is the distance between the mass centres of the i-th and k-th body.

Equations of motion (1.1) then read

$$M_S \ddot{\mathbf{R}}_S = - GM_S \left(M_\oplus \frac{\mathbf{R}_S - \mathbf{R}_\oplus}{\Delta_{S\oplus}^3} + M_\leftmoon \frac{\mathbf{R}_S - \mathbf{R}_\leftmoon}{\Delta_{S\leftmoon}^3} + M_\odot \frac{\mathbf{R}_S - \mathbf{R}_\odot}{\Delta_{S\odot}^3} \right)$$

$$+ \mathrm{grad}_{\mathbf{R}_S}(R_{S\oplus} + R_{S\leftmoon} + R_{S\odot}) ,$$

$$M_\oplus \ddot{\mathbf{R}}_\oplus = - GM_\oplus \left(M_\leftmoon \frac{\mathbf{R}_\oplus - \mathbf{R}_\leftmoon}{\Delta_{\oplus\leftmoon}^3} + M_\odot \frac{\mathbf{R}_\oplus - \mathbf{R}_\odot}{\Delta_{\oplus\odot}^3} \right)$$

$$+ \mathrm{grad}_{\mathbf{R}_\oplus}(R_{\oplus\leftmoon} + R_{\oplus\odot}) ,$$

$$M_\leftmoon \ddot{\mathbf{R}}_\leftmoon = - GM_\leftmoon \left(M_\oplus \frac{\mathbf{R}_\leftmoon - \mathbf{R}_\oplus}{\Delta_{\oplus\leftmoon}^3} + M_\odot \frac{\mathbf{R}_\leftmoon - \mathbf{R}_\odot}{\Delta_{\leftmoon\odot}^3} \right)$$

$$+ \mathrm{grad}_{\mathbf{R}_\leftmoon}(R_{\oplus\leftmoon} + R_{\leftmoon\odot}) ,$$

$$M_\odot \ddot{\mathbf{R}}_\odot = - GM_\odot \left(M_\oplus \frac{\mathbf{R}_\odot - \mathbf{R}_\oplus}{\varDelta_{\oplus\odot}^3} + M_{\text{☽}} \frac{\mathbf{R}_\odot - \mathbf{R}_{\text{☽}}}{\varDelta_{\text{☽}\odot}^3} \right)$$

$$+ \operatorname{grad}_{\mathbf{R}_\odot}(R_{\oplus\odot} + R_{\text{☽}\odot}) \, ; \tag{1.5}$$

they describe the motion in the so-called restricted four-body problem (Subbotin 1937, 1941; Brouwer and Clemence 1961). To be precise, the system of Eqs. (1.5) would need to be solved as a whole to derive the satellite's orbit exactly. However, in view of the facts already mentioned and to the general orientation of the subject, we shall now deal only with the first of Eqs. (1.5), and consider the positions of the three natural bodies as known. And since we are interested in the motion of the satellite relative to the mass centre of the Earth's body, we shall now translate system $\mathbf{R}\{X_j\}$ into the geocentric system $\mathbf{r}\{x_j\}, j = 1, 2, 3$;

$$\mathbf{r} = \mathbf{R} - \mathbf{R}_\oplus \, ; \tag{1.6}$$

thus

$$M_S \ddot{\mathbf{r}}_S = - GM_S \left[M_\oplus \frac{\mathbf{r}_S}{\varDelta_{S\oplus}^3} + M_{\text{☽}} \left(\frac{\mathbf{r}_S - \mathbf{r}_{\text{☽}}}{\varDelta_{S\text{☽}}^3} + \frac{\mathbf{r}_{\text{☽}}}{\varDelta_{\oplus\text{☽}}^3} \right) \right.$$

$$\left. + M_\odot \left(\frac{\mathbf{r}_S - \mathbf{r}_\odot}{\varDelta_{S\odot}^3} + \frac{\mathbf{r}_\odot}{\varDelta_{\oplus\odot}^3} \right) \right] + \operatorname{grad}_{\mathbf{R}_s}(R_{S\oplus} + R_{S\text{☽}} + R_{S\odot})$$

$$- \frac{M_S}{M_\oplus} \operatorname{grad}_{\mathbf{R}_\oplus}(R_{\oplus\text{☽}} + R_{\oplus\odot}) \, , \tag{1.7}$$

i.e.

$$M_S \ddot{\mathbf{r}}_S + GM_S M_\oplus \frac{\mathbf{r}_S}{\varDelta_{S\oplus}^3} = \operatorname{grad}_{r_s} R_{S\oplus} - GM_S M_{\text{☽}} \left(\frac{\mathbf{r}_S - \mathbf{r}_{\text{☽}}}{\varDelta_{S\text{☽}}^3} + \frac{\mathbf{r}_{\text{☽}}}{\varDelta_{\oplus\text{☽}}^3} \right)$$

$$- GM_S M_\odot \left(\frac{\mathbf{r}_S - \mathbf{r}_\odot}{\varDelta_{S\odot}^3} + \frac{\mathbf{r}_\odot}{\varDelta_{\oplus\odot}^3} \right) + \varDelta \ddot{\mathbf{r}}_S \, . \tag{1.8}$$

In this particular case in which the satellite is assumed to be close to the Earth, all the terms on the right-hand side (rhs) of Eq. (1.8) have the nature of perturbations. The first expresses the effect of the deviations of the Earth's gravitational field from a spherically symmetric field, being larger the closer the satellite is to the Earth. The second and third terms reflect the effect of the 'third bodies', i.e. the Moon and Sun, assuming, however, that their gravitational fields are spherically symmetrical. The last term, $\varDelta \ddot{\mathbf{r}}_S$, contains the effect of the deviations of gravitational fields of the third bodies from spherically symmetrical.

If the perturbations due to the third bodies are small enough to be neglected, the motion of the satellite would be described by the equation

$$M_S \ddot{\mathbf{r}}_S + GM_S M_\oplus \frac{\mathbf{r}_S}{\varDelta_{S\oplus}^3} = \operatorname{grad}_{r_s} R_{S\oplus} \, , \tag{1.9}$$

which solves the restricted two-body problem. And if the Earth's gravitational field were also spherically symmetrical, then $R_{S\oplus} = 0$, and the motion of the satellite would be described by the equation

$$\ddot{\mathbf{r}}_S + GM_\oplus \frac{\mathbf{r}_S}{\Delta_{S\oplus}^3} = 0 , \tag{1.10}$$

which solves the problem of two point masses, or of spheres of spherically symmetrical density, i.e. an ideal, unperturbed motion. In this case the force function reads $V = GM_S M_\oplus / \Delta_{S\oplus}$ and the gravitational potential is $GM_\oplus / \Delta_{S\oplus}$.

The elementary problem (1.10) has an exact solution. This is given by a system of three second-order differential equations, i.e. by six general integrals with six integration constants. These will be introduced as is usual in classical celestial mechanics (Subbotin 1937, 1941; Brouwer and Clemence 1961), where they represent six orbital elements. Hereinafter the magnitude of the geocentric radius-vector of the satellite will be designated ϱ ($\varrho = |\mathbf{r}_S| = \Delta_{S\oplus}$); the subscript 'S' will be omitted. We shall retain scalar r to designate the principal variable of the potential theory, i.e. to designate the distance between two mass elements.

System (1.10) has seven first integrals: three integrals of areas or angular momentum, one integral of energy and three Laplace integrals. However, only five of them are mutually independent. Equation (1.10) immediately yields

$$[\mathbf{r}_S \times \ddot{\mathbf{r}}_S] = 0 , \tag{1.11}$$

$$\frac{d}{dt}[\mathbf{r}_S \times \dot{\mathbf{r}}_S] = 0 , \tag{1.12}$$

and therefore

$$\mathbf{r}_S \times \dot{\mathbf{r}}_S = \mathbf{C} = \begin{vmatrix} \mathbf{e}_1 & \mathbf{e}_2 & \mathbf{e}_3 \\ x_1 & x_2 & x_3 \\ \dot{x}_1 & \dot{x}_2 & \dot{x}_3 \end{vmatrix} , \quad \mathbf{C} = \{C_j\} , \quad j = 1, 2, 3 , \tag{1.13}$$

$$\mathbf{C} \cdot \mathbf{r}_s = 0 , \quad \left(\sum_j C_j x_j = 0 \right) . \tag{1.14}$$

Equation (1.13) represents a vector integral of areas, and its components three integrals of areas, reflecting the law of conservation of angular momentum $\mathbf{L}_S = \mathbf{r}_S \times M_S \dot{\mathbf{r}}_S = \mathbf{r}_S \times M_S \mathbf{v}_S = \text{const}$. Equation (1.14) proves that the orbit is planar and that the orbital plane passes through the mass centre O of the ideal body, being perpendicular to vector $\mathbf{C}$ (Fig. 1.2). We assume $\mathbf{C} \neq 0$; the case of $\mathbf{C} = 0$ is not considered because it describes linear motion. Vector $\mathbf{C}$ is referred to as the vector constant of areas; it is a constant vector, parallel to the vector of normal $\mathbf{n}$ to the orbital plane, whose position in space it defines together with the given point O. Equation (1.11) also implies that the acting force always points to centre O. Modulus C has a dynamic significance: $C = 2\dot{P}$, where $\dot{P}$ is the areal velocity; consequently, with a view to (1.13)

$$C = [(x_2 \dot{x}_3 - x_3 \dot{x}_2)^2 + (x_3 \dot{x}_1 - x_1 \dot{x}_3)^2 + (x_1 \dot{x}_2 - x_2 \dot{x}_1)^2]^{1/2} . \tag{1.15}$$

It is therefore equal to the magnitude of the satellite's orbital angular momentum.

Components C_j read

$$C_j = 2\dot{P}\cos(\mathbf{n}, x_j), \quad j = 1, 2, 3 \tag{1.16}$$

and the direction cosines of vector $\mathbf{C}$ and normal $\mathbf{n}$ (Fig. 1.3)

$$\cos(\mathbf{n}, x_1) = \sin(\Omega - \Theta)\sin i,$$
$$\cos(\mathbf{n}, x_2) = -\cos(\Omega - \Theta)\sin i,$$
$$\cos(\mathbf{n}, x_3) = \cos i; \tag{1.17}$$

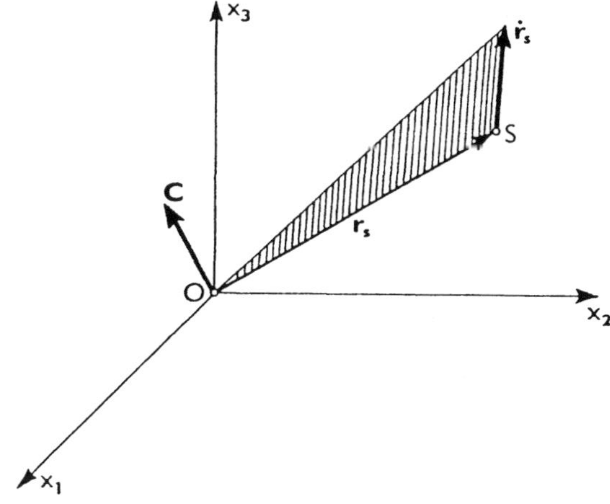

Fig. 1.2. Law of conservation of angular momentum (integral of areas)

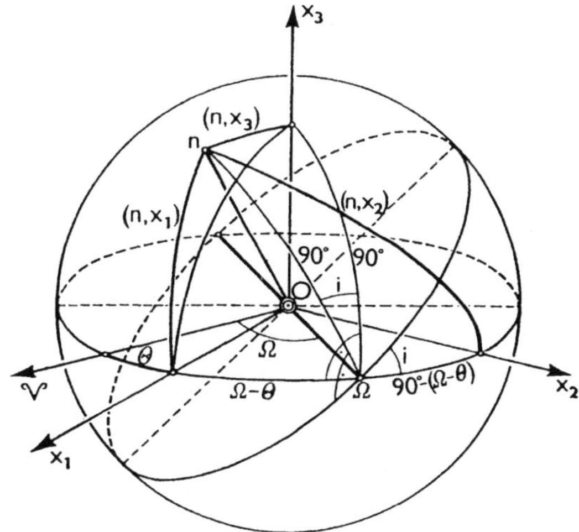

Fig. 1.3. Direction cosines of the normal to the plane of the satellite orbit

Ω is the right ascension of the ascending node Ω of the satellite's orbit, i the inclination of its orbital plane to the equatorial plane, Θ the hour angle of the vernal equinox Υ relative to the zero meridian $(x_1 x_3)$. We have so far considered the coordinate system x_j as fixed with the Earth, the x_3-axis as close to the axis of rotation and the Earth's body as perfectly rigid. However, the equations of motion, which from now on will be our point of departure, will be expressed in the inertial system; otherwise we would have to consider the Coriolis, centrifugal and possibly also the Euler force, and this would complicate the issue. It will, therefore, be simpler to transform the quantities expressed in the Earth's rotating coordinate system into the inertial system (rotation through angle Θ) when it becomes necessary to substitute them into the equations of motion; for example, we shall replace $\Omega - \Theta$ in (1.17) with Ω, and Λ in the geopotential with $\Lambda + \Theta$, etc.

To deal with the satellite orbit in Eq. (1.14) it is convenient to use the polar coordinate system ϱ, u; $\varrho = \overline{OS}$ is the geocentric radius-vector of satellite S, and u is the angle between ϱ and the nodal line $\overline{O\Omega}$ (Fig. 1.4). The specification of Lagrange's equations of the second kind is

$$\frac{\mathrm{d}}{\mathrm{d}t}\left(\frac{\partial L}{\partial \dot{q}_r}\right) - \frac{\partial L}{\partial q_r} = 0 . \tag{1.18}$$

The quantities $\dot{q}_r = \mathrm{d}q_r/\mathrm{d}t$ are generalized velocities, $\partial L/\partial q_r$ generalized forces and $\partial L/\partial \dot{q}_r$ generalized impulses; integral $S = \int_{t_1}^{t_2} L(q, \dot{q}, t)\,\mathrm{d}t$ is a functional (called action) and, according to Hamilton's principle, $S \to$ min. (least action) its first variation must read

$$\delta S = \int_{t_1}^{t_2}\left(\frac{\partial L}{\partial q}\,\delta q + \frac{\partial L}{\partial \dot{q}}\,\delta \dot{q}\right)\mathrm{d}t = 0 .$$

Let us specify (1.18) for generalized (Lagrange) coordinates $q_1 = \varrho$, $q_2 = u$, if the

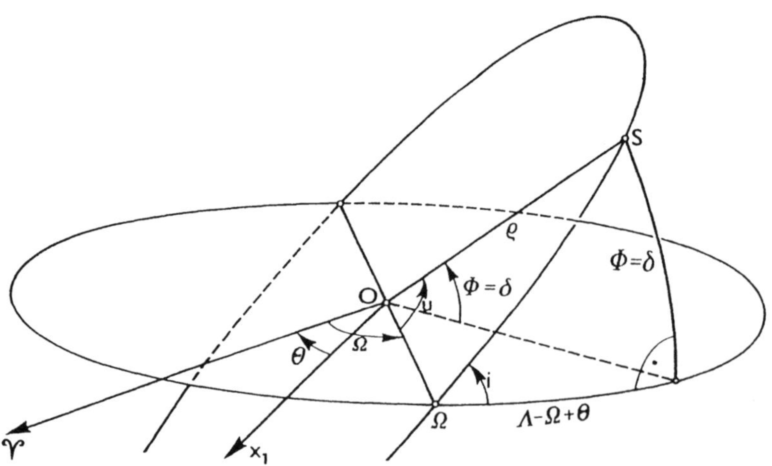

Fig. 1.4. Polar coordinates of the satellite, ϱ, u

Lagrangian function (the Lagrangian of the system) is $L = T - U$, where $T = \frac{1}{2}M_S(\dot{\varrho}^2 + \varrho^2\dot{u}^2)$ is the kinetic energy and $U = -GM_SM_\oplus/\varrho$ the potential energy. The equations of motion

$$\ddot{\varrho} - \varrho\dot{u}^2 + GM_\oplus\varrho^{-2} = 0 , \quad \frac{\mathrm{d}}{\mathrm{d}t}(\varrho^2\dot{u}) = 0 \tag{1.19}$$

follow immediately. Note that potential energy U differs from force function V in sign; the potential is numerically equal to the force function if the mass is unity ($M_S = 1$).

Since $\varrho^2\dot{u} = |\mathbf{r} \times \dot{\mathbf{r}}|$, the second equation of (1.19) again yields the integral of areas (1.13),

$$\varrho^2\dot{u} = C , \tag{1.20}$$

also equal to the angular momentum and generalized impulse, which follows directly from Kepler's second law. We shall modify the first equation in (1.19) with a view to (1.20), and after introducing a new variable $\varrho' = \varrho^{-1}$ as follows (Binet's formula),

$$\frac{\mathrm{d}^2\varrho'}{\mathrm{d}u^2} + \varrho' = \frac{GM_\oplus}{C^2} , \tag{1.21}$$

which is a linear inhomogeneous differential equation of the second order. The general solution of the homogeneous equation, without the rhs, reads

$$\varrho'_0 = k\cos(u - \omega) , \tag{1.22}$$

where k and ω are integration constants. The particular integral of the inhomogeneous equation (1.21) is clearly

$$\varphi(u) = \frac{GM_\oplus}{C^2} , \tag{1.23}$$

and, consequently, the general solution reads

$$\varrho' = k\cos(u - \omega) + \frac{GM_\oplus}{C^2}$$

or

$$\varrho = \frac{p}{1 + e\cos v} . \tag{1.24}$$

This is the focal equation of the conic section in polar coordinates ϱ and

$$v = u - \omega , \tag{1.25}$$

related to its focus and principal axis (Figs. 1.5 and 1.6): v is the true anomaly, ω the argument of the perigee,

$$p = \frac{C^2}{GM_\oplus} \tag{1.26}$$

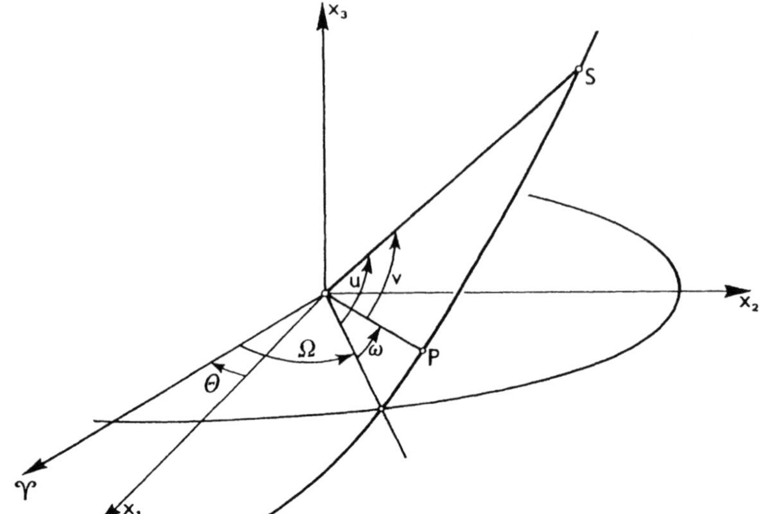

Fig. 1.5. True anomaly v

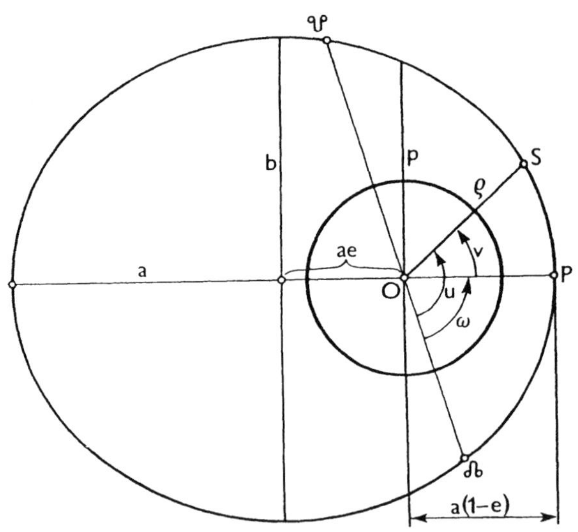

Fig. 1.6. Elliptical orbit

the parameter [the length of the radius-vector perpendicular to the semimajor axis ($v = 90°$, $270°$)],

$$e = k\,\frac{C^2}{GM_\oplus}$$

(1.27)

the numerical eccentricity;

$$k = \frac{e}{p}. \tag{1.28}$$

For elliptic motion, (1.26) yields the integral of areas

$$C = [GM_\oplus a(1 - e^2)]^{1/2}. \tag{1.29}$$

The energy (or vis viva) integral or Hamiltonian (Hamilton's function; total mechanical energy) reads

$$H = T_k + U = \tfrac{1}{2}M_S(\dot{\varrho}^2 + \varrho^2\dot{u}^2) - G\frac{M_S M_\oplus}{\varrho}$$

$$= \tfrac{1}{2}M_S\sum \dot{x}_j^2 - G\frac{M_S M_\oplus}{\varrho} = \text{const}. \tag{1.30}$$

Using the Hamiltonian, the equations of motion (1.10) can be expressed in canonical form; $M_S\dot{q}_j = \partial H/\partial p_j$, $M_S\dot{p}_j = -\partial H/\partial q_j$; if rectangular coordinates x_j are taken to be generalized Lagrange coordinates q_j and derivatives $\dot{x}_j$ to be the generalized impulses p_j, $M_S\dot{x}_j = \partial H/\partial \dot{x}_j$, $M_S\dfrac{\mathrm{d}}{\mathrm{d}t}\dot{x}_j = -\partial H/\partial x_j$, then the Hamilton–Jacobi integration method may be used.

The value of constant H can be derived, e.g. from the position in the perigee P, where $v = 0$, $\dot{\varrho} = 0$, $\varrho = p(1 + e)^{-1}$,

$$H = \tfrac{1}{2}M_S(GM_\oplus)^2(e^2 - 1)C^{-2}, \tag{1.31}$$

which translates into

$$e^2 = 1 + 2HM_S^{-1}C^2(GM_\oplus)^{-2}. \tag{1.32}$$

For an ellipse (and circle) $H < 0$, $0 < e < 1$ $(e = 0)$; for a hyperbola $H > 0$, $e > 1$; for a parabola $H = 0$, $e = 1$. If the ellipse's semimajor axis is a, $H = -\tfrac{1}{2}GM_S M_\oplus a^{-1}$ and, with a view to (1.30), we arrive at

$$a = -M_S\frac{GM_\oplus}{2H} = -\frac{GM_\oplus}{v^2 - 2GM_\oplus \varrho^{-1}} = \frac{\varrho}{2 - \dfrac{v^2}{v_k^2}}, \tag{1.33}$$

where v is the orbital velocity of the satellite (the symbol used for velocity is the same as that used for the true anomaly), and

$$v_k = \left(\frac{GM_\oplus}{\varrho}\right)^{1/2} \tag{1.34}$$

is the circular velocity. If $v = v_k\sqrt{2}$, $a \to \infty$, the orbit is a parabola, and

$$v_p = \left(\frac{2GM_\oplus}{\varrho}\right)^{1/2} \tag{1.35}$$

is the parabolic or escape velocity.

Laplace's vector integral reads

$$\dot{\mathbf{r}}_S \times \mathbf{C} - GM_\oplus \mathbf{r}_S/\varrho = \boldsymbol{\lambda}\,; \quad \lambda = \{\lambda_j\}, \quad j = 1, 2, 3\,; \tag{1.36}$$

$\boldsymbol{\lambda}$ is the vector constant (Laplace's constant vector). It follows from the relation

$$\ddot{\mathbf{r}}_S \times \mathbf{C} = \frac{\mathrm{d}}{\mathrm{d}t}(\dot{\mathbf{r}}_S \times \mathbf{C}) = \ddot{\mathbf{r}}_S \times [\mathbf{r}_S \times \dot{\mathbf{r}}_S] = -\frac{GM_\oplus}{\varrho^3}\,\mathbf{r}_S \times [\mathbf{r}_S \times \dot{\mathbf{r}}_S]$$

$$= -\frac{GM_\oplus}{\varrho^3}\,[(\mathbf{r}_S \cdot \dot{\mathbf{r}}_S)\mathbf{r}_S - \varrho^2 \dot{\mathbf{r}}_S]$$

$$= -\frac{GM_\oplus}{\varrho^3}\,(\varrho\dot{\varrho}\mathbf{r}_S - \varrho^2 \dot{\mathbf{r}}_S) = GM_\oplus \frac{\mathrm{d}}{\mathrm{d}t}\left(\frac{\mathbf{r}_S}{\varrho}\right), \tag{1.37}$$

to which (1.10), (1.13) and the relation between the vector and scalar product have been applied.

However, the first integral (1.36) is not independent of the integral of areas (1.13) and the integral of energy (1.30). Indeed,. the following holds:

$$\mathbf{C} \cdot \boldsymbol{\lambda} = 0\,, \quad \left(\sum_j C_j \lambda_j = 0\right), \tag{1.38}$$

(vector $\boldsymbol{\lambda}$ lies in the orbital plane and is thus perpendicular to vector $\mathbf{C}$) and, with a view to (1.31) and (1.36), also

$$\lambda^2 - 2HM_S^{-1}C^2 = (GM_\oplus)^2\,, \tag{1.39}$$

since

$$[\dot{\mathbf{r}}_S \times \mathbf{C} - GM_\oplus \mathbf{r}_S/\varrho]^2 = [\dot{\mathbf{r}}_S \times \mathbf{C}]^2 - 2GM_\oplus \mathbf{r}_S \cdot [\dot{\mathbf{r}}_S \times \mathbf{C}]/\varrho + (GM_\oplus \mathbf{r}_S/\varrho)^2$$

$$= \dot{\mathbf{r}}_S^2 C^2 - (\dot{\mathbf{r}}_S \cdot \mathbf{C})^2 - 2GM_\oplus [\mathbf{r}_S \times \dot{\mathbf{r}}_S] \cdot \mathbf{C}/\varrho + (GM_\oplus)^2$$

$$= (\dot{\mathbf{r}}_S^2 - 2GM_\oplus/\varrho)C^2 + (GM_\oplus)^2 = 2HM_S^{-1}C^2 + (GM_\oplus)^2\,.$$

The geocentric radius-vector of the satellite (1.24) can also be expressed in terms of Laplace's vector integral. Since

$$\boldsymbol{\lambda} \cdot \mathbf{r}_S = [\dot{\mathbf{r}}_S \times \mathbf{C}] \cdot \mathbf{r}_S - GM_\oplus \varrho = \mathbf{C} \cdot [\mathbf{r}_S \times \dot{\mathbf{r}}_S] - GM_\oplus \varrho$$

$$= C^2 - GM_\oplus \varrho = \lambda \varrho \cos(\mathbf{r}_S, \boldsymbol{\lambda})\,, \tag{1.40}$$

then

$$\varrho = \frac{C^2}{GM_\oplus + \lambda \cos(\mathbf{r}_S, \boldsymbol{\lambda})} = \frac{p}{1 + e \cos v}\,, \tag{1.41}$$

Eqs. (1.26) and (1.27) holding for parameter p and the eccentricity, respectively:

$$e = \frac{\lambda}{GM_\oplus} = \left[1 + 2\frac{H}{M_S}\left(\frac{C}{GM_\oplus}\right)^2\right]^{1/2}, \tag{1.42}$$

i.e. (1.32); $(\mathbf{r}_S, \boldsymbol{\lambda}) = v$, in other words Laplace's vector $\boldsymbol{\lambda}$ points along the axis of the orbit running through the perigee (Fig. 1.7).

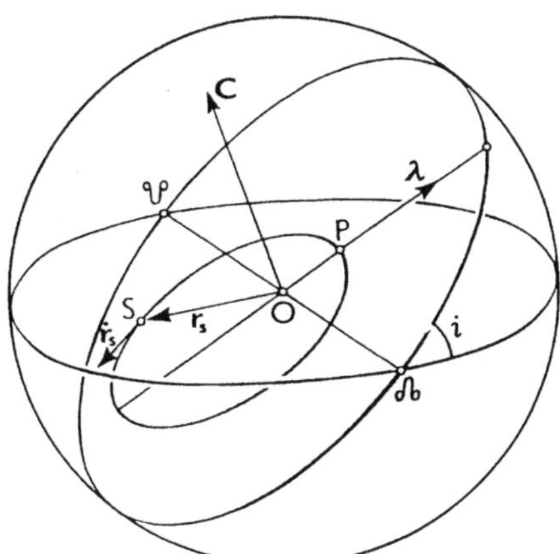

Fig. 1.7. Laplace's vector

There is a total of seven first integrals (if expressed in components) mutually linked by two relations, (1.38) and (1.39), and only five of them are independent. Any two of the set of seven constants $C_1, C_2, C_3, H, \lambda_1, \lambda_2, \lambda_3$ may be expressed in terms of the other five.

None of the first integrals contains the independent variable, time t. However, the general solution of Eq. (1.21), called the orbit integral and which describes the motion in the orbital plane already given, contains variable v, (1.25), defining the position of the satellite along its orbit. It is related to time t by the integral of areas (1.13) which, in view of (1.20), (1.25) and $\omega = $ const, also reads

$$\varrho^2 \dot{v} = C ,\tag{1.43}$$

and with a view to (1.24)

$$\frac{.p^2 \, \mathrm{d}v}{(1 + e \cos v)^2} = C \, \mathrm{d}t .\tag{1.44}$$

Integration yields

$$p^2 \int_0^v \frac{\mathrm{d}v}{(1 + e \cos v)^2} = C(t - t_0) ,\tag{1.45}$$

where t_0 is the integration constant, time, corresponding to the position in the perigee ($v = 0$).

The six integration constants which define the orbit and, together with the given time t, also define the satellite's position along it in the geocentric coordinate system in space uniquely, may be given various forms, e.g. that of one of the first integrals. However, in classical celestial mechanics six so-called orbital elements $\Omega, i, \omega, a, e, t_0$, which we have already defined, were introduced. They represent five integrals of motion and the instant from which time is reckoned. The instantaneous mean anomaly M is frequently introduced as an independent variable instead of time t; it also runs uniformly:

$$M = M_0 + n(t - \bar{t}_0) \,. \tag{1.46}$$

The sixth orbital element is then M_0 which represents the value of the mean anomaly at epoch $t = \bar{t}_0$; n is the mean motion (mean angular orbital velocity):

$$n = \frac{2\pi}{T} \,. \tag{1.47}$$

Since $\frac{1}{2}\varrho^2 \, dv$ is the area described by radius-vector ϱ as it changes its direction by dv, the orbital period T can be determined by integrating (1.43) over time interval T:

$$\int_0^{2\pi} \varrho^2 \, dv = C \int_0^T dt = 2\pi ab \,,$$

$$T = 2\pi abC^{-1} = 2\pi a^2(1 - e^2)^{1/2} C^{-1} \tag{1.48}$$

$[b = a(1 - e^2)^{1/2}$ is the semiminor axis of the ellipse$]$ and, with a view to (1.26), after substituting $p = a(1 - e^2)$

$$T = 2\pi a^{3/2}(GM_\oplus)^{-1/2} \,, \tag{1.49}$$

which also follows from Kepler's third law $(n^2 a^3 = GM_\oplus)$. The relation between T and a is illustrated by the data in Table 1.1 and in Fig. 1.8 for artificial Earth satellites, and for artificial satellites of the Moon and Mars for comparison. The mean and true anomalies are related by the eccentric anomaly E. Its definition for elliptic motion is illustrated in Fig. 1.9, which also clearly indicates that

$$\varrho \cos v = a(\cos E - e) \tag{1.50}$$

Table 1.1. Orbital period T of the satellite as a function of the semi-axis a of the osculating ellipse $\left(T^2 = \dfrac{4\pi^2}{GM} a^3 \right)$

	a (km)		
T (h)	Earth $(GM_\oplus = 398\,600.44 \times 10^9\,\mathrm{m^3 s^{-2}})$	Moon $(GM_\math{\leftmoon} = 4\,902.78 \times 10^9\,\mathrm{m^3 s^{-2}})$	Mars $(GM_\mars = 42\,828.4 \times 10^9\,\mathrm{m^3 s^{-2}})$
1.6	6 945	–	–
2.0	8 059	1860	3831
2.5	9 352	2159	4446
3.0	10 560	2438	5020
6.0	16 763	3870	7969
12.0	26 610	6143	12651
18.0	34 869	8049	16 577
24.0	42 241	9751	20 082

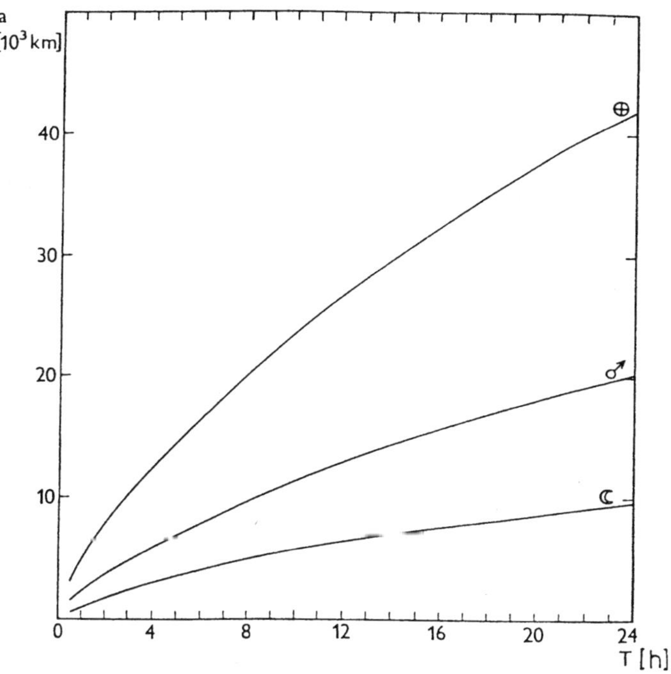

Fig. 1.8. The length of the semimajor axis a as a function of the orbital period T of artificial satellites of the Earth, Moon and Mars

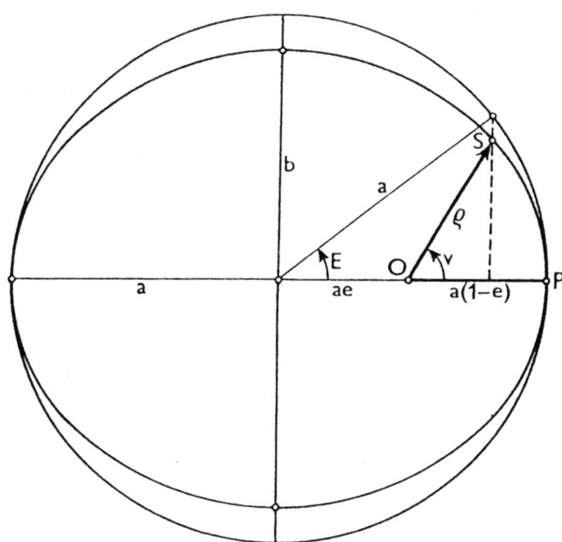

Fig. 1.9. Relation between the true (v) and eccentric (E) anomaly

and then

$$\varrho \sin v = a(1 - e^2)^{1/2} \sin E = b \sin E \,, \tag{1.51}$$

$$\varrho = a(1 - e \cos E) \,, \tag{1.52}$$

$$dv = \frac{C}{a^2(1 - e \cos E)^2} \, dt = \frac{[GM_\oplus(1 - e^2)]^{1/2}}{a^{3/2}(1 - e \cos E)^2} \, dt$$

$$= \frac{(1 - e^2)^{1/2}}{1 - e \cos E} \, dE \,; \tag{1.53}$$

finally, after integration, we arrive at

$$E - e \sin E = (GM_\oplus)^{1/2} a^{-3/2}(t - t_0) \,, \tag{1.54}$$

which is Kepler's equation; t_0 is the time at $E = 0$ (perigee passage). By comparing (1.54) and (1.49) and in view of (1.47) and (1.46) we obtain

$$E - e \sin E = M \,, \tag{1.55}$$

provided that the mean anomaly is reckoned from $t = t_0$.

If the orbital elements are known, the coordinates x_j (Fig. 1.10) and the velocity components $\dot{x}_j$ can be calculated for any instant of time. Figures 1.10 and 1.11, which show the direction cosines of the radius-vector, indicate that [Θ – hour angle of the vernal equinox for the zero meridian $(x_1 x_3)$, for which

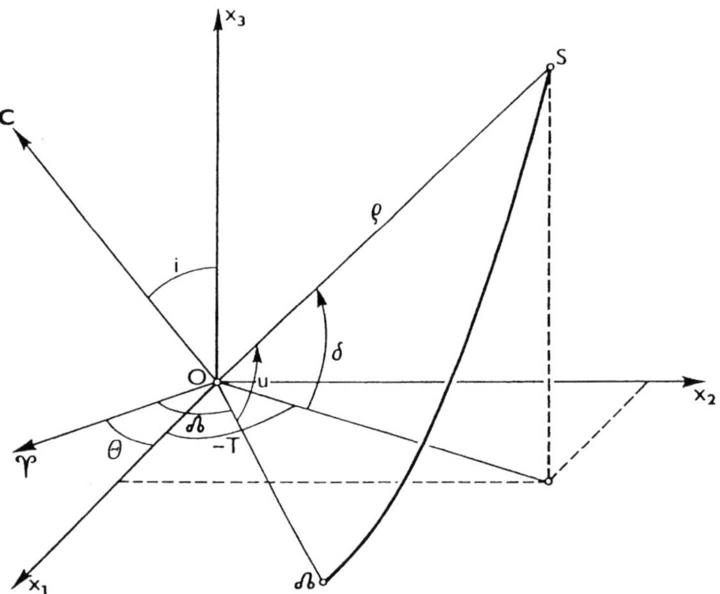

Fig. 1.10. Spatial coordinates of the satellite

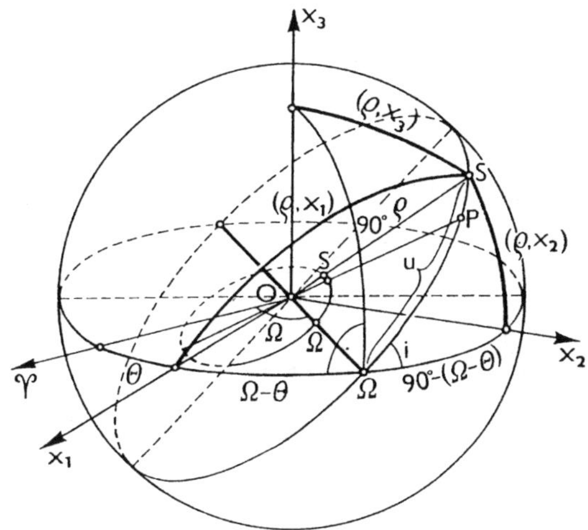

Fig. 1.11. Direction cosines of the satellite's radius-vector

also the hour angle T of the satellite is defined, δ – declination (Fig. 1.12)]

$$\dot{x}_1 = \varrho\cos(\varrho, x_1) = \varrho[\cos u\cos(\Omega - \Theta) - \sin u\sin(\Omega - \Theta)\cos i]\,,$$

$$= \varrho\cos\delta\cos T\,,$$

$$x_2 = \varrho\cos(\varrho, x_2) = \varrho[\cos u\sin(\Omega - \Theta) + \sin u\cos(\Omega - \Theta)\cos i]$$

$$= -\varrho\cos\delta\sin T\,,$$

$$x_3 = \varrho\sin u\sin i = \varrho\sin\delta\,, \qquad\qquad (1.56)$$

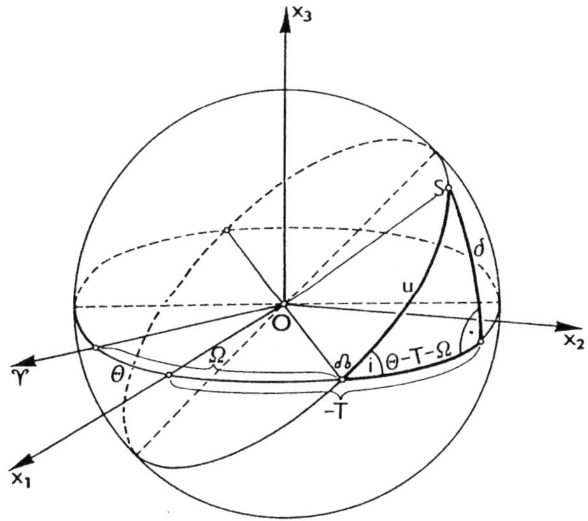

Fig. 1.12. Relation between the orbital elements and geocentric spherical coordinates of the satellite

and consequently

$$\dot{x}_1 = \dot{\varrho}\,\frac{x_1}{\varrho} - \varrho\,[\sin u\cos(\Omega - \Theta) + \cos u\sin(\Omega - \Theta)\cos i]\,\dot{v}\,,$$

$$\dot{x}_2 = \dot{\varrho}\,\frac{x_2}{\varrho} - \varrho\,[\sin u\sin(\Omega - \Theta) - \cos u\cos(\Omega - \Theta)\cos i]\,\dot{v}\,,$$

$$\dot{x}_3 = \dot{\varrho}\,\frac{x_3}{\varrho} + (\varrho\cos u\sin i)\dot{v}\,,$$

$$\sum_{j=1}^{3} \dot{x}_j^2 = 2\left(\frac{GM_{\oplus}}{\varrho} + \frac{H}{M_S}\right) = GM_{\oplus}\left(\frac{2}{\varrho} - \frac{1}{a}\right)\,, \tag{1.57}$$

in view of (1.43), (1.29), (1.41) and (1.53)

$$\dot{v} = \frac{C}{\varrho^2} = \frac{1}{\varrho^2}[GM_{\oplus}a(1 - e^2)]^{1/2}\,,$$

$$\dot{\varrho} = (GM_{\oplus})^{1/2}[a(1 - e^2)]^{-1/2}e\sin v = C^{-1}GM_{\oplus}e\sin v\,,$$

and also making use of the first integral (1.30)

$$\dot{\varrho} = \left[\frac{2H}{M_S} - \frac{C^2}{\varrho^2} + 2\frac{GM_{\oplus}}{\varrho}\right]^{1/2} = \left[\frac{2H}{M_S} - \frac{GM_{\oplus}(p - 2\varrho)}{\varrho^2}\right]^{1/2} = \left[\frac{2T_k}{M_S} - \frac{C^2}{\varrho^2}\right]^{1/2}\,,$$

$$\dot{v} = \frac{GM_{\oplus}}{\varrho^2}\left[\frac{M_S(e^2 - 1)}{2H}\right]^{1/2}\,.$$

If, however, the integration constants are not known, they can be calculated using the six orbital data at time $t = t_0$, e.g. the position $(x_j)_0$ and velocity components $(\dot{x}_j)_0$ at that particular time. The values of the first integrals are then given by (1.13) and (1.30) for $x_j = (x_j)_0$, $\dot{x}_j = (\dot{x}_j)_0$, $\varrho = \varrho_0 = \left[\sum_j (x_j)_0^2\right]^{1/2}$ and by (1.36) which yields

$$\lambda_j = (x_j)_0 \sum_j (\dot{x}_j)_0^2 - \varrho_0\dot{\varrho}_0(\dot{x}_j)_0 - GM_{\oplus}\frac{(x_j)_0}{\varrho_0}\,. \tag{1.58}$$

Assuming the validity of (1.13) and (1.17), the values of the orbital elements are determined by the following expressions:

$$\tan i = (C_1^2 + C_2^2)^{1/2}/C_3\,,$$

$$\tan \Omega = -C_1/C_2\,,$$

$$\cos \omega = (\lambda_1\cos\Omega + \lambda_2\sin\Omega)/\lambda\,,$$

$$a = \frac{C^2}{GM_{\oplus}(1 - e^2)} = -\frac{1}{2H}M_S GM_{\oplus}\,; \tag{1.59}$$

the numerical eccentricity is given by Eq. (1.42). The true anomaly can be determined from the scalar product of vector ϱ and Laplace's vector λ:

$$\cos v_0 = \frac{1}{\lambda \varrho_0} \sum_{j=1}^{3} \lambda_j (x_j)_0 \,. \tag{1.60}$$

Also

$$\sin E_0 = \frac{\varrho_0 \sin v_0}{a(1 - e^2)^{1/2}}$$

and, finally, the last orbital element

$$M_0 = E_0 - e \sin E_0 \,. \tag{1.61}$$

The six orbital elements Ω, i, ω, a, e, M_0 are only constants if the motion is unperturbed, i.e. if it takes place in a spherically symmetrical gravitational field. With regard to the Earth's artificial satellites, perturbations are mostly due to the deviations of the gravitational field of the real Earth from the ideal field. They are symbolically expressed by the perturbing function $R_{S\oplus}$; in this case the satellite's motion is described by Eq. (1.9). Also, the perturbations due to the Moon and Sun have an effect, as expressed by equations of motion (1.8). Before dealing with perturbed motion, we shall explain the perturbing function in its actual form.

1.3 Perturbing Function and Perturbing Potential

1.3.1 General Definitions

The perturbing function is part of the force function. Generally speaking, it may be any part of the latter not included in the homogeneous equations of motion and, therefore, constitutes the rhs of complete equations of motion. In the restricted four-body problem being considered, it could be the sum of all perturbations in the force function (1.4), i.e. the sum of all its terms except for the first term on its rhs. However, assuming that the satellite is close to the Earth and the effect of the Moon and Sun on its motion is relatively small, the effect of both third bodies as a whole could be included in the perturbing function. Moreover, with close Earth satellites perturbations due to the non-sphericity of the fields of both perturbing bodies, $R_{S\mathbb{D}}$, $R_{S\odot}$, need not be considered and, consequently, the gravitational fields of the Moon and Sun may be assumed to be spherically symmetrical. The error incurred thereby is of the order of 2×10^{-4} in the lunar field and 10^{-5} in the solar field; however, the perturbations themselves are of the order of 10^{-4} of the principal perturbing term in the Earth's gravitational potential due to its polar flattening. The error will thus be at least two orders of magnitude smaller than the accuracy with which the perturbing function as a whole can be currently expressed.

This means that $\Delta \ddot{\mathbf{r}}_S$ is made equal to zero in (1.8), and the aggregate perturbing function then reads

$$R = R_{S\oplus} + \Delta V_{S\mathrm{D}} + \Delta V_{S\odot} + \delta V_{S\mathrm{D}} + \delta V_{S\odot} \,, \tag{1.62}$$

where

$$R_{S\oplus} = GM_S \left(\int\limits_{M_\oplus} \frac{dm_\oplus}{r_{S\oplus}} - \frac{M_\oplus}{\varrho} \right). \tag{1.63}$$

Terms $\Delta V_{S\mathrm{D}}$ and $\Delta V_{S\odot}$ represent the first and second terms on the rhs of Eq. (1.8):

$$\mathrm{grad}\,\Delta V_{S\mathrm{D}} = - GM_{\mathrm{D}} \left(\frac{\mathbf{r}_S - \mathbf{r}_{\mathrm{D}}}{\Delta_{S\mathrm{D}}^3} + \frac{\mathbf{r}_{\mathrm{D}}}{\Delta_{\oplus\mathrm{D}}^3} \right), \tag{1.64}$$

$$\mathrm{grad}\,\Delta V_{S\odot} = - GM_\odot \left(\frac{\mathbf{r}_S - \mathbf{r}_\odot}{\Delta_{S\odot}^3} + \frac{\mathbf{r}_\odot}{\Delta_{\oplus\odot}^3} \right), \tag{1.65}$$

$$\Delta V_{S\mathrm{D}} = GM_{\mathrm{D}} \left(\frac{1}{\Delta_{S\mathrm{D}}} - \frac{\mathbf{r}_S \cdot \mathbf{r}_{\mathrm{D}}}{\Delta_{\oplus\mathrm{D}}^3} \right), \tag{1.66}$$

$$\Delta V_{S\odot} = GM_\odot \left(\frac{1}{\Delta_{S\odot}} - \frac{\mathbf{r}_S \cdot \mathbf{r}_\odot}{\Delta_{\oplus\odot}^3} \right). \tag{1.67}$$

Terms $\delta V_{S\mathrm{D}}$, $\delta V_{S\odot}$ represent perturbations due to tidal deformations caused by the Moon and Sun (discussed in Chap. 3).

If $M_S = 1$, the perturbing function is equal to the perturbing potential. This definition of the perturbing potential does not, of course, agree with its definition in the boundary-value problems of the theory of the gravity potential.

1.3.2 Perturbing Gravitational Potential of the Earth in Outer Space

All anomalies of the Earth's figure and density, i.e. its deviations from a spherically homogeneous sphere, generate perturbing gravitational potential. In outer space, where satellite M_S is orbiting, the perturbing potential is given by Eq. (1.63) with $M_S = 1$. In outer space the potential is a harmonic function. It can be expressed by expansion into a series of spherical harmonics which always converge uniformly outside the sphere of convergence surrounding all particles of the body and, therefore, in this part of space the series may always be integrated term by term.

P will denote the potential point (note that the same symbol is used for the perigee) in which the potential is being considered; its position will be expressed in coordinate system x_j, whose origin O is assumed to be located in the Earth's centre of gravity, axis x_3, close to the axis of rotation, and the system is fixed relative to the Earth. Positions on the surface of and within the body, as well as in its close vicinity, will be defined by the geocentric radius-vector ϱ, the geocentric latitude ϕ and longitude Λ (positive to the east); the coordinates within the body will be marked with a prime. The position of the satellite will, as a rule, be defined by coordinates ϱ, δ, T, with the index 'S' being omitted to save space; δ is the geocentric declination and T the geocentric hour angle relative to the prime meridian $(x_1 x_3)$, positive to the west.

The symbol ψ will be used to denote the angle between radius-vector ϱ' of variable element $dm_\oplus$ and radius-vector ϱ of potential point P (Fig. 1.13), i.e. the point at which satellite S is located at that particular moment. The reciprocal distance $r_{S\oplus} = \overline{P\,dm_\oplus}$ in (1.63) will be expressed as a series (see, e.g., Idel'son 1936),

$$\frac{1}{r_{S\oplus}} = \frac{1}{\varrho} \sum_{n=0}^{\infty} \left(\frac{\varrho'}{\varrho}\right)^n \mathrm{P}_n^{(0)}(\cos\psi);\tag{1.68}$$

in outer space, where the satellite is orbiting, $\varrho > \varrho'$ always, and it can easily be proved that series (1.68) converges uniformly there. According to Legendre's addition theorem (decomposition formula) of spherical harmonics

$$\mathrm{P}_n^{(0)}(\cos\psi) = \mathrm{P}_n^{(0)}(\sin\phi')\mathrm{P}_n^{(0)}(\sin\delta)$$

$$+ 2 \sum_{k=1}^{n} \frac{(n-k)!}{(n+k)!} \mathrm{P}_n^{(k)}(\sin\phi')\mathrm{P}_n^{(k)}(\sin\delta)\cos k(\Lambda' + T);\tag{1.69}$$

$\mathrm{P}_n^{(k)}$ is an associated Legendre function of degree n and order k:

$$\mathrm{P}_n^{(k)}(t) = \frac{1}{2^n n!}(1 - t^2)^{k/2}\frac{d^{n+k}}{dt^{n+k}}(t^2 - 1)^n.\tag{1.70}$$

By substituting (1.68) into the relation for the perturbing potential (1.63) and putting $M_S = 1$, we arrive at

$$R_{S\oplus} = \frac{GM_\oplus}{\varrho} \sum_{n=2}^{\infty} \sum_{k=0}^{n} \left(\frac{a_0}{\varrho}\right)^n (J_n^{(k)}\cos kT - S_n^{(k)}\sin kT)\mathrm{P}_n^{(k)}(\sin\delta),\tag{1.71}$$

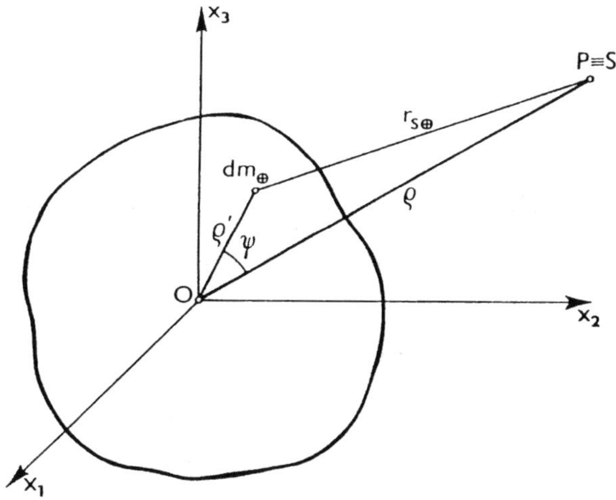

Fig. 1.13. External potential point P

where

$$\left.\begin{array}{c} J_n^{(k)} \\ S_n^{(k)} \end{array}\right\} = \frac{(2 - \delta_{0k})(n - k)!}{M_\oplus a_0^n (n + k)!} \int\limits_{M_\oplus} \varrho'^n P_n^{(k)}(\sin \phi') \begin{array}{c} \cos k \Lambda' \\ \sin k \Lambda' \end{array} \, dm_\oplus ; \qquad (1.72)$$

δ_{0k} is the Kronecker delta and a_0 an optional factor of length (the semimajor axis of the Earth's rotational ellipsoid as a rule) introduced to render $J_n^{(k)}$ and $S_n^{(k)}$ dimensionless. Series (1.71) does not contain terms with $n = 1$ if the coordinate system is strictly geocentric, so that $J_1^{(0)} = 0$, $J_1^{(1)} = 0$, $S_1^{(1)} = 0$.

Quantities $J_n^{(k)}$ and $S_n^{(k)}$, called dynamic or harmonic geopotential coefficients, are functions of the Earth's moments of inertia of degree n and order k. They belong to the class of so-called Stokes constants of the body, i.e. of integrals of the type

$$S = \int\limits_\tau U \sigma \, d\tau , \qquad (1.73)$$

in which U is any function harmonic within volume τ bounded by the body, and σ is the density in $d\tau$. Hereinafter we shall refer to them as Stokes parameters or geopotential coefficients. An excellent property of Stokes constants S is that they can be determined without knowing the density distribution within the body, provided that gravitational potential V and its normal derivative $\dfrac{\partial V}{\partial n}$ are known on surface Σ which bounds the body. This follows directly from Green's first theorem which, when applied to the functions mentioned, U, V, reads

$$\int\limits_\tau (U \Delta V - V \Delta U) \, d\tau = \int\limits_\Sigma (U \operatorname{grad} V - V \operatorname{grad} U) \cdot d\Sigma$$

$$= \int\limits_\Sigma \left(U \frac{\partial V}{\partial n} - V \frac{\partial U}{\partial n} \right) d\Sigma . \qquad (1.74)$$

Since U is a function harmonic within τ by definition, i.e. $\Delta U = 0$, and the gravitational potential satisfies Poisson's equation

$$\Delta V = - 4\pi G \sigma , \qquad (1.75)$$

inside τ we do in fact get

$$S = - \frac{1}{4\pi G} \int\limits_\Sigma \left(U \frac{\partial V}{\partial n} - V \frac{\partial U}{\partial n} \right) d\Sigma \qquad (1.76)$$

quite independently of the body's density structure.

In Eq. (1.72) $\varrho'^n P_n^{(k)} (\sin \phi') \cos k\Lambda' = {}_1 U_n^{(k)}$, $\varrho'^n P_n^{(k)} (\sin \phi') \sin k\Lambda' = {}_2 U_n^{(k)}$ are solid spherical harmonics, i.e. $\Delta_1 U_n^{(k)} = 0$, $\Delta_2 U_n^{(k)} = 0$; the fundamental condition has thus been satisfied and $J_n^{(k)}$ and $S_n^{(k)}$ are indeed Stokes constants which can be determined without knowing the internal structure of the body.

The fundamental property of the perturbing potential is that it is a function which is continuous and limited within space as a whole, and harmonic in outer space,

$$\Delta R_{S\oplus} = 0 , \qquad (1.77)$$

and regular in infinity:

$$\lim_{\varrho \to \infty} R_{S\oplus} = 0, \quad \lim_{\varrho \to \infty} \varrho R_{S\oplus} = GM_\oplus ; \tag{1.78}$$

all its derivatives are also continuous and limited in outer space.

1.3.3 Perturbations due to the Moon and the Sun

We shall consider the Earth to be perfectly rigid. The effect of elastic deformations of the Earth, due to the tidal effects of the Moon and the Sun, on the satellite's orbit will only be derived after dealing with the tide-forming potential in Chapter 4. Perturbations $\delta V_{S\mathbb{D}}$ and $\delta V_{S\odot}$ will not be considered at this point.

With this simplification and under the constraints formulated in Section 1.3.1, the perturbations due to the Moon and Sun will be expressed by the sum $\Delta V_{S\mathbb{D}} + \Delta V_{S\odot}$. The two terms are defined by Eqs. (1.66) and (1.67) which can be expressed as

$$\Delta V_{S\mathbb{D}} = GM_\mathbb{D} \left(\frac{1}{\varDelta_{S\mathbb{D}}} - \frac{\varrho}{\varDelta_{\oplus\mathbb{D}}^2} \cos S_\mathbb{D} \right), \tag{1.79}$$

$$\Delta V_{S\odot} = GM_\odot \left(\frac{1}{\varDelta_{S\odot}} - \frac{\varrho}{\varDelta_{\oplus S\odot}^2} \cos S_\odot \right), \tag{1.80}$$

where $S_\mathbb{D}$ $(S_\odot)$ are the angles made by the geocentric radius-vector of the satellite ϱ and the geocentric radius-vector $\varDelta_{\oplus\mathbb{D}}(\varDelta_{\oplus\odot})$ of the Moon's and (Sun's) centre of mass, respectively, as shown in Fig. 1.14. They are related (δ, T being equatorial coordinates already defined; $\delta_\mathbb{D}$, $T_\mathbb{D}$ and $\delta_\odot$, $T_\odot$ being equatorial coordinates

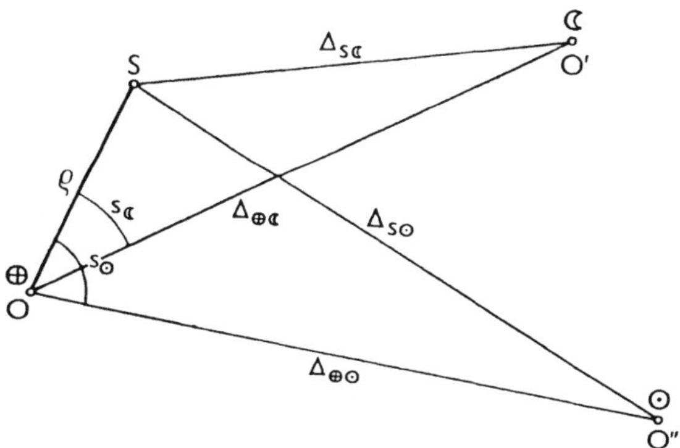

Fig. 1.14. Positions of the perturbing bodies (Moon and Sun) and of the Earth's artificial satellite

of the Moon's and Sun's centre of mass, respectively) as

$$\cos S_{\text{☽}} = \frac{\mathbf{r}_S \cdot \mathbf{r}_{\text{☽}}}{\varrho \Delta_{\oplus\text{☽}}} = \sin \delta \sin \delta_{\text{☽}} + \cos \delta \cos \delta_{\text{☽}} \cos (T - T_{\text{☽}}),\tag{1.81}$$

$$\cos S_{\odot} = \frac{\mathbf{r}_S \cdot \mathbf{r}_{\odot}}{\varrho \Delta_{\oplus\odot}} = \sin \delta \sin \delta_{\odot} + \cos \delta \cos \delta_{\odot} \cos (T - T_{\odot}).\tag{1.82}$$

Since

$$\frac{1}{\Delta_{S\text{☽}}} = \frac{1}{\Delta_{\oplus\text{☽}}} \sum_{n=0}^{\infty} \left(\frac{\varrho}{\Delta_{\oplus\text{☽}}} \right)^n P_n^{(0)}(\cos S_{\text{☽}}),\tag{1.83}$$

$$\frac{1}{\Delta_{S\odot}} = \frac{1}{\Delta_{\oplus\odot}} \sum_{n=0}^{\infty} \left(\frac{\varrho}{\Delta_{\oplus\odot}} \right)^n P_n^{(0)}(\cos S_{\odot}),\tag{1.84}$$

(1.79) and (1.80) can be modified to read

$$\Delta V_{S\text{☽}} = \frac{GM_{\text{☽}}}{\Delta_{\oplus\text{☽}}} \left[1 + \sum_{n=2}^{\infty} \left(\frac{\varrho}{\Delta_{\oplus\text{☽}}} \right)^n P_n^{(0)}(\cos S_{\text{☽}}) \right],\tag{1.85}$$

$$\Delta V_{S\odot} = \frac{GM_{\odot}}{\Delta_{\oplus\odot}} \left[1 + \sum_{n=2}^{\infty} \left(\frac{\varrho}{\Delta_{\oplus\odot}} \right)^n P_n^{(0)}(\cos S_{\odot}) \right],\tag{1.86}$$

so that both perturbing terms are harmonic functions.

1.4 Solution of the Perturbed Motion

The perturbed motion of an Earth artificial satellite is described by Eq. (1.8). If the perturbations due to the non-sphericity of the Moon's and Sun's gravitational fields are neglected, this equation becomes

$$M_S \ddot{\mathbf{r}}_S + G M_S M_{\oplus} \frac{\mathbf{r}_S}{\varrho^3} = \text{grad}_{\mathbf{r}_S} R,\tag{1.87}$$

where the perturbing function R is defined by Eq. (1.62). If $M_S = 1$, i.e. if R is the perturbing potential, in the inertial system Eq. (1.87) will read

$$\ddot{x}_j + G M_{\oplus} \frac{x_j}{\varrho^3} = \frac{\partial R}{\partial x_j}.\tag{1.88}$$

In terms of components ($j = 1, 2, 3$) Eq. (1.88) represents a system of three inhomogeneous differential equations of the second order. Since their rhs $\frac{\partial R}{\partial x_j}$, which give the satellite its perturbing acceleration, are several orders of magnitude smaller than terms $GM_{\oplus} x_j \varrho^{-3}$ on the lhs, generated by the spherically symmetrical Earth, Lagrange's method of variation of constant should evidently be used to produce a solution of the form

$$x_j = x_j [t; \Omega(t), i(t), \omega(t), a(t), e(t), M_0(t)].\tag{1.89}$$

The orbital elements $\Omega, i, \omega, a, e\,M_0$, which were integration constants in the system of homogeneous Eqs. (1.10), have now become functions of time. They have to satisfy six conditions. Since [putting $\Omega(T) = e_1$, $i(t) = e_2$, $\omega(t) = e_3$, $a(t) = e_4$, $e(t) = e_5$, $M_0(t) = e_6$]

$$\dot{x}_j = \frac{\partial x_j}{\partial t} + Q_j, \quad j = 1, 2, 3, \tag{1.90}$$

$$Q_j = \sum_{i=1}^{6} \dot{e}_i \frac{\partial x_j}{\partial e_i}, \quad i = 1, 2, \ldots, 6, \tag{1.91}$$

$$\ddot{x}_j = \frac{\partial^2 x_j}{\partial t^2} + P_j \tag{1.92}$$

and the derivative of (1.90), in view of (1.92), comes out as

$$P_j = \sum_{i=1}^{6} \dot{e}_i \frac{\partial}{\partial e_i} \left(\frac{\partial x_j}{\partial t} \right) + \dot{Q}_j, \tag{1.93}$$

after substituting into Eq. (1.88)

$$\frac{\partial^2 x_j}{\partial t^2} + GM_\oplus \frac{x_j}{\varrho^3} + P_j = \frac{\partial R}{\partial x_j}. \tag{1.94}$$

However, if the position of the satellite and its velocity are now to be determined at a particular time by means of the integrals of homogeneous equations (of unperturbed motion), the orbital elements e_i now being functions of time, these first three natural conditions must be satisfied:

$$P_j = \frac{\partial R}{\partial x_j}, \quad j = 1, 2, 3. \tag{1.95}$$

The following three conditions must also be satisfied,

$$Q_j = 0, \tag{1.96}$$

because only then, in view of (1.90), does

$$\dot{x}_j = \frac{\partial x_j}{\partial t}. \tag{1.97}$$

The first three conditions (1.95) then simplify to

$$\frac{\partial R}{\partial x_j} = \sum_{i=1}^{6} \dot{e}_i \frac{\partial}{\partial e_i} \left(\frac{\partial x_j}{\partial t} \right). \tag{1.98}$$

The orbital elements e_i, satisfying the first three conditions (1.95), are referred to as instantaneous; they are not defined uniquely. If the other three conditions (1.96) are also satisfied, i.e. a total of six conditions (1.96) and (1.98), the orbital elements defined in this manner are referred to as osculating elements. By applying the relations describing unperturbed motion ($R = 0$), these elements yield not only the actual position but also the satellite's velocity at the given time (osculation epoch).

There are various ways of expressing the time variations of orbital elements $\dot{e}$ explicitly. We shall reproduce Lagrange's solution here. Starting with relation

$$\frac{\partial R}{\partial e_k} = \sum_{j=1}^{3} \frac{\partial R}{\partial x_j} \frac{\partial x_j}{\partial e_k}, \tag{1.99}$$

which is evident and which, in view of (1.98) and (1.96) and after some algebra, can be expressed as

$$\frac{\partial R}{\partial e_k} = \sum_{i=1}^{6} \dot{e}_i [e_k, e_i], \qquad k = 1, 2, \ldots, 6; \quad i \neq k, \tag{1.100}$$

where

$$[e_k, e_i] = \sum_{j=1}^{3} \left[\frac{\partial x_j}{\partial e_k} \frac{\partial}{\partial e_i} \left(\frac{\partial x_j}{\partial t} \right) - \frac{\partial x_j}{\partial e_i} \frac{\partial}{\partial e_k} \left(\frac{\partial x_j}{\partial t} \right) \right] \tag{1.101}$$

are referred to as Lagrangian brackets.

Relation (1.100) represents a system of six linear equations in six unknowns $\dot{e}_i$, containing a total of 36 Lagrangian brackets, related as

$$[e_k, e_i] = - [e_i, e_k], \quad [e_k, e_k] = 0, \tag{1.102}$$

by definition (1.101). An excellent property of Lagrangian brackets is that they are time-invariant, i.e.

$$\frac{\partial}{\partial t} [e_k, e_i] = 0. \tag{1.103}$$

They are thus constant for any point of the satellite's orbit and may be used to advantage to simplify a number of relations substantially by computing them, e.g. for the perigee position where $v = 0$, $M = 0$, $\dot{\varrho} = 0$, $\varrho = a(1 - e)$ and $u = \omega$. Nevertheless, the computation, described in detail in Burša (1970) for example, is tedious. The expressions for Lagrangian brackets read:

$$[\Omega, i] = - (GM_\oplus)^{1/2} [a(1 - e^2)]^{1/2} \sin i,$$

$$[\Omega, \omega] = 0,$$

$$[\Omega, a] = \tfrac{1}{2} a^{-1/2} (GM_\oplus)^{1/2} (1 - e^2)^{1/2} \cos i,$$

$$[\Omega, e] = - (GM_\oplus) e a^{1/2} (1 - e^2)^{-1/2} \cos i,$$

$$[\Omega, M_0] = 0, \; [i, \omega] = 0, \quad [i, a] = 0, \quad [i, e] = 0, \; [i, M_0] = 0,$$

$$[\omega, a] = \tfrac{1}{2} a^{-1/2} (GM_\oplus)^{1/2} (1 - e^2)^{1/2},$$

$$[\omega, e] = - (GM_\oplus)^{1/2} e a^{1/2} (1 - e^2)^{-1/2},$$

$$[\omega, M_0] = 0, \; [a, e] = 0,$$

$$[a, M_0] = - \tfrac{1}{2} a^{-1/2} (GM_\oplus)^{1/2}, \quad [e, M_0] = 0. \tag{1.104}$$

On substituting them in (1.100) we obtain the expressions for derivatives $\frac{\partial R}{\partial \Omega}, \frac{\partial R}{\partial i}, \ldots, \frac{\partial R}{\partial M_0}$ as functions of the time derivatives of orbital elements $d\Omega/dt$, $di/dt, \ldots, dM_0/dt$. However, we shall be more interested in the inverse relations, i.e. the Lagrange equations for the osculating orbital elements (also referred to as 'planetary equations')

$$\operatorname{cosec} i (GM_\oplus)^{-1/2} [a(1-e^2)]^{-1/2} \frac{\partial R}{\partial i} = \frac{d\Omega}{dt}, \tag{1.105}$$

$$-\operatorname{cosec} i (GM_\oplus)^{-1/2} [a(1-e^2)]^{-1/2} \frac{\partial R}{\partial \Omega}$$

$$+ \cot i (GM_\oplus)^{-1/2} [a(1-e^2)]^{-1/2} \frac{\partial R}{\partial \omega} = \frac{di}{dt}, \tag{1.106}$$

$$-\cot i (GM_\oplus)^{-1/2} [a(1-e^2)]^{-1/2} \frac{\partial R}{\partial i}$$

$$+ (1-e^2)^{1/2} e^{-1} (GM_0)^{-1/2} a^{-1/2} \frac{\partial R}{\partial e} = \frac{d\omega}{dt}, \tag{1.107}$$

$$2a^{1/2} (GM_\oplus)^{-1/2} \frac{\partial R}{\partial M_0} = \frac{da}{dt}, \tag{1.108}$$

$$-(1-e^2)^{1/2} e^{-1} (GM_\oplus)^{-1/2} a^{-1/2} \frac{\partial R}{\partial \omega}$$

$$+ (1-e^2) e^{-1} (GM_\oplus)^{-1/2} a^{-1/2} \frac{\partial R}{\partial M_0} = \frac{de}{dt}, \tag{1.109}$$

$$-2a^{1/2} (GM_\oplus)^{-1/2} \frac{\partial R}{\partial a} - (1-e^2) e^{-1} (GM_\oplus)^{-1/2} a^{-1/2} \frac{\partial R}{\partial e} = \frac{dM_0}{dt}. \tag{1.110}$$

This form (1.105) of Lagrange equations cannot be used directly if the osculating plane is in the plane of the equator ($i = 0$), or if the osculating orbit is circular ($e = 0$). In the former case it is sufficient to introduce a new variable, $J = \cos i$; in the latter, for example, Delaunay variables (orbital elements) can be used (Delaunay 1860):

$$L = (GM_\oplus a)^{1/2}, \quad G = [GM_\oplus a(1-e^2)]^{1/2},$$

$$H = [GM_\oplus a(1-e^2)]^{1/2} \cos i. \tag{1.111, 1.112, 1.113}$$

The absolute values of all non-zero Lagrange brackets are then equal to unity. Denoting the other three elements $M = l$, $\omega = g$, $\Omega = h$, Eqs. (1.105)–(1.110) formally simplify to read:

$$\frac{\partial R}{\partial l} = \dot{L}, \quad -\frac{\partial R}{\partial L} = \dot{l}, \tag{1.114}$$

$$\frac{\partial R}{\partial g} = \dot{G}, \qquad -\frac{\partial R}{\partial G} = \dot{g}, \tag{1.115}$$

$$\frac{\partial R}{\partial h} = \dot{H}, \qquad -\frac{\partial R}{\partial H} = \dot{h}. \tag{1.116}$$

If $e \doteq 0$, it is sufficient to substitute, e.g., $e \sin \omega = w$ and $e \cos \omega = q$.

The Lagrange equations solve the problem of perturbed motion. Their application, however, requires perturbing function R to be transformed so that its arguments are orbital elements, with respect to which it then becomes easy to compute their derivatives $(\partial R/\partial e_i)$. These will be dealt with in the next chapter.

1.5 Transformation of the Perturbing Gravitational Potential into the Function of the Satellite's Orbital Elements

1.5.1 Transformation of Potential $R_{S\oplus}$

The perturbing potential $R_{S\oplus}$ is a function of the geocentric coordinates of the satellite S (Fig. 1.12), i.e. of radius-vector ϱ and angular coordinates $\vartheta = \frac{1}{2}\pi - \delta$, $\Lambda = -T$. This coordinate system is transformed by rotation about the origin so that axes x_1 and x_2 of the new Cartesian system lie in the plane of the satellite's unperturbed motion in the field of the central forces, with axis x_1 pointing towards the perigee. This transformation can be effected by means of Euler angles α, β, γ (Fig. 1.15); $\alpha = \Omega - \frac{1}{2}\pi, \beta = i, \gamma = \omega + \frac{1}{2}\pi$. By introducing spherical

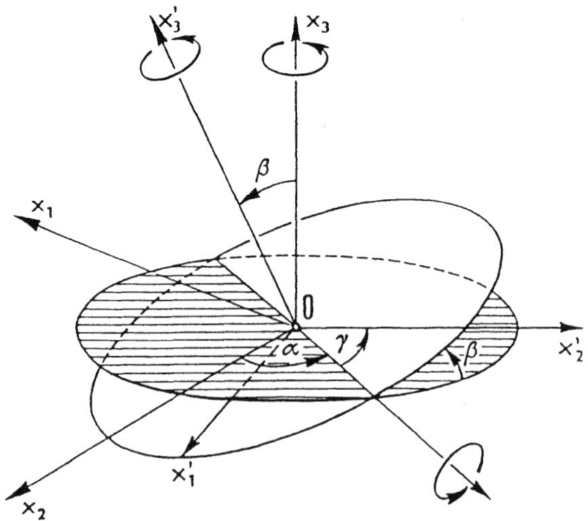

Fig. 1.15. Angular elements of transformation

harmonics $Y_{nk}(\vartheta, \Lambda)$ (2.4) and complex geopotential coefficients A_{nk}, which are related to $J_n^{(k)}$ and $S_n^{(k)}$ by (2.12), the perturbing potential can be modified to read

$$R_{S\oplus} = \frac{GM_\oplus}{\varrho} \sum_{n=2}^{\infty} \left(\frac{a_0}{\varrho}\right)^n \sum_{k=-n}^{n} A_{nk}\, Y_{nk}(\vartheta, \Lambda) . \tag{1.117}$$

The rotation described transforms coefficients A_{nk} according to formula (2.28), and $Y_{nk}(\cos\vartheta)$ can be expressed in terms of associated Legendre functions $P_n^{(k)}(\cos\vartheta)$, (2.4). Figure 1.12 can be used to prove the following for argument $\cos\vartheta$:

$$\cos\vartheta = \sin\delta = \sin i \sin u , \tag{1.118}$$

$$u = \omega + v . \tag{1.119}$$

The expansion for the associated Legendre polynomials is now substituted for $\cos\vartheta$. It is then necessary to express quantities ϱ and v in terms of orbital elements a, M, e. The procedure is tedious and therefore the reader is referred to the results of Kaula (1966) who modified the perturbing potential to read

$$R_{S\oplus} = \frac{GM_\oplus}{a} \sum_{n=1}^{\infty} \sum_{k=0}^{n} \left(\frac{a_0}{a}\right)^n \sum_{r=0}^{n} F_{nkr}(i)$$

$$\times \sum_{s=-\infty}^{\infty} G_{nrs}(e) S_{nkrs}(\omega, M, \Omega, \Theta) ; \tag{1.120}$$

function $F_{nkr}(i)$ is a trigonometrical polynomial in powers of $\sin i$, and $G_{nrs}(e)$ is a function of eccentricity e only,

$$S_{nkrs} = \left.\begin{array}{l} J_n^{(k)} \\ -S_n^{(k)} \end{array}\right\} \begin{array}{l}(n-k)\text{ even} \\ (n-k)\text{ odd}\end{array} \cos\left[(n-2r)\omega + (n-2r+s)M + k(\Omega - \Theta)\right]$$

$$+ \left.\begin{array}{l} S_n^{(k)} \\ J_n^{(k)} \end{array}\right\} \begin{array}{l}(n-k)\text{ even} \\ (n-k)\text{ odd}\end{array} \sin\left[(n-2r)\omega + (n-2r+s)M + k(\Omega - \Theta)\right] ; \tag{1.121}$$

$$R_{S\oplus} = \frac{GM_\oplus}{\varrho} \sum_{n=1}^{\infty} \sum_{k=0}^{n} \left(\frac{a_0}{\varrho}\right)^n J_{n,k} \left\{ \sum_{r=0}^{n} F_{nkr}(i) \sum_{s} G_{nrs} \cos\left[(n-2r)\omega \right.\right.$$

$$\left.\left. + (n-2r+s)M + k(\Omega - \Theta - \Lambda_{n,k})\right] \right\} ; \tag{1.122}$$

$J_{n,k}$ and $\Lambda_{n,k}$ are the amplitude and phase $(J_n^{(k)} + iS_n^{(k)})$. Equation (1.122) indicates that each perturbing harmonic term has the form

$$\delta R_{n,k} = A\left(\frac{a_0}{\varrho}\right)^n J_{n,k} \cos(\dot\varphi t + \varphi_0)$$

$$= A\left(\frac{a_0}{\varrho}\right)^n J_{n,k} \cos(2\pi ft + \varphi_0) ; \tag{1.123}$$

φ_0 stands for the phase angle, f for frequency

$$f = \frac{1}{2\pi}\left[(n-2r)\frac{d\omega}{dt} + (n-2r+s)\frac{dM}{dt} + k\left(\frac{d\Omega}{dt} - \frac{d\Theta}{dt} - \Lambda_{n,k}\right)\right], \qquad (1.124)$$

in which velocities $d\Theta/dt = 2\pi/\text{day}$ in units of mean solar time (angular velocity of the Earth's rotation) and $dM/dt = (2\pi/\text{day})$ times the number of the satellite's orbital periods dominate. Velocities $d\Omega/dt$ and $d\omega/dt$ are smaller in order of magnitude. Polynomials F_{nkr} and G_{nrs} have been tabulated by Kaula up to $n = 4$.

1.5.2 Transformation of Potentials $\Delta V_{S\math{D}}$, $\Delta V_{S\odot}$

The two perturbing potentials are defined by Eqs. (1.85) and (1.86). Since $\varrho/\Delta_{\oplus\mathrm{D}}$ with close Earth's satellites $\sim 1/60$, and $\varrho/\Delta_{\oplus\odot} \sim 4 \times 10^{-5}$, $GM_{\mathrm{D}} = 5 \times 10^{12}\,\mathrm{m}^3\,\mathrm{s}^{-2}$, $GM_\odot = 1.3 \times 10^{20}\,\mathrm{m}^3\,\mathrm{s}^{-2}$, it is sufficient to put $n = 3$ in (1.85) and $n = 2$ in (1.86). We shall simplify them accordingly, omitting the constant spherical terms which become zero when differentiated with respect to e_i, and denote the remaining parts $\Delta R_{S\mathrm{D}}$ and $\Delta R_{S\odot}$:

$$\Delta R_{S\mathrm{D}} = \frac{GM_{\mathrm{D}}}{\Delta_{\oplus\mathrm{D}}}\left[\left(\frac{\varrho}{\Delta_{\oplus\mathrm{D}}}\right)^2 (\tfrac{3}{2}\cos^2 S_{\mathrm{D}} - \tfrac{1}{2})\right.$$

$$\left. + \left(\frac{\varrho}{\Delta_{\oplus\mathrm{D}}}\right)^3 (\tfrac{5}{2}\cos^3 S_{\mathrm{D}} - \tfrac{3}{2}\cos S_{\mathrm{D}})\right], \qquad (1.125)$$

$$\Delta R_{S\odot} = \frac{GM_\odot}{\Delta_{\oplus\odot}}\left(\frac{\varrho}{\Delta_{\oplus\odot}}\right)^2 (\tfrac{3}{2}\cos^2 S_\odot - \tfrac{1}{2}). \qquad (1.126)$$

The error incurred by the simplification is, in the case of (1.125), of the order of 10^{-6} of the principal perturbing term due to the Earth's flattening, i.e. about 10^{-2} of the effect of gravity anomalies which can still be distinguished in satellite orbital dynamics. For (1.126) the corresponding error is three orders of magnitude smaller.

The use of Lagrange equations requires Eqs. (1.125) and (1.126) to be transformed into functions of orbital elements, i.e. transformation of functions $\cos S_{\mathrm{D}}$ and $\cos S_\odot$. They are defined by Eqs. (1.81) and (1.82) which can be altered to read

$$\cos S_{\mathrm{D}} = \sin \delta_{\mathrm{D}}\, \mathrm{P}_1^{(0)}(\sin \delta) + \cos \delta_{\mathrm{D}} \cos T_{\mathrm{D}}\, \mathrm{P}_1^{(1)}(\sin \delta) \cos T$$

$$+ \cos \delta_{\mathrm{D}} \sin T_{\mathrm{D}}\, \mathrm{P}_1^{(1)}(\sin \delta) \sin T, \qquad (1.127)$$

$$\cos S_\odot = \sin \delta_\odot \sin i \sin(\omega + v) + \cos \delta_\odot \cos T_\odot\, \mathrm{P}_1^{(1)}(\sin \delta) \cos T$$

$$+ \cos \delta_\odot \sin T_\odot\, \mathrm{P}_1^{(1)}(\sin \delta) \sin T. \qquad (1.128)$$

To transform the spherical harmonics $P_1^{(1)} (\sin \delta) \cos T$ and $P_1^{(1)} (\sin \delta) \sin T$, it is possible to use directly the transformation relation which follows from (1.56):

$$P_1^{(1)}(\sin \delta) \cos T = \cos(\omega + v) \cos(\Omega - \Theta) - \sin(\omega + v) \sin(\Omega - \Theta) \cos i,$$

$$P_1^{(1)}(\sin \delta) \sin T = - \cos(\omega + v) \sin(\Omega - \Theta) - \sin(\omega + v) \cos(\Omega - \Theta) \cos i.$$

$$(1.129)$$

The effect of the perturbations due to the gravitational potential of the Moon and Sun on the motion of artificial satellites close to the Earth is relatively small. For example, with satellite $1958\,\beta2$ (Kozai 1959) the mean diurnal motion of the nodal line $\Omega = -3.01507° \pm 0.00004°$, the part due to the Moon being $0.00028°$ and that due to the Sun $0.00013°$. Analogously, the mean diurnal motion of the perigee $\dot\omega = 4.40462° \pm 0.00010°$, the perturbations due to the Moon and Sun being $0.00039°$ and $0.00018°$, respectively.

As regards the effects of other bodies of the Solar System, they are at least another three orders of magnitude smaller with artificial Earth satellites and practically need not be considered.

1.6 Fundamentals of the Theory of Determining the Parameters of the Earth's Gravitational Potential by Satellite Methods

1.6.1 Motion of the Nodal Line due to the Earth's Polar and Equatorial Flattening

Once the perturbing potential has been transformed into the function of osculating orbital elements, it is easy to compute the derivatives $\dfrac{\partial R}{\partial e_i}$, $i = 1, 2, \ldots, 6$, and after substituting them into the Lagrange equations, also to compute the time derivatives $\dot e_i$. This cannot be done here in full. We shall only outline the method by computing perturbation $\dot\Omega$, i.e. the motion of the nodal line $\overline{O\Omega}$, but only that part of it generated by geopotential coefficients of the second degree $(J_2^{(0)}, J_2^{(2)}, S_2^{(2)})$, i.e. by the polar and equatorial flattening of the Earth. We shall denote the perturbation being derived $\dot\Omega_{n=2}$ and the appropriate part of the perturbing potential $R_{n=2}$. In view of (1.117)

$$R_{n=2} = \frac{GM_\oplus}{a} \left(\frac{a_0}{a}\right)^2 \left[\frac{1 + e \cos v}{1 - e^2}\right]^3 (F_2^{(0)} + F_2^{(2)}),$$

$$(1.130)$$

$$F_2^{(0)} = J_2^{(0)}\{ -\tfrac{1}{2} + \tfrac{3}{4} \sin^2 i[1 + \cos 2(\omega + v)]\},$$

$$(1.131)$$

$$F_2^{(2)} = \tfrac{3}{4} J_2^{(2)} \{\cos 2[(\omega + v) + (\Omega - \Theta)](1 + \cos i)^2$$

$$+ 2\cos 2(\Omega - \Theta)\sin^2 i + \cos 2[(\omega + v) - (\Omega - \Theta)](1 - \cos i)^2\}$$

$$+ \tfrac{3}{4} S_2^{(2)} \{\sin 2[(\omega + v) + (\Omega - \Theta)](1 + \cos i)^2$$

$$+ 2\sin 2(\Omega - \Theta)\sin^2 i$$

$$- \sin 2[(\omega + v) - (\Omega - \Theta)](1 - \cos i)^2\}$$

$$= \tfrac{3}{2} J_2^{(2)}[\cos 2(\omega + v)\cos 2(\Omega - \Theta)(1 + \cos^2 i)$$

$$+ \cos 2(\Omega - \Theta)\sin^2 i - 2\sin 2(\omega + v)\sin 2(\Omega - \Theta)\cos i]$$

$$+ \tfrac{3}{2} S_2^{(2)}[\cos 2(\omega + v)\sin 2(\Omega - \Theta)(1 + \cos^2 i)$$

$$+ \sin 2(\Omega - \Theta)\sin^2 i + 2\sin 2(\omega + v)\cos 2(\Omega - \Theta)\cos i]\,, \qquad (1.132)$$

and, therefore, with regard to (1.105)

$$\dot{\Omega}_{n=2} = \frac{3}{2}\frac{a_0^2 (GM_\oplus)^{1/2}}{[a(1 - e^2)]^{7/2}}(1 + e\cos v)^3\, F_2(i, \omega, \Omega - \Theta, v)\,, \qquad (1.133)$$

$$F_2(i, \omega, \Omega - \Theta, v) = J_2^{(0)}\cos i[1 + \cos 2(\omega + v)]$$

$$+ J_2^{(2)}\{-\cos[2(\omega + v) + 2(\Omega - \Theta)](1 + \cos i)$$

$$+ 2\cos 2(\Omega - \Theta)\cos i + \cos[-2(\omega + v) + 2(\Omega - \Theta)]$$

$$\times (1 - \cos i)\} + S_2^{(2)}\{-\sin[2(\omega + v) + 2(\Omega - \Theta)](1 + \cos i)$$

$$+ 2\sin 2(\Omega - \Theta)\cos i + \sin[-2(\omega + v) + 2(\Omega - \Theta)]$$

$$\times (1 - \cos i)\}\,. \qquad (1.134)$$

Function (1.134) contains variations of a different nature (Fig. 1.16):
(a) secular, (b) long-period and (c) short-period. The secular vary linearly with

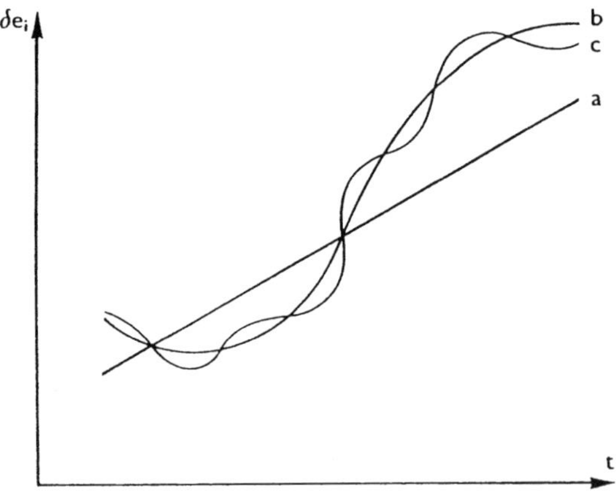

Fig. 1.16. Schematic representation of the character of variations of satellite orbital elements. *a* secular; *b* long-period; *c* short-period

time and can be separated, for example, by integrating (1.134) within the interval of one orbit, i.e. $[v = 0, v = 2\pi]$:

$$\frac{1}{2\pi} \int_0^{2\pi} \frac{d\Omega_{n=2}}{dv}\, dv$$

$$= \frac{1}{2\pi} \int_0^{2\pi} \dot{\Omega}_{n=2} \frac{dt}{dv}\, dv$$

$$= \frac{1}{2\pi} (GM_\oplus)^{-1/2} \int_0^{2\pi} \dot{\Omega}_{n=2} [a(1 - e^2)]^{3/2} (1 + e\cos v)^{-2}\, dv$$

$$= \tfrac{3}{2} J_2^{(0)} \left(\frac{a_0}{a}\right)^2 (1 - e^2)^{-2} \cos i . \tag{1.135}$$

Since $J_2^{(0)} = -1082.63 \times 10^{-6} < 0$, the secular term (given $i < 90°$) in $\dot{\Omega}$ is negative, and the nodal line moves from east to west, i.e. against the direction of the Earth's rotation (Fig. 1.17). If $i > 90°$ the sense of rotation is the opposite. As regards satellites close to the Earth, the motion of the nodal line amounts to $\sim 4°/\text{day}$ given an orbital inclination $i \sim 60°$; if $i \sim 0°$ it is largest, as much as $\sim 8°/\text{day}$; it is zero when $i = 90°$.

The secular variations due to geopotential coefficients, given any other zonal spherical harmonics, can be derived in a similar manner. With tesseral and sectorial functions geopotential coefficients do not generate any secular motion, but they do generate variations of a periodic nature. For their description and

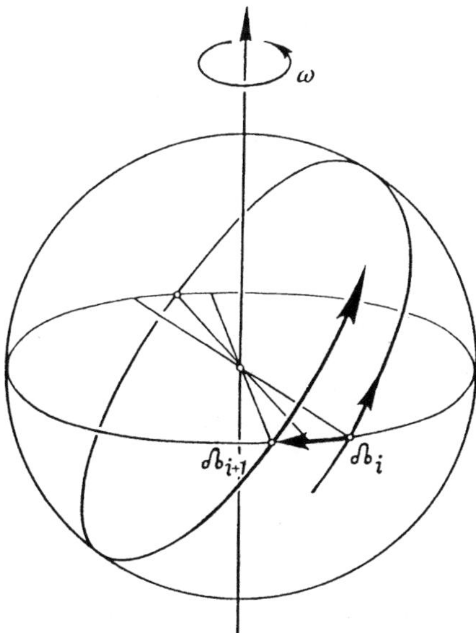

Fig. 1.17. Schematic representation of the regression of the nodal line of the satellite's orbital plane

the way in which the time variations $\dot{e}_i$ of the other orbital elements are derived, we refer the reader to the literature (e.g. Aksenov 1977).

As indicated by (1.121), reliable determination of the amplitude and phase shift in $\dot{e}_i$ requires observations which are distributed well with respect to argument $[(n - 2r)\omega + k(\Omega - \Theta)]$. This distribution requires, above all, suitable localization of satellite stations in geographic latitude.

1.6.2 Geopotential Coefficients Determined from the Variation in Satellite Orbital Elements – An Outline. Numerical Results

A thorough and quite general analysis of the time variations in satellite orbital elements from a uniform point of view was made by Kaula, to whose solution we refer the reader (Kaula 1966).

If the positions of a satellite are observed for a sufficiently long time, so that the variations of orbital elements $\dot{e}_i$ can be determined numerically, equations of the type

$$\dot{e}_i = \sum_{n=2}^{\bar{n}} K_n^{(0)} J_n^{(0)} + \sum_{n=2}^{\bar{n}} \sum_{k=1}^{n} K_n^{(k)} J_n^{(k)} + \sum_{n=2}^{\bar{n}} \sum_{k=1}^{n} L_n^{(k)} S_n^{(k)} , \qquad (1.136)$$

which determine the set of geopotential coefficients $J_n^{(0)}$, $J_n^{(k)}$, $S_n^{(k)}$ up to degree $\bar{n}$, can be formed. Coefficients $K_n^{(0)}$, $K_n^{(k)}$, $L_n^{(k)}$ are functions of the orbital elements. In general, it is also necessary to consider terms $\sum K_{n,m} J_n^{(k)} J_m^{(j)}$, i.e. the terms containing the squares or products of geopotential coefficients. These terms appeared in the expressions for e_i due to iterative integration. With bodies of the Solar System the largest perturbation is $J_2^{(0)}$, generated by the polar flattening of the body, and of the quadratic terms it is necessary to consider $(J_2^{(0)})^2$.

In general, the separate harmonic terms in the expression for $\dot{e}$ contain factor $\dfrac{GM_\oplus}{a} \left(\dfrac{a_0}{a} \right)^n$, which indicates that a more accurate determination of the geopotential coefficients would require the semi-axis a of the satellite to be small. However, if the central body which the satellite is orbiting has an atmosphere, the length of the perigee radius-vector $a(1 - e)$ has to be larger in view of the perturbing decelerating effects of the lower layers of the atmosphere, in other words, a larger semi-axis and smaller eccentricity are required. Generally speaking, a perigee height at which the decelerating effects of the atmosphere would be smaller than the effect of solar radiation, i.e. in the interval $\sim 800\text{--}1100$ km, would be optimal. However, it is now possible to introduce corrections for the effect of solar radiation with relatively high accuracy, and geodynamic satellites, whose principal programme is to determine the geopotential coefficients of the Earth's gravitational potential, now orbit at distances of several thousands of kilometres. For example, the height of the perigee of the geodynamic satellite LAGEOS (launched 4 May 1976; a sphere of diameter 60 cm; relatively large weight of 411 kg; 426 corner reflectors for laser ranging on the surface) was 5800 km at the time of launching. Its orbit is known with an accuracy of a few centimetres.

This method was used to derive the geopotential coefficients not only of the Earth ($\bar{n} = 180$), but also of the Moon ($\bar{n} = 16$), Mars ($\bar{n} = 18$), Jupiter ($\bar{n} = 6$) and Venus ($\bar{n} = 18$). Priority in deriving coefficient $J_2^{(0)}$ is due to Buchar who analysed the orbits of the first two satellites (Sputniks I and II) (Buchar 1958). The motion of the nodal lines of these satellites was faster than theoretically assumed, because the internationally accepted value of the polar flattening (and coefficient $J_2^{(0)}$ derived from it) included an error of about 0.35%.

Coefficient $J_3^{(0)}$ and, consequently, also the equatorial asymmetry of the geoid were first derived in analysing the orbit of the Vanguard satellite (O'Keefe et al. 1959). Table 1.2 shows that the amplitude of the third zonal harmonic term $J_3^{(0)}$ is about 16 m and that it is commensurable with the amplitudes of some of the other terms. Therefore, to speak of the Earth as being 'pear-shaped', which was the case after the orbit analysis of the Vanguard 1 satellite (O'Keefe et al. 1959), is inappropriate even if all odd harmonic terms and the fact that the geoid surface has a depression in the region of Antarctica are considered (see Sect. 2.10 and Figs. 2.10 and 2.11). In Chapter 2 the fully normalized coefficients are denoted $\bar{C}_{nk}, \bar{S}_{nk}$; therefore, $\bar{J}_n^{(k)} = \bar{C}_{nk}$ and $\bar{S}_n^{(k)} = \bar{S}_{nk}$.

Tables 1.2 and 1.3 give the numerical values of the geopotential coefficient up to $\bar{n} = 12$, appropriate to model GEM-T2 ($\bar{n} = 36$, $k = 36$) (Marsh et al. 1990). This model was derived from laser tracking of 11 satellites, especially geodynamic satellites, whose orbits are determined with a radial accuracy of ± 10 cm. Doppler observations and other techniques were also used. A total of 2 386 000 observations from 1130 arcs were processed. The error in the derived geopotential coefficients ranges from $\pm 5 \times 10^{-10}$ for low harmonics ($n = 2$), where it is the smallest, to $\pm 65 \times 10^{-10}$ for the higher harmonics ($n = 20$); once

Table 1.2. GEM-T2 geopotential (Stokes') zonal parameters: conventional $J_n^{(0)}$; fully normalized $\bar{J}_n^{(0)} = \bar{C}_n$ (rounded values)

n	$J_n^{(0)}$ (10^{-6})	$\bar{J}_n^{(0)}$ (10^{-6})
2	-1082.6265	-484.1653
3	2.5321	0.9570
4	1.6197	0.5399
5	0.2278	0.0687
6	-0.5394	-0.1496
7	0.3489	0.0901
8	0.1995	0.0484
9	0.1240	0.0284
10	0.2519	0.0550
11	-0.2491	-0.0519
12	0.1705	0.0341

Table 1.3. GEM-T2 geopotential (Stokes') tesseral and sectorial parameters: fully normalized $\bar{J}_n^{(k)} = \bar{C}_{l,m}$, $\bar{S}_n^{(k)} = \bar{S}_{l,m}$ (rounded values)

n	k	$\bar{J}_n^{(k)}\ (10^{-6})$	$\bar{S}_n^{(k)}\ (10^{-6})$	n	k	$\bar{J}_n^{(k)}\ (10^{-6})$	$\bar{S}_n^{(k)}\ (10^{-6})$
2	1	− 0.0017	0.0012	9	3	− 0.1613	− 0.0857
	2	2.4390	− 1.4001		4	− 0.0122	0.0259
					5	− 0.0241	− 0.0575
3	1	2.0308	0.2496		6	0.0667	0.2234
	2	0.9035	− 0.6190		7	− 0.1229	− 0.0951
	3	0.7215	1.4137		8	0.1882	− 0.0037
					9	− 0.0613	0.0970
4	1	− 0.5353	− 0.4741				
	2	0.3483	0.6640	10	1	0.0832	− 0.1356
	3	0.9913	− 0.2014		2	− 0.0819	− 0.0501
	4	− 0.1894	0.3090		3	− 0.0041	− 0.1604
					4	− 0.0939	− 0.0688
5	1	− 0.0608	− 0.0950		5	− 0.0489	− 0.0458
	2	0.6561	− 0.3241		6	− 0.0345	− 0.0784
	3	− 0.4519	− 0.2171		7	0.0090	− 0.0022
	4	− 0.2950	0.0514		8	0.0422	− 0.0929
	5	0.1719	− 0.6691		9	0.1244	− 0.0389
					10	0.0966	− 0.0189
6	1	− 0.0771	0.0253				
	2	0.0524	− 0.3752	11	1	0.0188	− 0.0302
	3	0.0584	0.0069		2	0.0126	− 0.0920
	4	− 0.0888	− 0.4711		3	− 0.0310	− 0.1318
	5	− 0.2661	− 0.5369		4	− 0.0363	− 0.0702
	6	0.0097	− 0.2370		5	0.0405	0.0584
					6	− 0.0022	0.0280
7	1	0.2819	0.0962		7	0.0033	− 0.0874
	2	0.3210	0.0957		8	− 0.0059	0.0238
	3	0.2524	− 0.2096		9	− 0.0401	0.0433
	4	− 0.2742	− 0.1242		10	− 0.0530	− 0.0214
	5	− 0.0001	0.0197		11	0.0455	− 0.0646
	6	− 0.3586	0.1516				
	7	− 0.0015	0.0253	12	1	− 0.0543	− 0.0442
					2	0.0067	0.0318
8	1	0.0246	0.0584		3	0.0404	0.0176
	2	0.0695	0.0673		4	− 0.0633	− 0.0046
	3	− 0.0166	− 0.0871		5	0.0373	0.0044
	4	− 0.2431	0.0670		6	− 0.0022	0.0428
	5	− 0.0236	0.0871		7	− 0.0160	0.0349
	6	− 0.0649	0.3102		8	− 0.0234	0.0148
	7	0.0689	0.0747		9	0.0422	0.0237
	8	− 0.1214	0.1207		10	− 0.0084	0.0320
					11	0.0098	− 0.0084
9	1	0.1421	0.0254		12	− 0.0052	− 0.0111
	2	0.0285	− 0.0350				

$n > 20$ the error decreases slightly, and for $\bar{n} = 36$ it is about $\pm 40 \times 10^{-10}$ (Fig. 1.18). The GEM-T2 model enables the external equipotential surface of geopotential to be expressed globally with an accuracy of about ± 1 m. Table 1.4 illustrates the orbit accuracy of geodetic and geodynamic satellites (Marsh et al. 1990). The geopotential coefficients of the first degree here are a priori zero since the origin has been located in the Earth's centre of mass: $J_0^{(0)} = 1$.

The coefficients marked with a bar correspond to fully normalized spherical harmonics $\bar{P}_n^{(k)}(\sin\phi)\cos k\Lambda$, $\bar{P}_n^{(k)}(\sin\phi)\sin\Lambda$, for which the following holds:

$$\frac{1}{4\pi} \int\limits_0^{2\pi} \int\limits_{-\pi/2}^{\pi/2} \left[\bar{P}_n^{(k)}(\sin\phi) \frac{\cos k\Lambda}{\sin k\Lambda} \right]^2 \cos\phi \, d\phi \, d\Lambda = 1 \, ,$$

$$\bar{P}_n^{(k)}(\sin\phi) = N_n^{(k)} P_n^{(k)}(\sin\phi) \, , \tag{1.137}$$

$$N_n^{(k)} = \left[\frac{[2 - \delta(0, k)](2n + 1)(n - k)!}{(n + k)!} \right]^{1/2} \, . \tag{1.138}$$

Table 1.4. Orbit accuracy of satellites used for geodetic purposes. (Marsh et al. 1990)

Geopotential model	LAGEOS (m)	AJISAI (m)	STRLT (m)	BB–C (m)	GEOS1 (m)	GEOS2 (m)	GEOS3 (m)	NOVA (cm/s)
GEM-9 (1979)	0.333	0.951	1.16	0.873	1.26	1.18	1.72	0.95
GEM-L2 (1981)	0.199	0.797	1.00	0.893	1.07	1.09	1.87	0.79
GEM-T1 (1987)	0.069	0.181	0.172	0.396	0.387	0.655	0.693	0.44
GEM-T2 (1989)	0.066	0.151	1.102	0.334	0.316	0.667	0.249	0.37
Noise floor	0.038	0.038	0.040	0.102	0.206	0.343	0.101	0.335

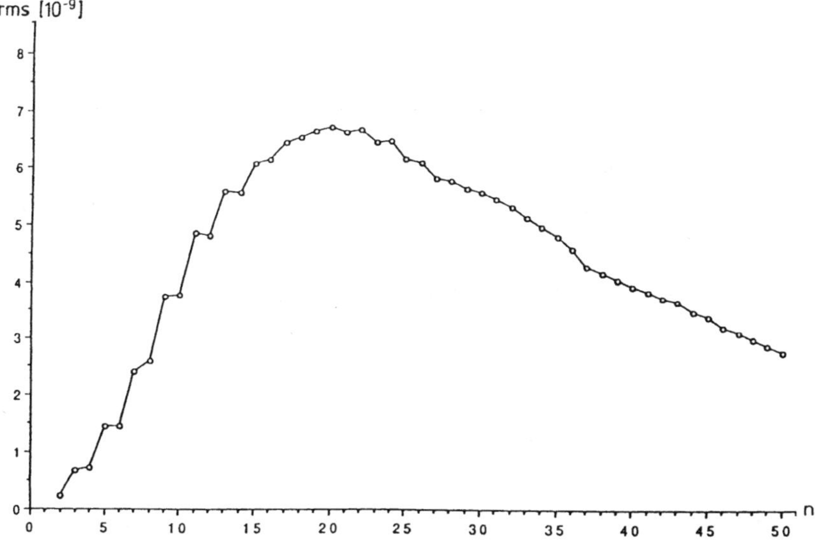

Fig. 1.18. Root-mean-square error of GEM-T2 geopotential coefficients as a function of harmonics degree n. (Marsh et al. 1990)

1.7 The Geocentric Gravitational Constant

In the presatellite era, the geocentric gravitational constant, the product of Newton's universal constant of gravitation G and the Earth's mass $M_\oplus$, was known with relatively very low accuracy; its current determinations from orbit analyses of space probes, artificial Earth satellites and lunar laser ranging enable it to be used to define the linear dimension of the Earth. The practically unsurmountable complications of the presatellite theory of the Earth's gravitational and gravity field were, to a large extent, caused by the inaccuracy in $GM_\oplus$, which had to be 'evaded'. However, Newton's constant G has still not been determined accurately by laboratory measurements. Its most probable value now is (IERS Standards 1989)

$$G = (6.672\,59 \pm 0.00030) \times 10^{-11}\,\mathrm{m}^3\,\mathrm{s}^{-2}\,\mathrm{kg}^{-1}\,. \tag{1.139}$$

In the celestial mechanics of natural celestial bodies the Gaussian gravitational constant k is used traditionally:

$$k = 0.017\,202\,098\,95$$

$$(k/86\,400 = 1.990\,983\,675 \times 10^{-7})\,; \tag{1.140}$$

it is given by definition and relates the heliocentric gravitational constant of the Sun $GM_\odot$ (in $\mathrm{m}^3\,\mathrm{s}^{-2}$), the astronomical unit of length $A_\odot$ (in m) and the unit of time as

$$GM_\odot = A_\odot^3 \left(\frac{k}{86\,400} \right)^2 , \tag{1.141}$$

which follows from Kepler's third law:

$$\left(\frac{2\pi}{T} \right)^2 A_\odot^3 = GM_\odot\,. \tag{1.142}$$

In (1.142) $GM_\odot$ is in $\mathrm{m}^3\,\mathrm{s}^{-2}$, $A_\odot$ in m and T in s;

$$GM_\odot = 13\,271\,244.0 \times 10^{13}\,\mathrm{m}^3\,\mathrm{s}^{-2}\,, \tag{1.143}$$

$$A_\odot = 1.495\,978\,70 \times 10^{11}\,\mathrm{m}\,. \tag{1.144}$$

If T is the length of year expressed in days (T_d),

$$\left(\frac{2\pi}{T_\mathrm{d}\,86\,400} \right)^2 A_\odot^3 = GM_\odot\,, \tag{1.145}$$

and by comparing (1.145) with (1.141)

$$k = \frac{2\pi}{T_\mathrm{d}}\,. \tag{1.146}$$

The Gaussian constant thus represents the mean diurnal motion (in radians) of a particle orbiting the Sun at distance $A_\odot$ (1 AU).

The heliocentric gravitational constant (1.143) has been determined from the orbit dynamics of planets and space probes. Given the Gaussian constant (1.140) the astronomical unit $A_\odot$ can be determined from (1.141), and its accuracy can be improved as the accuracy of the dynamic determination of the heliocentric gravitational constant (1.143) increases.

The Gaussian constant, which Newton used (with a smaller number of decimal places: 0.017 202 12) before Gauss, has an advantage if the units in astronomical computations are $A_\odot = 1$ (length), the length of day $= 1$ (time), and $M_\odot = 1$ (mass). Then

$$\left(\frac{2\pi}{T_\mathrm{d}}\right)^2 A_\odot^3 = k^2 M_\odot \,, \tag{1.147}$$

i.e. (1.146) holds with $A_\odot = 1$ and $M_\odot = 1$.

The geocentric gravitational constant occurs in equations of motion (1.88), which yield it in the form

$$GM_\oplus = -\frac{\varrho^3}{x_j}\left(\ddot{x}_j - \frac{\partial R}{\partial x_j}\right). \tag{1.148}$$

However, it also occurs in perturbing function R, in its part $R_{S\oplus}$, generated by geopotential coefficients. In the interest of achieving high accuracy, therefore, it is desirable to consider the effects of these perturbations and the effect of perturbations of non-gravitational origin as precisely as possible. This is easier with satellites and space probes moving above the denser layers of the atmosphere and at distances from the Earth at which the effects of perturbations (geopotential coefficients) of a higher degree are practically zero, or relatively very small. However, perturbations due to third bodies must be taken into account. In this respect, lunar laser ranging can be used to advantage.

$GM_\oplus$ can also be derived from Kepler's third law, according to which [see (1.49)]

$$GM_\oplus = \left(\frac{2\pi}{T}\right)^2 a^3 = n^2 a^3 \,, \tag{1.149}$$

but this requires the instantaneous semi-axis of the satellite's osculating ellipse and its period T to be known precisely.

Besides Kepler's third law, also the energy integral (1.30), generalized for the real, i.e. spherically asymmetrical gravitational field of the Earth, can be used for this purpose. It is then necessary to measure the length of geocentric radius-vector ϱ and velocity $v = \left(\sum_j \dot{x}_j^2\right)^{1/2}$ of the space probe at a distance sufficiently far from the Earth and for at least two positions ϱ_1, δ_1, T_1; ϱ_2, δ_2, T_2. If no other perturbations, whether of gravitational or non-gravitational origin, exist besides the perturbations due to the Earth's gravitational field, then the following must hold at both positions ($i = 1, 2$):

$$\frac{1}{2} M_S v_i^2 - G \frac{M_S M_\oplus}{\varrho_i} \left[1 + \sum_{n=2}^{\infty} \sum_{k=0}^{n} \left(\frac{\varrho_0}{\varrho_i} \right)^n \right.$$

$$\left. \times (J_n^{(k)} \cos kT_i - S_n^{(k)} \sin kT_i) P_n^{(k)} (\sin \delta_i) \right] = H . \tag{1.150}$$

The difference then yields this relation for the geocentric gravitational constant:

$$GM_\oplus = \frac{1}{2} (v_2^2 - v_1^2) \left\{ \frac{1}{\varrho_2} \left[1 + \sum_{n=2}^{\infty} \sum_{k=0}^{n} \left(\frac{a_0}{\varrho_2} \right)^n \right. \right.$$

$$\left. \times (J_n^{(k)} \cos kT_2 - S_n^{(k)} \sin kT_2) P_n^{(k)} (\sin \delta_2) \right]$$

$$- \frac{1}{\varrho_1} \left[1 + \sum_{n=2}^{\infty} \sum_{k=0}^{n} \left(\frac{a_0}{\varrho_1} \right)^n \right.$$

$$\left. \left. \times (J_n^{(k)} \cos kT_1 - S_n^{(k)} \sin kT_1) P_n^{(k)} (\sin \delta_1) \right] \right\}^{-1} . \tag{1.151}$$

If the values of ϱ are large, the effect of geopotential coefficients $J_n^{(k)}$ and $S_n^{(k)}$ will be weaker the further the probe is from the Earth and the higher the degree n of the harmonic terms in the expansion of the gravitational potential.

The most probable value of the geocentric gravitational constant now is (Ries et al. 1989)

$$GM_\oplus = (398\,600.440 \pm 0.001) \times 10^9 \ \mathrm{m^3 \, s^{-2}} . \tag{1.152}$$

This is a weighted mean of a number of values derived by different satellite methods (Table 1.5).

Table 1.5. Values of the geocentric gravitational constant determined by methods of space geodesy

Method	Authors	Year	GM $(10^9\,\mathrm{m^3\,s^{-2}})$	Standard error $(10^9\,\mathrm{m^3\,s^{-2}})$
Satellite; series of satellites	Lerch et al.	1978	398 600.5	$\pm\,0.04$
LAGEOS satellite	Smith et al.	1985	398 600.434	$\pm\,0.002$
Satellite; series of satellites	Marsh et al.	1985	398 600.434	$\pm\,0.005$
LAGEOS satellite	Tapley et al.	1985	398 600.440	$\pm\,0.002$
Lunar laser ranging	Newhall et al.	1987	398 600.443	$\pm\,0.006$
Satellite; series of satellites	Ries et al.	1989	398 600.4409	$\pm\,0.0009$
LAGEOS satellite	Ries et al.	1989	398 600.4405	$\pm\,0.0011$
Adopted value	Ries et al.	1989	398 600.4405	$\pm\,0.001$

For comparison we shall also give the values of the heliocentric $GM_\odot$, selenocentric GM_{D}, areocentric $GM_{\mathrm{\delta}}$, and afroditocentric GM_{Q} gravitational constants:

$$GM_\odot = 13\,271\,244.0 \times 10^{13}\ \mathrm{m}^3\,\mathrm{s}^{-2}\,,$$

$$GM_{\mathrm{D}} = 4902.799 \times 10^9\ \mathrm{m}^3\,\mathrm{s}^{-2}\,,$$

$$GM_{\mathrm{\delta}} = 42\,828.2 \times 10^9\ \mathrm{m}^3\,\mathrm{s}^{-2}\,,$$

$$GM_{\mathrm{Q}} = 324\,858.8 \times 10^9\ \mathrm{m}^3\,\mathrm{s}^{-2}\,. \tag{1.153}$$

These, together with (1.152), yield relatively very accurate mass ratios:

$$\frac{M_\odot}{M_\oplus} = 332\,946.04 \pm 0.01\,,$$

$$\frac{M_\odot}{M_\oplus + M_{\mathrm{D}}} = 328\,900.55 \pm 0.01\,,$$

$$\frac{M_\odot}{M_{\mathrm{\delta}}} = 3\,098\,716 \pm 5\,,$$

$$\frac{M_\oplus}{M_{\mathrm{D}}} = 81.300\,57 \pm 0.000\,02\,,$$

$$\frac{M_\odot}{M_{\mathrm{Q}}} = 408\,523.5 \pm 0.1\,. \tag{1.154}$$

1.8 Resonance Phenomena

As satellites orbit, resonance phenomena may occur due to some of the harmonic terms in the expansion of the perturbing gravitational potential (1.71). These phenomena may in turn be used to derive these terms or their functions.

The problem was solved quite generally by Kaula (1966) and developed in detail by Allan (1971); the occurrence of a resonance phenomenon can be determined immediately by function (1.120). In general, this phenomenon will occur if the arguments in function (1.121) satisfy the condition

$$(n - 2r)\frac{\mathrm{d}\omega}{\mathrm{d}t} + (n - 2r + s)\frac{\mathrm{d}M}{\mathrm{d}t} + k\frac{\mathrm{d}(\Omega - \Theta)}{\mathrm{d}t} = 0 \tag{1.155}$$

or, to simplify matters, because variations $\mathrm{d}\omega/\mathrm{d}t$ and $\mathrm{d}\Omega/\mathrm{d}t$ are small compared with $\mathrm{d}M/\mathrm{d}t$ and $\omega_\oplus$,

$$(n - 2r + s)\frac{\mathrm{d}M}{\mathrm{d}t} - k\omega_\oplus \doteq 0\,; \tag{1.156}$$

$\dfrac{\mathrm{d}\Theta}{\mathrm{d}t} = \omega_\oplus$ is the angular velocity of the Earth's rotation and [see (1.46)]

$$\frac{\mathrm{d}M}{\mathrm{d}t} = \frac{2\pi}{T} = n = \left[\frac{GM_\oplus}{a^3}\right]^{1/2}. \tag{1.157}$$

The large time variation in the mean anomaly is, in general, approximately compensated in this case by the term which depends on the Earth's angular velocity of rotation, and which, with regard to determining geopotential coefficients, is advantageous. For example, $\dfrac{\mathrm{d}M}{\mathrm{d}t} : \dfrac{\mathrm{d}\Theta}{\mathrm{d}t} = 13$ for altitudes of 800 to 1200 km and small eccentricity and, if $n - 2r + s = 1$, the resonating harmonic terms are of order $k = 13$. This applies, for example, to geopotential coefficients $J_{13}^{(13)}$, $S_{13}^{(13)}$, $J_{15}^{(13)}$, $S_{15}^{(13)}$, $J_{17}^{(13)}$, $S_{17}^{(13)}$, etc. Analogously, if the Earth rotates through an angle of $24°$ during one satellite orbital period, terms with $k = 15$, i.e. of the 15th order, will resonate.

If the satellite is more remote, the ratio $\dfrac{\mathrm{d}M}{\mathrm{d}t} : \dfrac{\mathrm{d}\Theta}{\mathrm{d}t}$ is much smaller; however, the attenuation factor, $\dfrac{GM_\oplus}{a}\left(\dfrac{a_0}{a}\right)^n$ diminishes the zone in which the resonance phenomenon may occur. The special case in which this ratio is equal to unity refers to a geostationary satellite with a period of 24 h (see Sect. 1.9).

Of special significance is the case in which

$$\frac{\mathrm{d}\omega}{\mathrm{d}t} + \frac{\mathrm{d}M}{\mathrm{d}t} + \frac{\mathrm{d}\Omega}{\mathrm{d}t} - \frac{\mathrm{d}\Theta}{\mathrm{d}t} = 0, \tag{1.158}$$

since resonance then occurs for every harmonic term of degree n and order k for which the argument in (1.121) represents an integral multiple of the sum $(\omega + M + \Omega - \Theta)$; this occurs if

$$s = 0, \quad k = n - 2r, \tag{1.159}$$

i.e. if the difference between the $(n - k)$th degree and order of the spherical harmonic is an even number, and if $r = \tfrac{1}{2}(n - k)$.

Under conditions (1.159) Eq. (1.120) will yield an expression for the part of the perturbing potential $(R_{S\oplus})_{\mathrm{reson}}$ whose harmonic terms may generate a resonance phenomenon:

$$(R_{S\oplus})_{\mathrm{reson}} = \frac{GM_\oplus}{a} \sum_{\substack{n=1 \\ (n-k)\ \mathrm{even}}}^{\infty} \sum_{k=0}^{n} \left(\frac{a_0}{a}\right)^n F_{nk\frac{n-k}{2}}(i)\, G_{n\frac{n-k}{2}0}(e)\, \phi_{n,k}, \tag{1.160}$$

$$\phi_{n,k} = [J_n^{(k)} \cos k(\omega + M + \Omega - \Theta) + S_n^{(k)} \sin k(\omega + M + \Omega - \Theta)]$$

$$= J_{n,k} \cos k(\omega + M + \Omega - \Theta - \Lambda_{n,k}). \tag{1.161}$$

On substituting this into the Lagrange equations, we obtain the perturbations of orbital elements quite easily and from there on the procedure is similar to that in Section 1.6.2.

We would have to deal with a more general case if only part of condition (1.159) were satisfied: $s = 0$; then

$$(n - 2r)\left(\frac{d\omega}{dt} + \frac{dM}{dt}\right) = k\,\frac{d(\Theta - \Omega)}{dt}.$$ (1.162)

This means that $(n - 2r)$ nodal periods of the satellite are equal to k rotational periods of the Earth, reckoned with respect to the perturbed nodal line of the satellite orbit. Quantity $\alpha(\dot\omega + \dot M) + \beta(\dot\Omega - \dot\Theta)$ is called the 'resonance variable'; $\alpha = n - 2r$, $\beta = k$. From a practical point of view the case, defined by (1.162), is more significant, because (1.158) practically defines the geostationary orbit where the effect of most geopotential coefficients is subject to considerable attenuation.

The behaviour of a satellite close to resonance can be described by the equation of motion of a mathematical pendulum in which the role of the amplitude is played by the longitude of the stroboscopic mean node,

$$\lambda_N - \frac{\alpha}{\beta}\left[(M + \omega) + (\Omega - \Theta)\right].$$

A long-period perturbation is generated, and its period is longer the more accurately the resonance condition is satisfied.

Although the 'resonance methods' of determining geopotential coefficients are in fact restricted considerably, they can contribute to partial assessment of the accuracy of the sets of geopotential coefficients $J_n^{(k)}, S_n^{(k)}$, derived from the variations in orbital elements.

1.9 Geostationary Satellites

Geosynchronous satellites, whose orbital period T_S is such that

$$\frac{2\pi}{T_S} = \omega_\oplus,$$ (1.163)

where $\omega_\oplus = d\Theta/dt$ is the angular velocity of the Earth's rotation, have a quite special status with regard to geodynamic problems.

A special case of geosynchronous satellites are geostationary satellites whose orbit, in the ideal case, is circular and lies in the plane of the Earth's equator; they orbit in the direction of the Earth's rotation and their sidereal rotation period is equal to $(86\,164.0989 + 0.0015\,T)\,s$, T being expressed in centuries as of epoch 1900.0.

It is convenient to describe their motion in the geocentric spherical system, ϱ, ϕ, Λ, fixed with the Earth, i.e. in a non-inertial system. It is then, of course, necessary to consider not only the force of gravitation but also the centrifugal force and Coriolis force, i.e. acceleration $[2\boldsymbol{\omega}_\oplus \times \mathbf{v}]$, where $\boldsymbol{\omega}_\oplus$ is the vector of instantaneous rotation of the Earth and $\mathbf{v}$ the satellite velocity vector.

The generalized coordinates q_r in the Lagrange equations of motion of the second kind, (1.18), are now replaced by quantities ϱ, ϕ, Λ. The Lagrangian of the problem in this system now reads

$$L(\varrho, \phi, \Lambda) = T(\varrho, \phi, \Lambda) + W(\varrho, \phi, \Lambda) + \varrho^2 \dot{\Lambda} \omega_\oplus \cos^2 \phi \,, \tag{1.164}$$

where $T(\varrho, \phi, \Lambda)$ is the kinetic energy of the satellite ($M_S = 1$).

$$T(\varrho, \phi, \Lambda) = \tfrac{1}{2}(\dot{\varrho}^2 + \varrho^2 \dot{\phi}^2 + \varrho^2 \dot{\Lambda}^2 \cos^2 \phi) \,, \tag{1.165}$$

and $W(\varrho, \phi, \Lambda)$ is the Earth's gravity potential. This yields

$$\ddot{\varrho} - \varrho \dot{\phi}^2 - \varrho \dot{\Lambda}^2 \cos^2 \phi = \frac{\partial W(\varrho, \phi, \Lambda)}{\partial \varrho} + 2\varrho \dot{\Lambda} \omega_\oplus \cos^2 \phi \,,$$

$$\ddot{\phi} \varrho^2 + 2\dot{\phi} \varrho \dot{\varrho} + \varrho^2 \dot{\Lambda}^2 \sin \phi \cos \phi = \frac{\partial W(\varrho, \phi, \Lambda)}{\partial \phi} - 2\varrho^2 \omega_\oplus \dot{\Lambda} \sin \phi \cos \phi \,,$$

$$\ddot{\Lambda} \varrho^2 \cos^2 \phi + 2\dot{\Lambda} \varrho \dot{\varrho} \cos^2 \phi - \dot{\Lambda} \varrho^2 \dot{\phi} \sin 2\phi$$

$$= \frac{\partial W(\varrho, \phi, \Lambda)}{\partial \Lambda} - \frac{\mathrm{d}}{\mathrm{d}t}(\varrho^2 \omega_\oplus \cos^2 \phi) \,, \tag{1.166}$$

where

$$\frac{\mathrm{d}}{\mathrm{d}t}(\varrho^2 \omega_\oplus \cos^2 \phi) = 2\varrho \dot{\varrho} \omega_\oplus \cos^2 \phi - \omega_\oplus \varrho^2 \dot{\phi} \sin 2\phi \,, \tag{1.167}$$

provided that the angular velocity of the Earth's rotation $\omega_\oplus$ is considered to be constant in time. In the ideal case, if the geostationary satellite is orbiting in the plane of the Earth's equator, $\phi = 0$, $\dot{\phi} = 0$, and if $\partial W(\varrho, \phi, \Lambda)/\partial \phi = 0$, system (1.166) reduces to two equations. If, in addition, its radial motion may be neglected and $\dot{\varrho}$ made equal to zero, we are left with one equation:

$$\ddot{\Lambda} \varrho^2 = \frac{\partial W(\varrho, \phi, \Lambda)}{\partial \Lambda} \,. \tag{1.168}$$

The gravity potential

$$W(\varrho, \phi, \Lambda) = V(\varrho, \phi, \Lambda) + Q(\varrho, \phi, \Lambda), \tag{1.169}$$

and, with a view to (1.63) and (1.71),

$$V(\varrho, \phi, \Lambda) = \frac{GM_\oplus}{\varrho} \sum_{n=0}^{\infty} \sum_{k=0}^{n} \left(\frac{a_0}{\varrho}\right)^n (J_n^{(k)} \cos k\Lambda + S_n^{(k)} \sin k\Lambda) \, \mathrm{P}_n^{(k)}(\sin \phi) \,, \tag{1.170}$$

Q being the potential of centrifugal forces.

In this ideal case we shall explain the principle of using geostationary satellites to derive the geopotential coefficients of some lower degrees, and to verify the position of the instantaneous Earth's centre of mass relative to the instantaneous axis of rotation. If $\delta x_1''$ and $\delta x_2''$ are the geocentric coordinates of the point of intersection O'' of the instantaneous axis of rotation with the plane of the geocentric equator, the potential of centrifugal forces in system $(\varrho, \phi, \varLambda)$ will read

$$Q = \tfrac{1}{2} \omega_\oplus^2 \varrho^2 \cos^2 \phi - \delta x_1'' \omega_\oplus^2 \varrho \cos \phi \cos \varLambda - \delta x_2'' \omega_\oplus^2 \varrho \cos \phi \sin \varLambda, \qquad (1.171)$$

where $\delta x_1''$ and $\delta x_2''$ are considered as very small quantities.

In view of (1.170) and (1.171), and after substituting $\phi = 0$, the equation of motion (1.168) becomes

$$\ddot{\varLambda}\varrho^2 + 2\varrho\dot{\varrho}(\dot{\varLambda} + \omega_\oplus) = \frac{GM_\oplus}{\varrho} \sum_{n=2}^{\infty} \left[\left(\frac{a_0}{\varrho}\right)^n \sum_{k=1}^{n} k(-J_n^{(k)} \sin k\varLambda$$

$$+ S_n^{(k)} \cos k\varLambda) P_n^{(k)}(0) \right] + \omega_\oplus^2 \varrho(\delta x_1'' \sin \varLambda - \delta x_2'' \cos \varLambda) . \qquad (1.172)$$

However, in this illustrative and extremely simplified solution radial motion has not been considered (we have put $\dot{\varrho} = 0$) and, consequently, the second term on the lhs of Eq. (1.172) is equal to zero. In addition, all harmonic terms on the rhs for which $(n - k)$ is an odd number are equal to zero, because then $P_n^{(k)}(0) = 0$. Equation (1.172) may now be expressed as

$$\ddot{\varLambda}\varrho^2 = \frac{GM_\oplus}{\varrho} \left[\sum_{\substack{n=2 \\ (n-k)}}^{\infty} \sum_{\substack{k=0 \\ \text{even or } 0}}^{n} k \left(\frac{a_0}{\varrho}\right)^n (-J_n^{(k)} \sin k\varLambda + S_n^{(k)} \cos k\varLambda) P_n^{(k)}(0) \right.$$

$$\left. + q \left(\frac{a_0}{\varrho}\right)^{-3} \left(\frac{\delta x_1''}{\varrho} \sin \varLambda - \frac{\delta x_2''}{\varrho} \cos \varLambda\right) \right], \quad q = \frac{\omega_\oplus^2 a_0^3}{GM_\oplus} . \qquad (1.173)$$

This equation describes the libration of the geostationary satellite in longitude but in the idealized case ($\dot{\varrho} = 0$, $\phi = 0$). It indicates that this method can be used to derive geopotential coefficients with which difference $(n - k)$ is an even number or zero and, in particular, geopotential coefficients $J_2^{(2)}$ and $S_2^{(2)}$ which are dominant in magnitude and have a decisive effect on the shape and orientation of the equatorial ellipse substituting for the equatorial section of the geoid.

However, even if the density distribution inside the Earth were spherically symmetrical, i.e. all geopotential coefficients $J_n^{(k)}$ and $S_n^{(k)}$ were equal to zero, $n \neq 0$, libration of the satellite could occur provided that $\delta x_1'' \neq 0$ or $\delta x_2'' \neq 0$. In this case the libration of the satellite in longitude would be described by the equation

$$\ddot{\varLambda} = -k_1 \sin(\varLambda - \varLambda_{1,1}), \qquad (1.174)$$

which is the equation of pendulum motion, where

$$k_1 = \frac{\omega_\oplus^2}{\varrho} [(\delta x_1'')^2 + (\delta x_2'')^2]^{1/2} = \frac{GM_\oplus}{\varrho} q a_0^{-3} [(\delta x_1'')^2 + (\delta x_2'')^2]^{1/2}, \quad k_1 > 0; \qquad (1.175)$$

$$\cos \varLambda_{1,1} = -\delta x_1'' [(\delta x_1'')^2 + (\delta x_2'')^2]^{-1/2},$$

$$\sin \varLambda_{1,1} = -\delta x_2'' [(\delta x_1'')^2 + (\delta x_2'')^2]^{-1/2} . \qquad (1.176)$$

Equation (1.174) yields the relation for the square of the velocity of an ideal geostationary satellite in longitude,

$$\dot{\Lambda}^2 = 2k_1 \cos(\Lambda - \Lambda_{1,1}) + C_{1,1} , \tag{1.177}$$

where

$$C_{1,1} = -2k_1 \cos(\bar{\Lambda} - \Lambda_{1,1}) \tag{1.178}$$

is an integration constant defined by the initial conditions, and $\bar{\Lambda}$ is the longitude of the point of the satellite's orbit at which $\dot{\Lambda} = 0$, i.e. longitude of maximum elongation. Only if $\cos(\bar{\Lambda} - \Lambda_{1,1}) \leqq \cos(\Lambda - \Lambda_{1,1})$, i.e. $|\Lambda - \Lambda_{1,1}| \leqq |\bar{\Lambda} - \Lambda_{1,1}|$, does the solution have a physical meaning and reduce to the well-known elliptical integral.

It can be proved that the displacements of the position of the instantaneous axis of rotation and instantaneous centre of mass of the Earth, a few centimetres in magnitude, could be determined, and the basic assumption in the theory of rotation of an elastic Earth could be verified using this method.

2 The Earth's Gravity Field and Its Sources

2.1 Introduction

The properties of the gravity field are closely related to the figure of the Earth. The problem of determining the figure and dimensions of the Earth is of scientific and practical significance. Initially, based on experience from its immediate vicinity, the Earth was considered to be a planar plate. The first concept of a spherical Earth came from Pythagoras (sixth century B.C.); he arrived at this from his conviction of a perfectly organized universe. He considered the sphere to be a perfect geometrical shape from which he deduced the spherical shape of the Earth. Eratosthenes (third century B.C.) made the first scientific attempt at measuring the Earth's circumference. The origin of the dynamic concept of the figure of the Earth and the idea of an ellipsoidal Earth was due to Newton who formulated the law of universal gravitation and supposed that the rotating Earth should be flatter at the poles.

The geometrical problem of determining the figure of the Earth has thus been treated for more than two millenia, and the dynamic problem is more than three centuries old. The subsequent development of understanding the Earth's gravity field was influenced by the works of Clairaut, Legendre, Laplace, Stokes, Bruns, Helmert, Jeffreys, Molodenskij and many other scientists. In the era of satellite observations some new aspects connected with new ways of studying the Earth's gravity field have emerged. First of all, the Earth's gravitational field is the most significant force affecting satellite motion and, consequently, a detailed knowledge of it is important with regard to the orbiting of satellites about the planet. Conversely, the observation of satellite orbits provides detailed data on the structure of the field, data which cannot be obtained by classical ground-based methods to such a large extent and in such great detail. The existence of these data has opened up new possibilities in the theory of the Earth's gravity field. In the first instance this concerns the study of the relations between the detailed structure of the Earth's gravity field and the distribution of density within it, namely the study of lateral variations in density. A detailed knowledge of the density distribution, i.e. also the knowledge of deviations from hydrostatic equilibrium, is of fundamental importance for studying the dynamic processes within the Earth and of the non-hydrostatic stresses associated with them.

In this chapter we shall deal with the properties of the Earth's gravity field directly connected with the field's description in terms of geopotential coefficients, whose determination using satellite methods was discussed in the preceding chapter. We shall intentionally omit some classical parts of the theory of the Earth's gravity field which are not related directly to satellite methods, e.g. Stokes' theorem and the isostasy hypothesis, which have been treated in a number of monographs and textbooks (Pick et al. 1973). In contrast to the usual method of treatment we shall devote more attention to describing the properties of equipotential surfaces, especially of the geoid, which can be derived from the expansion of the radius-vector of the equipotential surface into a series of spherical harmonics. Considerable attention will be devoted to investigating the limitations for the three-dimensional density distribution within the Earth; this follows from our knowledge of geopotential coefficients up to a high degree (Moritz 1973; Pěč and Martinec 1984).

Although we are aware that the Earth's gravity field varies with time and that geopotential coefficients are not constants, but that they also depend on time, we shall ignore this fact in this chapter. Other chapters will be devoted to the time-variable components of the field related to the Earth's rotation and tides.

2.2 Gravitational and Gravity Potentials

The gravity field is a conservative field of force and its potential W (geopotential, gravity potential) at external point P is represented by the sum

$$W(P) = V(P) + Q(P) + \delta W(P), \tag{2.1}$$

where V is the gravitational potential, Q the potential of centrifugal forces, and δW its variable part due to free nutation (motion of the Earth's poles) and tidal effects of the Moon and Sun. The tidal component contains a constant part which will also be included in δW. The variable component will be treated in Chapters 3–5. The gravitational potential is a harmonic function of coordinates, i.e. it satisfies Laplace's equation

$$\Delta V(P) = 0. \tag{2.2}$$

In the system of spherical coordinates $(\varrho, \vartheta = 90° - \phi, \Lambda)$ with its origin at the Earth's centre of mass, there are two independent partial solutions of Eq. (2.2): $\varrho^j Y_{jm}(\vartheta, \Lambda)$ and $\varrho^{-j-1} Y_{jm}(\vartheta, \Lambda)$, where $Y_{jm}(\vartheta, \Lambda)$ are spherical harmonics. The first solution has a singularity in infinity and, therefore, is unsatisfactory at an external point. The general solution of (2.2) will be expressed as

$$V(\varrho, \vartheta, \Lambda) = \frac{GM}{\varrho} \sum_{j=0}^{\infty} \left(\frac{a_0}{\varrho}\right)^j \sum_{m=-j}^{j} A_{jm} Y_{jm}(\vartheta, \Lambda), \tag{2.3}$$

where A_{jm} are complex geopotential coefficients and the other symbols have their usual meaning. Spherical harmonics $Y_{jm}(\vartheta, \Lambda)$ are related to Legendre functions $P_j^{(m)}(\cos \vartheta)$, introduced by Eq. (1.70), as follows ($-j \le m \le j$):

$$Y_{jm}(\vartheta, \Lambda) = \exp(im\Lambda) \left\{ (-1)^m \sqrt{\left(\frac{2j+1}{4\pi} \frac{(j-m)!}{(j+m)!} \right)} \, P_j^{(m)}(\cos \vartheta) \right\}. \tag{2.4}$$

Spherical harmonics $Y_{jm}(\vartheta, \Lambda)$ are fully normalized, i.e.

$$\int_0^{2\pi} d\Lambda \int_0^\pi \sin \vartheta Y_{jm}(\vartheta, \Lambda) Y_{j'm'}^*(\vartheta, \Lambda) d\vartheta = \delta_{jj'} \delta_{mm'} . \tag{2.5}$$

It also holds that

$$Y_{jm}^*(\vartheta, \Lambda) = Y_{jm}(\vartheta, -\Lambda) = (-1)^m Y_{j-m}(\vartheta, \Lambda) . \tag{2.6}$$

By using the above relations, series (2.3) for the real potential $V(P)$ can be arranged to read

$$V(\varrho, \vartheta, \Lambda) = \frac{GM}{\varrho} \left[1 + \sum_{j=1}^\infty \sum_{m=0}^j \left(\frac{a_0}{\varrho} \right)^j (J_j^{(m)} \cos m\Lambda \right.$$

$$\left. + S_j^{(m)} \sin m\Lambda) P_j^{(m)}(\cos \vartheta) \right]. \tag{2.7}$$

Quantities $J_j^{(m)}$ and $S_j^{(m)}$ are the geopotential coefficients introduced in Chapter 1; they are related to complex constants A_{jm} by the following transformation relations:

$$A_{j0} = \frac{2\sqrt{\pi}}{\sqrt{(2j+1)}} J_j^{(0)} = -\frac{2\sqrt{\pi}}{\sqrt{(2j+1)}} J_j ,$$

$$A_{jm} = (-1)^m \sqrt{(2\pi)} [J_j^{(m)} - iS_j^{(m)}]/N_j^{(m)} , \quad m > 0$$

$$A_{j-m} = \sqrt{(2\pi)} [J_j^{(m)} + iS_j^{(m)}]/N_j^{(m)} , \quad m > 0 . \tag{2.8}$$

Quantity $N_j^{(m)}$ was introduced by (1.138).

Apart from expressions (2.3) and (2.7) the following expansion is also used for the external gravitational potential:

$$V(\varrho, \vartheta, \Lambda) = \frac{GM}{\varrho} \sum_{j=0}^\infty \left(\frac{a_0}{\varrho} \right)^j \sum_{m=0}^j [\bar{C}_{jm} \cos m\Lambda + \bar{S}_{jm} \sin m\Lambda] \bar{P}_j^m(\cos \vartheta) , \tag{2.9}$$

where the associated Legendre functions $\bar{P}_j^{(m)}(\cos \vartheta)$ are related to functions $P_j^{(m)}(\cos \vartheta)$, introduced earlier by (1.70), as follows:

$$\bar{P}_j^m(\cos \vartheta) = N_j^{(m)} P_j^{(m)}(\cos \vartheta); \tag{2.10}$$

coefficients $\bar{C}_{jm}, \bar{S}_{jm}$ are thus related to coefficients $J_j^{(m)}, S_j^{(m)}$ and/or $\bar{J}_j^{(m)}, \bar{S}_j^{(m)}$ as

$$\bar{C}_{jm} = J_j^{(m)}/N_j^{(m)} = \bar{J}_j^{(m)} ; \quad \bar{S}_{jm} = S_j^{(m)}/N_j^{(m)} = \bar{S}_j^{(m)} . \tag{2.11}$$

Quantities $\bar{J}_j^{(m)}$ and $\bar{S}_j^{(m)}$ are fully normalized potential coefficients and can be expressed in terms of complex geopotential coefficients as follows:

$$A_{j0} = 2\sqrt{(\pi)}\,\bar{J}_j^{(0)},$$

$$A_{jm} = \sqrt{(2\pi)}(-1)^m(\bar{J}_j^{(m)} - i\bar{S}_j^{(m)}), \quad m > 0$$

$$A_{j-m} = \sqrt{(2\pi)}(\bar{J}_j^{(m)} + i\bar{S}_j^{(m)}), \qquad m > 0. \tag{2.12}$$

To simplify the record we shall introduce symbol Ω for angular spherical coordinates (ϑ, Λ), i.e. $\Omega = (\vartheta, \Lambda)$, and symbol $d\Omega$ will represent an element of solid angle:

$$d\Omega = \sin\vartheta\,d\vartheta\,d\Lambda. \tag{2.13}$$

We shall write out spherical harmonics $Y_{jm}(\Omega)$ up to degree $j = 2$ explicitly for later use:

$$Y_{00}(\Omega) = 1/(2\sqrt{\pi}); \quad Y_{10}(\Omega) = \tfrac{1}{2}\sqrt{\left(\frac{3}{\pi}\right)}\cos\vartheta;$$

$$Y_{20}(\Omega) = \tfrac{1}{4}\sqrt{\left(\frac{5}{\pi}\right)}(3\cos^2\vartheta - 1),$$

$$Y_{1\pm1}(\Omega) = \mp\tfrac{1}{2}\sqrt{\left(\frac{3}{2\pi}\right)}\sin\vartheta\exp(\pm i\Lambda),$$

$$Y_{2\pm1}(\Omega) = \mp\tfrac{1}{2}\sqrt{\left(\frac{3\cdot5}{2\pi}\right)}\cos\vartheta\sin\vartheta\exp(\pm i\Lambda),$$

$$Y_{2\pm2}(\Omega) = \tfrac{1}{4}\sqrt{\left(\frac{3\cdot5}{2\pi}\right)}\sin^2\vartheta\exp(\pm 2i\Lambda). \tag{2.14}$$

The potential of centrifugal forces can be expressed in terms of spherical harmonics $Y_{00}(\Omega)$, $Y_{20}(\Omega)$ and Helmert's parameter q:

$$Q(P) = \tfrac{1}{2}\omega^2\varrho^2\sin^2\vartheta = \frac{GM}{\varrho}q\left(\frac{\varrho}{a_0}\right)^3\frac{2\sqrt{\pi}}{3}\left[Y_{00}(\Omega) - \frac{1}{\sqrt{5}}Y_{20}(\Omega)\right]$$

$$= \tfrac{1}{3}\frac{GM}{\varrho}q\left(\frac{\varrho}{a_0}\right)^3[1 - P_2^{(0)}(\cos\vartheta)]. \tag{2.15}$$

Note that the potential of centrifugal forces, as opposed to the gravitational potential, is not harmonic; it satisfies the equation

$$\Delta Q = 2\omega^2 \neq 0. \tag{2.16}$$

Parameter q in (2.15) is defined as

$$q = \omega^2 a_0^3/(GM). \tag{2.17}$$

The geometrical locus of points satisfying condition

$$W = W_0, \tag{2.18}$$

where W_0 is a constant, represents a closed surface of constant geopotential, i.e. an equipotential surface. Constant W_0 defines this surface in terms of dimension and the set of geopotential coefficients in terms of shape. Instead of constant W_0 we shall introduce the ratio

$$GM/W_0 = R_0 , \tag{2.19}$$

which we shall refer to as the geopotential scale factor. If R_0 and/or W_0 are considered to be parameters, we arrive at a system of closed equipotential surfaces which do not intersect anywhere. One of them has an exceptional status: in the region of oceans and seas it identifies with their mean undisturbed surfaces and, therefore, represents the surface of the Earth as a whole. Listing (1873) called this surface the geoid. We shall denote its geopotential W_0 and its scale factor R_0. Quantity W_0, or rather R_0, is the fundamental quantity which defines the Earth's dimension directly and uniquely. Other quantities defining the dimensions of the body, e.g. the semimajor axis of the Earth ellipsoid, are parametric and depend on additional conditions which have to be formulated.

2.3 Transformation of the Gravitational Potential and Potential of Centrifugal Forces Under Rotation of the Coordinate System. Transformation of Geopotential Coefficients

We shall consider the gravitational potential $V(P)$ at external point $P(\varrho, \Omega)$ in the form of (2.3), and denote the spherical coordinate system in which it is to be expressed $S(\varrho, \vartheta, \Lambda)$. We now change to another spherical coordinate system, $S'(\varrho', \vartheta', \Lambda')$, which is created by rotating the original system S about the origin of coordinates. We shall describe the rotation by means of angles α, β, γ. Potential $V(P)$ at the same point, but now expressed in system S', will read

$$V(P) = \frac{GM}{\varrho'} \sum_{j=0}^{\infty} \left(\frac{a_0}{\varrho'}\right)^j \sum_{M=-j}^{j} A'_{jM} \, Y_{jm}(\vartheta, \Lambda') . \tag{2.20}$$

Symbol $\hat{\mathbf{D}}(\alpha, \beta, \gamma)$ has been used to denote the operator of the system's rotation. Since spherical harmonics $Y_{jm}(\vartheta, \Lambda)$ are covariant irreducible tensors, the effect of the rotation operator can be expressed as follows:

$$Y_{jm}(\vartheta', \Lambda') = \hat{\mathbf{D}}(\alpha, \beta, \gamma) Y_{jm}(\vartheta, \Lambda) = \sum_{m=-j}^{j} Y_{jm}(\vartheta, \Lambda) D^j_{mM}(\alpha, \beta, \gamma) , \tag{2.21}$$

where $D^j_{mM}(\alpha, \beta, \gamma)$ are Wigner functions.

Angles ϑ, Λ and ϑ', Λ', defining the vector's direction in the original and rotated system, are related as

$$\cos \vartheta = \cos \vartheta' \cos \Lambda - \sin \vartheta' \sin \Lambda \cos(\Lambda' + \gamma) ,$$

$$\cot(\Lambda - \alpha) = \cot(\Lambda + \gamma)\cos \beta + \frac{\cot \vartheta' \sin \beta}{\sin(\Lambda' + \gamma)} \tag{2.22}$$

and, conversely,

$$\cos \vartheta' = \cos \vartheta \cos \beta + \sin \vartheta \sin \beta \cos(\Lambda - \alpha) ,$$

$$\cot(\Lambda' + \alpha) = \cot(\Lambda - \alpha) \cos \beta - \frac{\cot \vartheta \sin \beta}{\sin(\Lambda - \alpha)} . \tag{2.23}$$

By substituting (2.21) into the formula for potential $V(P)$ we get

$$V(P) = \frac{GM}{\varrho'} \sum_{j=0}^{\infty} \left(\frac{a_0}{\varrho'}\right)^j \sum_{m=-j}^{j} Y_{jm}(\vartheta, \Lambda) \sum_{M=-j}^{j} A'_{jM} D^j_{mM}(\alpha, \beta, \gamma) . \tag{2.24}$$

Considering that the rotation of the coordinate system about its origin has no effect on the length of the radius-vector, i.e. $\varrho = \varrho'$, the comparison of (2.24) with (2.3) will yield the relation between the complex geopotential coefficients in both systems:

$$A_{jm} = \sum_{M=-j}^{j} A'_{jM} D^j_{mM}(\alpha, \beta, \gamma) . \tag{2.25}$$

Since the rotation operator is a unitary matrix operator, i.e. the inverse operator $\hat{\mathbf{D}}^{-1}(\alpha, \beta, \gamma) = \hat{\mathbf{D}}^*(\alpha, \beta, \gamma)$, it follows that

$$\sum_{m=-j}^{j} D^j_{mM}(\alpha, \beta, \gamma) D^{j*}_{mM'}(\alpha, \beta, \gamma) = \delta_{MM'} . \tag{2.26}$$

Multiplying (2.25) by $D^{j*}_{mM'}(\alpha, \beta, \gamma)$ and summing over index m yields

$$\sum_{m=-j}^{j} A_{jm} D^{j*}_{mM'}(\alpha, \beta, \gamma) = \sum_{m=-j}^{j} A_{jm} \sum_{m=-j}^{j} D^j_{mM}(\alpha, \beta, \gamma) D^{j*}_{mM'}(\alpha, \beta, \gamma) . \tag{2.27}$$

In virtue of (2.26), we arrive at the relation inverse to (2.25):

$$A'_{jM} = \sum_{m=-j}^{j} A_{jm} D^{j*}_{mM}(\alpha, \beta, \gamma) . \tag{2.28}$$

Wigner functions $D^j_{mM}(\alpha, \beta, \gamma)$ are defined as

$$D^j_{mM}(\alpha, \beta, \gamma) = \exp(-im\alpha) d^j_{mM}(\beta) \exp(-iM\gamma) , \tag{2.29}$$

where

$$d^j_{mM}(\beta) = (-1)^{j-M} [(j+m)!(j-m)!(j+M)!(j-M)!]^{1/2}$$

$$\times \sum_k (-1)^k \frac{\left(\cos \frac{\beta}{2}\right)^{m+M+2k} \left(\sin \frac{\beta}{2}\right)^{2j-m-M-2k}}{k!(j-m-k)!(j-M-k)!(m+M+k)!} . \tag{2.30}$$

The summation in Eq. (2.30) is over natural numbers k for which all factorials in (2.30) are non-negative.

Due to the significance of this transformation matrices $d^j_{mM}(\beta)$ are given in Tables 2.1 and 2.2 for $j = 1$ and $j = 2$, respectively.

Table 2.1. Matrix $d^1_{mM}(\beta)$

m $\quad$ M	1	0	-1
1	$(1 + \cos\beta)/2$	$-\sin\beta/\sqrt{2}$	$(1 - \cos\beta)/2$
0	$\sin\beta/\sqrt{2}$	$\cos\beta$	$-\sin\beta/\sqrt{2}$
-1	$(1 - \cos\beta)/2$	$\sin\beta/\sqrt{2}$	$(1 + \cos\beta)/2$

Table 2.2. Matrix $d^2_{mM}(\beta)$

m	M	Matrix elements	m	M	Matrix elements
2	2	$(1 + \cos\beta)^2/4$	0	0	$(3\cos^2\beta - 1)/2$
2	1	$-\sin\beta(1 + \cos\beta)/2$	0	-1	$-\sqrt{(3)}\sin\beta\cos\beta/\sqrt{2}$
2	0	$\sqrt{(3)}\sin^2\beta/(2\sqrt{2})$	0	-2	$\sqrt{(3)}\sin^2\beta/(2\sqrt{2})$
2	-1	$-\sin\beta(1 - \cos\beta)/2$			
2	-2	$(1 - \cos\beta)^2/4$	-1	2	$\sin\beta(1 - \cos\beta)/2)$
			-1	1	$-(2\cos^2\beta - \cos\beta - 1)/2$
1	2	$\sin\beta(1 + \cos\beta)/2$	-1	0	$\sqrt{(3)}\sin\beta\cos\beta/\sqrt{2}$
1	1	$(2\cos^2\beta + \cos\beta - 1)/2$	-1	-1	$(2\cos^2\beta + \cos\beta - 1)/2$
1	0	$-\sqrt{(3)}\sin\beta\cos\beta/\sqrt{2}$	-1	-2	$-\sin\beta(1 + \cos\beta)/2$
1	-1	$-(2\cos^2\beta - \cos\beta - 1)/2$			
1	-2	$-\sin\beta(1 - \cos\beta)/2$	-2	2	$(1 - \cos\beta)^2/4$
			-2	1	$\sin\beta(1 - \cos\beta)/2$
0	2	$\sqrt{(3)}\sin^2\beta/(2\sqrt{2})$	-2	0	$\sqrt{(3)}\sin^2\beta/(2\sqrt{2})$
0	1	$\sqrt{(3)}\sin\beta\cos\beta/2$	-2	-1	$\sin\beta(1 + \cos\beta)/2$
			-2	-2	$(1 + \cos\beta)^2/4$

2.4 Gravity in Outer Space

Gravity $\mathbf{g}(\varrho, \Omega)$ at external point $P(\varrho, \Omega)$ is perpendicular to equipotential surface W passing through point P, and is defined as

$$\mathbf{g}(\varrho, \Omega) = -\nabla W(\varrho, \Omega) = -\nabla V(\varrho, \Omega) - \nabla Q(\varrho, \Omega) \,. \tag{2.31}$$

To be able to express the acceleration of gravity explicitly, operation gradient must be applied to $\varrho^{-j-1} Y_{jm}(\Omega)$ and to $\varrho^2 Y_{00}(\Omega) - 5^{-1/2} Y_{20}(\Omega)$. The gradient operation will transform scalar spherical harmonics $Y_{jm}(\Omega)$ into spherical vectors $\mathbf{Y}^{(\mu)}_{jm}(\Omega)$, $(\mu = 0, \pm 1)$.

Differential operator ∇ is expressed in the spherical system,

$$\nabla = \mathbf{e}_\varrho \nabla_\varrho + \mathbf{e}_\vartheta \nabla_\vartheta + \mathbf{e}_\Lambda \nabla_\Lambda \,, \tag{2.32}$$

where

$$\mathbf{V}_\varrho = \frac{\partial}{\partial \varrho}, \quad \mathbf{V}_\vartheta = \frac{1}{\varrho}\frac{\partial}{\partial \vartheta}, \quad \mathbf{V}_\varLambda = \frac{1}{\varrho \sin \vartheta}\frac{\partial}{\partial \varLambda}. \tag{2.33}$$

If unit normal ϱ/ϱ is denoted $\mathbf{n}$, (2.32) can be expressed as

$$\mathbf{V} = \mathbf{n}\frac{\partial}{\partial \varrho} + \frac{1}{\varrho}\mathbf{V}_\varOmega, \tag{2.34}$$

where $\mathbf{V}_\varOmega$ is the angular part of operator $\mathbf{V}$, i.e.

$$(\mathbf{V}_\varOmega)_\vartheta = \frac{\partial}{\partial \varrho}, \quad (\mathbf{V}_\varOmega)_\varLambda = \frac{1}{\sin \vartheta}\frac{\partial}{\partial \varLambda}. \tag{2.35}$$

Vectors $\mathbf{e}_\varrho$, $\mathbf{e}_\vartheta$, $\mathbf{e}_\varLambda$ are base vectors which form a right-handed system, i.e.

$$\mathbf{e}_\varrho \times \mathbf{e}_\vartheta = \mathbf{e}_\varLambda; \quad \mathbf{e}_\vartheta \times \mathbf{e}_\varLambda = \mathbf{e}_\varrho; \quad \mathbf{e}_\varLambda \times \mathbf{e}_\varrho = \mathbf{e}_\vartheta. \tag{2.36}$$

Spherical base vectors $\mathbf{e}_\varrho$, $\mathbf{e}_\vartheta$, $\mathbf{e}_\varLambda$ depend on the angular coordinates, and

$$\frac{\partial \mathbf{e}_\varrho}{\partial \varrho} = \frac{\partial \mathbf{e}_\vartheta}{\partial \varrho} = \frac{\partial \mathbf{e}_\varLambda}{\partial \varrho} = 0; \quad \frac{\partial \mathbf{e}_\varrho}{\partial \vartheta} = \mathbf{e}_\vartheta; \quad \frac{\partial \mathbf{e}_\vartheta}{\partial \vartheta} = -\mathbf{e}_\varrho; \quad \frac{\partial}{\partial \vartheta}\mathbf{e}_\varLambda = 0;$$

$$\frac{\partial}{\partial \varLambda}\mathbf{e}_\varrho = \mathbf{e}_\varLambda \sin \vartheta; \quad \frac{\partial}{\partial \varLambda}\mathbf{e}_\vartheta = \mathbf{e}_\varLambda \cos \vartheta; \quad \frac{\partial}{\partial \varLambda}\mathbf{e}_\varLambda = -\mathbf{e}_\varrho \sin \vartheta - \mathbf{e}_\vartheta \cos \vartheta \tag{2.37}$$

and their divergences and rotations are

$$\mathbf{V}\cdot\mathbf{e}_\varrho = \frac{2}{\varrho}; \quad \mathbf{V}\cdot\mathbf{e}_\vartheta = \frac{1}{\varrho}\cot \vartheta; \quad \mathbf{V}\cdot\mathbf{e}_\varLambda = 0;$$

$$\mathbf{V}\times\mathbf{e}_\varrho = 0; \quad \mathbf{V}\times\mathbf{e}_\vartheta = \frac{1}{\varrho}\mathbf{e}_\varLambda; \quad \mathbf{V}\times\mathbf{e}_\varLambda = \frac{1}{\varrho}\cot \vartheta \mathbf{e}_\varrho - \frac{1}{\varrho}\mathbf{e}_\vartheta. \tag{2.38}$$

Spherical vectors $\mathbf{Y}_{jm}^{(\lambda)}(\vartheta, \varLambda)$ are defined as follows:

$$\mathbf{Y}_{jm}^{(1)}(\varOmega) = \frac{1}{\sqrt{[j(j+1)]}}\,\mathbf{V}_\varOmega Y_{jm}(\varOmega),$$

$$\mathbf{Y}_{km}^{(0)}(\varOmega) = \frac{-1}{\sqrt{[j(j+1)]}}\,(\mathbf{n}\times\mathbf{V}_\varOmega)Y_{jm}(\varOmega),$$

$$\mathbf{Y}_{jm}^{(-1)}(\varOmega) = \mathbf{n}Y_{jm}(\varOmega). \tag{2.39}$$

Spherical vectors $\mathbf{Y}_{jm}^{(\lambda)}(\varOmega)$ ($\lambda = 0, \pm 1$) have the advantage that they are conveniently oriented with respect to the direction of radius-vector $\mathbf{n} = \varrho/\varrho$, i.e. $\mathbf{Y}_{jm}^{(-1)}(\varOmega)$ (longitudinal vector) is parallel with $\mathbf{n}$ and vectors $\mathbf{Y}_{jm}^{(0)}(\varOmega)$ and $\mathbf{Y}_{jm}^{(1)}(\varOmega)$ (transverse vectors) are perpendicular to $\mathbf{n}$, i.e.

$$\mathbf{n}\times\mathbf{Y}_{jm}^{(-1)}(\varOmega) = 0; \quad \mathbf{n}\cdot\mathbf{Y}_{jm}^{(1)}(\varOmega) = \mathbf{n}\cdot\mathbf{Y}_{jm}^{(0)}(\varOmega) = 0. \tag{2.40}$$

These relations indicate that the gradient can be expressed as

$$\mathbf{V}[\varrho^{-j-1}\mathbf{Y}_{jm}(\Omega)] = \frac{d\varrho^{-j-1}}{d\varrho}\,\mathbf{Y}_{jm}^{(-1)}(\Omega)$$
$$+ \sqrt{[j(j+1)]}\,\varrho^{-j-2}\mathbf{Y}_{jm}^{(1)}(\Omega)\,. \tag{2.41}$$

The spherical vector reads

$$\mathbf{Y}_{jm}^{(\lambda)}(\Omega) = \mathbf{Y}_{jm}^{(\lambda)}(\Omega)|_{\varrho}\mathbf{e}_{\varrho} + \mathbf{Y}_{jm}^{(\lambda)}(\Omega)|_{\vartheta}\mathbf{e}_{\vartheta} + \mathbf{Y}_{jm}^{(\lambda)}(\Omega)|_{\Lambda}\mathbf{e}_{\Lambda}\,, \tag{2.42}$$

and its components can be expressed in terms of spherical harmonics:

$$\mathbf{Y}_{jm}^{(1)}(\Omega)|_{\varrho} = \mathbf{Y}_{jm}^{(0)}(\varrho)|_{\varrho} = \mathbf{Y}_{jm}^{(-1)}(\Omega)|_{\vartheta} = \mathbf{Y}_{jm}^{(-1)}(\Omega)|_{\Lambda} = 0\,,$$

$$\mathbf{Y}_{jm}^{(1)}(\Omega)|_{\vartheta} = \frac{1}{\sqrt{[j(j+1)]}}\,\frac{\partial}{\partial\vartheta}\,Y_{jm}(\Omega)\,,$$

$$\mathbf{Y}_{jm}^{(1)}(\Omega)|_{\Lambda} = \frac{im}{\sqrt{[j(j+1)]}}\,\frac{1}{\sin\vartheta}\,Y_{jm}(\Omega)\,,$$

$$\mathbf{Y}_{jm}^{(0)}(\Omega)|_{\vartheta} = \frac{-m}{\sqrt{[j(j+1)]}}\,\frac{1}{\sin\vartheta}\,Y_{jm}(\Omega)\,,$$

$$\mathbf{Y}_{jm}^{(0)}(\Omega)|_{\Lambda} = -\frac{i}{\sqrt{[j(j+1)]}}\,\frac{\partial}{\partial\vartheta}\,Y_{jm}(\Omega)\,,$$

$$\mathbf{Y}_{jm}^{(-1)}(\Omega)|_{\varrho} = Y_{jm}(\Omega)\,, \tag{2.43}$$

where

$$\frac{\partial Y_{jm}(\Omega)}{\partial\vartheta} = \tfrac{1}{2}\{\sqrt{[(j-m)(j+m+1)]}\,\exp(-i\Lambda)\,Y_{jm+1}(\Omega)$$
$$- \sqrt{[(j+m)(j-m+1)]}\,\exp(i\Lambda)\,Y_{jm-1}(\Omega)\}$$
$$= -m\cot\vartheta\,Y_{jm}(\Omega) - \sqrt{[(j+m)(j-m+1)]}$$
$$\times Y_{jm-1}(\Omega)\exp(i\Lambda)\,. \tag{2.44}$$

Using these relations and in view of (2.31), (2.3) and (2.15), intensity $\mathbf{g}$ can be expressed in terms of spherical vectors as:

$$\mathbf{g}(\varrho,\vartheta,\Lambda) = \frac{GM}{\varrho^2}\sum_{jm}^{\infty}\left(\frac{a_0}{\varrho}\right)^j A_{jm}\{(j+1)\,\mathbf{Y}_{jm}^{(-1)}(\Omega) - \sqrt{[j(j+1)]}\,\mathbf{Y}_{jm}^{(1)}(\Omega)\}$$
$$- \frac{GM}{\varrho^2}q\left(\frac{\varrho}{a_0}\right)^3\frac{4\sqrt{\pi}}{3}\left\{\mathbf{Y}_{00}^{(-1)}(\Omega) - \frac{1}{\sqrt{5}}\left[\mathbf{Y}_{20}^{(-1)}(\Omega)\right.\right.$$
$$\left.\left. + \sqrt{\left(\frac{3}{2}\right)}\mathbf{Y}_{20}^{(1)}(\Omega)\right]\right\}\,. \tag{2.45}$$

The acceleration of gravity, expressed in components, now reads

$$\mathbf{g}|_\varrho = \frac{GM}{\varrho^2} \sum_{jm}^{\infty} (j+1) A_{jm} \left(\frac{a_0}{\varrho}\right)^j Y_{jm}(\Omega) - \omega^2 \varrho \sin^2 \vartheta \,, \tag{2.46}$$

$$\mathbf{g}|_\vartheta = -\frac{GM}{2\varrho^2} \sum_{jm}^{\infty} A_{jm} \left(\frac{a_0}{\varrho}\right)^j \{[(j-m)(j+m+1)] \exp(-i\Lambda) Y_{jm+1}(\Omega)$$

$$- \sqrt{[(j+m)(j-m+1)]} \exp(i\Lambda) Y_{jm-1}(\Omega)\}$$

$$+ \varrho\omega^2 \cos\vartheta \sin\vartheta \,, \tag{2.47}$$

$$\mathbf{g}|_\Lambda = -i\frac{GM}{\varrho^2} \sum_{jm} m A_{jm} \left(\frac{a_0}{\varrho}\right)^j \frac{1}{\sin\vartheta} A_{jm} Y_{jm}(\Omega) \,. \tag{2.48}$$

The components of centrifugal accelerations in (2.46) and (2.47) can be expressed in terms of spherical harmonics:

$$\mathbf{g}_Q|_\varrho = -\omega^2 \varrho \sin^2 \vartheta = -\frac{4\sqrt{\pi}}{3} \omega^2 \varrho \left[Y_{00}(\Omega) - \frac{1}{\sqrt{5}} Y_{20}(\Omega) \right], \tag{2.49}$$

$$\mathbf{g}_Q|_\vartheta = \omega^2 \varrho \cos\vartheta \sin\vartheta = \frac{2\sqrt{(2\pi)}}{\sqrt{(3\cdot 5)}} \omega^2 \varrho \exp(-i\Lambda) Y_{21}(\Omega)$$

$$= \frac{GM}{\varrho^2} \frac{\sqrt{(2\pi)}}{\sqrt{(3\cdot 5)}} \left(\frac{\varrho}{a_0}\right)^3 [\exp(-i\Lambda)Y_{21}(\Omega) - \exp(i\Lambda)Y_{2-1}(\Omega)] \,. \tag{2.50}$$

Denoting the direction cosines of the normal to the equipotential surface $\cos\alpha_\varrho = \sin\varphi$, $\cos\alpha_\vartheta = \cos\varphi \cos\lambda$, $\cos\alpha_\Lambda = \cos\varphi \sin\lambda$ (see 1.2),

$$\cos\alpha_\varrho : \cos\alpha_\vartheta : \cos\alpha_\Lambda = \mathbf{g}|_\varrho : \mathbf{g}|_\vartheta : \mathbf{g}|_\Lambda \,. \tag{2.51}$$

Spherical vectors $\mathbf{Y}_{jm}^{(\lambda)}(\Omega)$, $(\lambda = 0, \pm 1)$ are not irreducible spherical tensors of rank 1 and, consequently, they do not transform under rotation of the coordinate system as (2.29) using the D-matrices. Irreducible tensors of rank 1 are spherical vectors $\mathbf{Y}_{jm}^k(\Omega)$, $(k = j-1, j, j+1)$, which are eigenfunctions of the angular part of the Laplace operator Δ_Ω with eigenvalues $k(k+1)$, i.e. they satisfy the equation

$$[\Delta_\Omega + k(k+1)] \mathbf{Y}_{jm}^k(\Omega) = 0 \,. \tag{2.52}$$

Spherical vectors $\mathbf{Y}_{jm}^k(\Omega)$ are expressed in terms of spherical vectors $\mathbf{Y}_{jm}^{(\lambda)}(\Omega)$, $(\lambda = 0, \pm 1)$ as follows:

$$\mathbf{Y}_{jm}^{j+1}(\Omega) = \frac{1}{\sqrt{(2j+1)}} [\sqrt{(j)}\mathbf{Y}_{jm}^{(1)}(\Omega) - \sqrt{(j+1)} \mathbf{Y}_{jm}^{(-1)}(\Omega)] \,,$$

$$\mathbf{Y}_{jm}^{j}(\Omega) = \mathbf{Y}_{jm}^{(0)}(\Omega) \,,$$

$$\mathbf{Y}_{jm}^{j-1}(\Omega) = \frac{1}{\sqrt{(2j+1)}} [\sqrt{(j+1)} \mathbf{Y}_{jm}^{(1)}(\Omega) + \sqrt{(j)} \mathbf{Y}_{jm}^{(-1)}(\Omega)] \,. \tag{2.53}$$

Substituting spherical vectors $\mathbf{Y}_{jm}^k(\Omega)$ for vectors $\mathbf{Y}_{jm}^{(\lambda)}(\Omega)$ in Eq. (2.45) for the acceleration of gravity yields

$$\mathbf{g}(\varrho, \Omega) = -\frac{GM}{\varrho^2} \sum_{jm} \left(\frac{a_0}{\varrho}\right)^j \sqrt{[(j+1)(2j+1)]}\, A_{jm} \mathbf{Y}_{jm}^{j+1}(\Omega)$$

$$+ \frac{GM}{\varrho^2} q \left(\frac{\varrho}{a_0}\right)^3 \frac{4\sqrt{\pi}}{3} [\mathbf{Y}_{00}^1(\Omega) + \frac{1}{\sqrt{2}} \mathbf{Y}_{20}^1(\Omega)]. \tag{2.54}$$

If the coordinate system is rotated through Euler angles α, β, γ and the coordinates in the rotated system marked with a prime ($\varrho = \varrho'$), the transformation formulae for spherical vectors $\mathbf{Y}_{jm}^k(\Omega)$ become

$$\mathbf{Y}_{jm}^L(\Omega') = \sum_{m=-j}^{j} D_{mM}^j(\alpha, \beta, \gamma) \mathbf{Y}_{jm}^L(\Omega) \tag{2.55}$$

and the inverse transformation formulae, in view of the D-functions being unitary, now read

$$\mathbf{Y}_{jm}^L(\Omega) = \sum_{M=-j}^{j} D_{mM}^{j*}(\alpha, \beta, \gamma) \mathbf{Y}_{jm}^L(\Omega'). \tag{2.56}$$

In particular,

$$\mathbf{Y}_{00}^1(\Omega) = D_{00}^{0*}(\alpha, \beta, \gamma) \mathbf{Y}_{00}^1(\Omega'),$$

$$\mathbf{Y}_{20}^1(\Omega) = \sum_{M=-2}^{2} D_{0M}^{2*}(\alpha, \beta, \gamma) \mathbf{Y}_{2M}^1(\Omega'). \tag{2.57}$$

As regards the D-functions, in particular,

$$D_{00}^L(\alpha, \beta, \gamma) = P_L(\cos \beta),$$

$$D_{0m}^L(\alpha, \beta, \gamma) = \sqrt{\left(\frac{4\pi}{2L+1}\right)} Y_{L-m}(\beta, \gamma) = (-1)^m \sqrt{\left(\frac{4\pi}{2L+1}\right)} Y_{Lm}^*(\beta, \gamma). \tag{2.58}$$

Substituting for $\mathbf{Y}_{jM}^L(\Omega)$ in Eqs. (2.56) and (2.58), the formula for the acceleration of gravity in the rotated system becomes

$$\mathbf{g}(\varrho, \Omega') = -\frac{GM}{\varrho^2} \sum_{jmM} \left(\frac{a_0}{\varrho}\right)^j \sqrt{[(j+1)(2j+1)]}\, A_{jm} D_{mM}^{j*}(\alpha, \beta, \gamma)$$

$$\times \mathbf{Y}_{jM}^{j+1}(\Omega') + \frac{GM}{\varrho^2} \left(\frac{\varrho}{a_0}\right)^3 \frac{4\sqrt{\pi}}{3} q [D_{00}^{0*}(\alpha, \beta, \gamma) \mathbf{Y}_{00}^1(\Omega')$$

$$+ \frac{1}{\sqrt{2}} \sum_{M=-2}^{2} D_{0M}^{2*}(\alpha, \beta, \gamma) \mathbf{Y}_{2M}^1(\Omega')]$$

$$= -\frac{GM}{\varrho^2} \sum_{jM} \left(\frac{a_0}{\varrho}\right)^j \sqrt{[(j+1)(2j+1)]}\, A'_{jM} \mathbf{Y}_{jM}^{j+1}(\Omega')$$

$$+ \frac{GM}{\varrho^2} \left(\frac{\varrho}{a_0}\right)^3 \frac{4\sqrt{\pi}}{3} q \left[\mathbf{Y}_{00}^1(\Omega') + \frac{\sqrt{(2\pi)}}{\sqrt{5}} \right.$$

$$\left. \times \sum_{M=-2}^{2} (-1)^M \mathbf{Y}_{2M}^1(\Omega') Y_{2M}^*(\beta, \gamma) \right]. \tag{2.59}$$

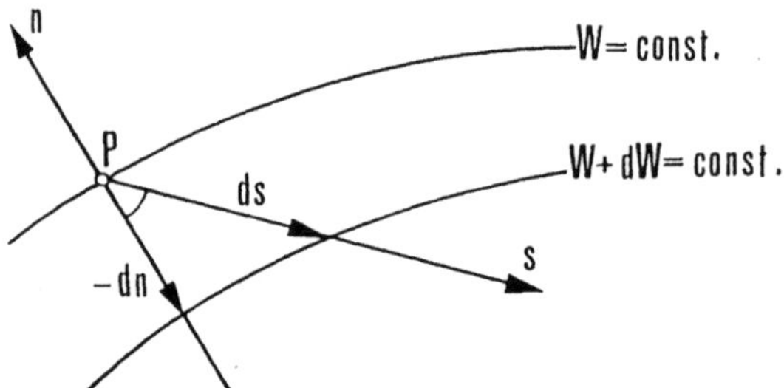

Fig. 2.1. Change of geopotential generated by displacement along orbit *ds*

By using Eqs. (2.53) and (2.43) the components of the acceleration of gravity can be expressed analogously to (2.46)–(2.48).

The infinitesimal change in geopotential (Fig. 2.1),

$$dW = \mathbf{g} \cdot d\mathbf{s} \tag{2.60}$$

corresponds to the displacement of potential point P along ds.

2.5 Listing's Geoid

2.5.1 Monge's Figure of the Geoid

In view of (2.18) Listing's geoid is defined as

$$W(\varrho,\ \vartheta,\ \Lambda) = W_0 = GM/R_0\,, \tag{2.61}$$

which represents the equation of a surface in coordinates $(\varrho,\ \vartheta,\ \Lambda)$. For some purposes it is convenient to express the geocentric radius-vector of the geoid point explicitly as a function of angular coordinates $(\vartheta,\ \Lambda)$, which is referred to as Listing's geoid. Hereafter we shall use two equivalent expressions for the geocentric radius-vector of the geoid:

$$\varrho = a_1 \sum_{j=0}^{\infty} \sum_{m=-j}^{j} E_{jm}\, Y_{jm}(\vartheta,\ \Lambda) \tag{2.62}$$

and

$$\varrho = R_0 \left[1 + A_0^{(0)} + \sum_{n=2}^{\infty} \sum_{k=0}^{n} (A_n^{(k)} \cos k\Lambda + B_n^{(k)} \sin k\Lambda)\, P_n^{(k)}(\cos \vartheta) \right]. \tag{2.63}$$

The coefficients in the expansions, E_{jm} and $A_n^{(k)}$, $B_n^{(k)}$ (Tables 2.3 and 2.4, respectively) describe the figure of the geoid in full and, consequently, may be expressed in terms of geopotential coefficients A_{jm} and $J_k^{(m)}$, $S_k^{(m)}$. The relevant relations are derived in Section 2.5.2.

Table 2.3. Coefficients of the zonal terms in the expansion of the geoid's radius-vector

n	$A_n^{(0)}$ (10^{-9})	$R_0 A_n^{(0)}$ (m)
2	$-2\,241\,889$	$-14\,266.65$
3	$2\,553$	16.25
4	$3\,117$	19.84
5	228	1.46
6	-553	-3.52
7	369	2.35
8	211	1.35
9	119	0.76
10	247	1.57
11	-234	-1.49
12	200	1.28

The following transformation relations between complex coefficients E_{jm} and real coefficients $A_j^{(m)}$, $B_j^{(m)}$ apply:

$$E_{00} = 2\sqrt{(\pi)}\,R_0(1 + A_0^{(0)})/a_1\,,$$

$$E_{j0} = 2\sqrt{(\pi)}(R_0/a_1)A_j^{(0)}/\sqrt{(2j+1)}\,, \qquad\qquad j > 0$$

$$E_{jm} = \sqrt{(2\pi)}(-1)^m(R_0/a_1)(A_j^{(m)} - iB_j^{(m)})/N_j^{(m)}\,, \quad j > 0,\ m > 0$$

$$E_{j-m} = \sqrt{(2\pi)}(R_0/a_1)(A_j^{(m)} + iB_j^{(m)})/N_j^{(m)}\,, \qquad j > 0,\ m > 0. \qquad (2.64)$$

2.5.2 Geometrical Properties of the Geoid

We shall study the properties of the geoid as a geometrical surface; the properties are inherent to the surface and can be described without referring to the ambient space. In other words, we shall be involved with the properties of the geoid which can be expressed by two coordinates u^1, u^2 on a surface. Every point on the geoid is uniquely defined by two parameters, u^1 and u^2, which represent its coordinates. Every point on the surface is defined as the point of intersection of the u^1-curve and the u^2-curve. In our particular case we shall adopt the curvilinear coordinates ϑ, Λ. This means that the ϑ-curves are lines of intersection of the geoid with a cone, apex angle 2ϑ, and the Λ-curves are lines of intersection of the geoid with meridional planes. To preserve the necessary symmetry and briefness in the symbols used, in this section we shall denote the space coordinates and coordinates on the geoid surface x^i ($i = 1, 2, 3$) and u^α ($\alpha = 1, 2$). In this section we shall also observe the rule that the lower-case Roman indices may take values of 1, 2, 3, and the Greek indices values of 1, 2.

Table 2.4. Coefficients of the tesseral and sectorial terms in the expansion of the geoid's radius-vector (fully normalized)

n	k	$\bar{A}_n^{(k)}$ (10^{-9})	$R_0\bar{A}_n^{(k)}$ (m)	$\bar{B}_n^{(k)}$ (10^{-9})	$R_0\bar{B}_n^{(k)}$ (m)	n	k	$\bar{A}_n^{(k)}$ (10^{-9})	$R_0\bar{A}_n^{(k)}$ (m)	$\bar{B}_n^{(k)}$ (10^{-9})	$R_0\bar{B}_n^{(k)}$ (m)
2	1	1	0.00	0	0.00	9	3	− 156	− 1.00	− 89	− 0.57
	2	2443	15.55	− 14.3	− 8.93		4	− 20	− 0.13	11	0.07
3	1	2044	13.01	255	1.62		5	− 18	− 0.12	− 53	− 0.34
	2	900	5.73	− 625	− 3.98		6	56	0.36	221	1.41
	3	718	4.57	1428	9.09		7	− 111	0.71	− 91	− 0.58
4	1	− 538	− 3.42	− 477	− 3.04		8	196	1.25	− 8	− 0.05
	2	355	2.26	675	4.30		9	− 50	− 0.32	98	0.63
	3	999	6.36	− 206	− 1.32	10	1	81	0.52	− 132	− 0.84
	4	− 195	− 1.24	309	1.97		2	− 90	− 0.58	− 31	− 0.20
5	1	− 55	− 0.35	− 97	− 0.62		3	− 10	− 0.06	− 161	− 1.02
	2	659	4.20	− 328	− 2.09		4	− 94	− 0.60	− 72	− 0.46
	3	− 466	− 2.97	− 218	− 1.39		5	− 57	− 0.36	− 43	− 0.28
	4	− 304	− 1.94	46	0.30		6	− 40	− 0.26	− 79	− 0.51
	5	159	1.01	− 672	− 4.28		7	8	0.05	6	0.04
6	1	− 78	− 0.50	29	0.18		8	41	0.27	− 84	− 0.53
	2	49	0.31	− 364	− 2.32		9	129	0.82	− 47	− 0.30
	3	61	0.39	4	0.03		10	104	0.66	− 30	− 0.19
	4	− 99	− 0.63	− 470	− 2.99	11	1	6	0.04	− 16	− 0.10
	5	− 267	− 1.70	− 546	− 3.48		2	27	0.17	− 91	− 0.58
	6	5	0.03	− 236	− 1.50		3	− 42	− 0.27	− 138	− 0.88
							4	− 47	− 0.30	− 64	− 0.41

n	m				
7	1	278	1.77	99	0.63
	2	330	2.10	106	0.68
	3	250	1.60	− 216	− 1.38
	4	− 283	− 1.80	− 127	− 0.81
	5	6	0.04	27	0.17
	6	− 368	− 2.35	141	0.90
	7	− 4	− 0.03	22	0.15
8	1	28	0.18	51	0.33
	2	73	0.47	72	0.46
	3	− 17	− 0.11	− 95	− 0.61
	4	− 251	− 1.60	77	0.49
	5	− 20	− 0.13	86	0.55
	6	− 68	− 0.44	318	2.02
	7	67	0.43	73	0.47
	8	− 127	− 0.81	21	0.14
9	1	155	0.99	14	0.09
	2	24	0.15	− 28	− 0.18

n	m				
	5	43	0.28	58	0.37
	6	− 10	− 0.07	35	0.23
	7	4	0.03	− 96	− 0.61
	8	− 3	− 0.02	24	− 0.15
	9	− 28	− 0.18	36	0.23
	10	− 4	− 0.03	− 13	− 0.09
	11	49	0.32	− 60	− 0.38
12	1	− 58	− 0.37	− 50	− 0.32
	2	4	0.03	23	0.15
	3	48	0.31	17	0.11
	4	− 74	− 0.47	− 0	− 0.01
	5	43	0.28	11	0.07
	6	8	0.06	39	0.25
	7	− 18	− 0.12	37	0.24
	8	− 28	− 0.18	21	0.14
	9	39	0.25	15	0.10
	10	− 1	− 0.01	30	0.19
	11	14	0.09	− 3	− 0.02
	12	0	0.00	− 10	− 0.06

For both types of indices we shall keep to the summation convention that the indices to which the summation refers appear twice in the relation.

Summarizing the coordinate notation,

$$x^1 = \varrho = a\,E_{jm}\,Y_{jm}(\Omega)\,,$$

$$x^2 = \vartheta = u^1\,,$$

$$x^3 = \Lambda = u^2\,. \tag{2.65}$$

The equation of the geoid surface may formally be expressed as

$$x^1 = x^1(u^1, u^2)\,;\quad x^2 = u^1\,;\quad x^3 = u^2\,. \tag{2.66}$$

We shall denote the derivative

$$\frac{\partial x^i}{\partial u^\alpha} = x^i_\alpha\,;\quad i = 1, 2, 3\,;\quad \alpha = 1, 2\,, \tag{2.67}$$

i.e. $x^1_1 = \partial\varrho/\partial\vartheta$, $x^1_2 = \partial\varrho/\partial\Lambda$, $x^2_1 = 1$, $x^2_2 = 0$, $x^3_1 = 0$, $x^3_2 = 1$.

The metric tensor $a_{\alpha\beta}$, which is used to define the distance between two points on a surface, is of fundamental importance in the geometry of surfaces. The square of the element of distance is expressed in space as $ds^2 = g_{mn}dx^m dx^n$ and on a surface as $ds^2 = a_{\alpha\beta}\,du^\alpha du^\beta$. Symbols g_{mn} stand for components of the spatial covariant metric tensor, and in this particular spherical system are expressed as

$$g_{11} = 1\,;\quad g_{22} = (x^1)^2 = \varrho^2\,;\quad g_{33} = (x^1)^2 \sin^2 x^2 = \varrho^2 \sin^2 \vartheta\,;$$

$$g_{ij} = 0\,;\ i \neq j\,;\quad G = \det|g_{ij}| = (x^1)^4 \sin^2 x^2 = \varrho^4 \sin^2 \vartheta\,. \tag{2.68}$$

For points on the surface both infinitesimal distances are equal; hence,

$$g_{mn}\,dx^m\,dx^n = g_{mn}\,x^m_\alpha\,du^\alpha\,x^n_\beta\,du^\beta = a_{\alpha\beta}\,du^\alpha\,du^\beta\,.$$

This yields the relation between the metric tensors:

$$a_{\alpha\beta} = g_{mn}\,x^m_\alpha\,x^n_\beta\,. \tag{2.69}$$

We shall denote the determinant of tensor $a_{\alpha\beta}$ $A = \det|a_{\alpha\beta}|$. Metric tensor $a_{\alpha\beta}$ reads

$$(a_{\alpha\beta}) = \begin{bmatrix} (x^1)^2 + (x^1_1)^2 & x^1_1 x^1_2 \\ x^1_1 x^1_2 & (x^1_2)^2 + (x^1 \sin x^2)^2 \end{bmatrix}$$

$$= \begin{bmatrix} \varrho^2 + \left(\dfrac{\partial\varrho}{\partial\vartheta}\right)^2 & \dfrac{\partial\varrho}{\partial\vartheta}\dfrac{\partial\varrho}{\partial\Lambda} \\ \dfrac{\partial\varrho}{\partial\vartheta}\dfrac{\partial\varrho}{\partial\Lambda} & (\varrho \sin\vartheta)^2 + \left(\dfrac{\partial\varrho}{\partial\Lambda}\right)^2 \end{bmatrix}\,. \tag{2.70}$$

Its determinant

$$A = a_{11}a_{22} - (a_{12})^2 = \varrho^4 \sin^2\vartheta\,. A_0 = GA_0\,, \tag{2.71}$$

where

$$A_0 = 1 + \left(\frac{1}{\varrho \sin \vartheta}\right)^2 \left(\frac{\partial \varrho}{\partial \Lambda}\right)^2 + \frac{1}{\varrho^2}\left(\frac{\partial \varrho}{\partial \vartheta}\right)^2 . \tag{2.72}$$

The contravariant components of the metric tensor are

$$a^{11} = a_{22}/A; \quad a^{22} = a_{11}/A; \quad a^{12} = -a_{12}/A; \quad a^{21} = -a_{21}/A . \tag{2.73}$$

The square of the element of length ds^2 can be expressed in terms of metric tensor $a_{\alpha\beta}$ as follows:

$$ds^2 = a_{\alpha\beta}\, du^\alpha\, du^\beta. \tag{2.74}$$

Element dS of the geoid surface then reads

$$dS = \sqrt{(A)}\, du^1\, du^2 = \varrho^2 \sqrt{(A_0)} \sin \vartheta\, d\vartheta\, d\Lambda . \tag{2.75}$$

In the geometry of the geoid quantity A_0 represents a norm. Indeed, it occurs not only in formula (2.75) for the surface element but also in the formulae for the geoid normal (2.95)–(2.97). We shall, therefore, derive the formulae for expressing it in space. To this end the following formulae for the products of the spherical harmonics and their derivatives will be required (Varshalovich et al. 1975):

$$Y_{j_1 m_1}(\Omega)\, Y_{j_2 m_2}(\Omega) = Q^{jm}_{j_1 m_1 j_2 m_2}\, Y_{jm}(\Omega) , \tag{2.76}$$

where

$$Q^{jm}_{j_1 m_1 j_2 m_2} = \{(2_{j_1} + 1)(2_{j_2} + 1)/[4\pi(2_j + 1)]\}^{1/2}$$
$$\times C^{j0}_{j_1 0 j_2 0}\, C^{jm}_{j_1 m_1 j_2 m_2} . \tag{2.77}$$

Quantities $C^{jm}_{j_1 m_1 j_2 m_2}$ are Clebsch–Gordan coefficients.

$$C^{c\gamma}_{a\alpha b\beta} = \delta^{\alpha+\beta}_{\gamma} \Delta(abc)\left[\frac{(c+\gamma)!(c-\gamma)!(2c+1)}{(a+\alpha)!(a-\alpha)!(b+\beta)!(b-\beta)!}\right]^{1/2}$$
$$\times \sum_k \frac{(-1)^{b+\beta+k}(c+b+\alpha-k)!(a-\alpha+k)!}{k!(c-a+b-k)!(c+\gamma-k)!(a-b-\gamma+k)!}, \tag{2.78}$$

where

$$\Delta(abc) = \left[\frac{(a+b-c)!(a-b+c)!(-a+b+c)!}{(a+b+c+1)!}\right]^{1/2} . \tag{2.79}$$

The summation in (2.78) is over all values of k for which all factorials are non-negative. The reader is reminded that the numerical computation of coefficients $Q^{jm}_{j_1 m_1 j_2 m_2}$ using Eqs. (2.77)–(2.79) is very demanding on computer time. It is much more convenient to use the fast numerical algorithm based on integration over the surface of the sphere and on fast Fourier transform (Martinec 1989b). The products of the derivatives of the spherical harmonics are

$$\frac{\partial Y_{j_1 m_1}(\Omega)}{\partial \vartheta}\frac{\partial Y_{j_2 m_2}(\Omega)}{\partial \vartheta} = \frac{1}{2}\Bigg[j_1(j_1+1) + j_2(j_2+1) $$

$$ -j(j+1) + 2\frac{m_1 m_2}{\sin^2\vartheta}\Bigg] Q^{jm}_{j_1 m_1 j_2 m_2} Y_{jm}(\Omega)\,, \tag{2.80}$$

$$\frac{\partial Y_{j_1 m_1}(\Omega)}{\partial \Lambda}\frac{\partial Y_{j_2 m_2}(\Omega)}{\partial \Lambda} = -m_1 m_2\, Q^{jm}_{j_1 m_1 j_2 m_2} Y_{jm}\,. \tag{2.81}$$

Using these relations A_0 is developed into a series of spherical harmonics:

$$A_0 = 1 + (a^2/2\varrho^2)\, E_{j_1 m_1} E_{j_2 m_2}[\,j_1(j_1+1)$$

$$+ j_2(j_2+1) - j(j+1)]\, Q^{jm}_{j_1 m_1 j_2 m_2}\, Y_{jm}(\Omega)\,. \tag{2.82}$$

In view of (2.159) and (2.164), by substituting the expansion for $1/\varrho^2$, A_0 can be expressed exclusively in terms of angular coordinates Ω and radius-vector coefficients E_{jm}:

$$A_0 = 1 + \tfrac{1}{2} E_{j_1 m_1} E_{j_2 m_2} E^{(-2)}_{j_3 m_3}\ \ [\,j_1(j_1+1) + j_2(j_2+1) - j_{12}(j_{12}+1)]$$

$$\times Q^{j_{12} m_{12}}_{j_1 m_1 j_2 m_2}\, Q^{jm}_{j_{12} m_{12} j_3 m_3}\, Y_{jm}(\Omega)\,. \tag{2.83}$$

The vectors tangent to the coordinate curves on surface $u^1 = \text{const}$ and $u^2 = \text{const}$ are expressed in spatial components as $x^r_1/\sqrt{a_{11}}$ and $x^r_2/\sqrt{a_{22}}$ ($r = 1,$ 2, 3), respectively. The cosine of the angle between the coordinate curves is $a_{12}/\sqrt{(a_{11}a_{22})}$; more generally, the cosine of angle Θ between two unit surface vectors λ^α and μ^α

$$\cos\Theta = a_{\alpha\beta}\lambda^\alpha\mu^\beta\,. \tag{2.84}$$

Tensor $a_{\alpha\beta}$ is also used to express the differential equation of the geodesic which represents the shortest distance between two points A and B on a surface,

$$L = \int_A^B \frac{\mathrm{d}s}{\mathrm{d}t}\,\mathrm{d}t = \int_A^B \left(a_{\alpha\beta}\frac{\mathrm{d}u^\alpha}{\mathrm{d}t}\frac{\mathrm{d}u^\beta}{\mathrm{d}t}\right)^{1/2}\mathrm{d}t = \min\,. \tag{2.85}$$

Using variation calculus the above condition of the differential equation of the geodesic can be expressed as (Sokolnikoff 1971):

$$\frac{\mathrm{d}^2 u^\alpha}{\mathrm{d}s^2} + \left\{\begin{matrix} \alpha \\ \beta\ \ \gamma \end{matrix}\right\}\frac{\mathrm{d}u^\beta}{\mathrm{d}s}\frac{\mathrm{d}u^\gamma}{\mathrm{d}s} = 0\,. \tag{2.86}$$

Symbols $\left\{\begin{matrix} \alpha \\ \beta\ \ \gamma \end{matrix}\right\}$ are Christoffel symbols of the second kind which, in terms of tensor components $a_{\alpha\beta}$, read

$$\left\{\begin{matrix} \alpha \\ \beta\ \ \gamma \end{matrix}\right\} = a^{\alpha\delta}[\beta\gamma,\,\delta]\,, \tag{2.87}$$

where $a^{\alpha\delta}$ are contravariant components of tensor $a_{\alpha\delta}$, and

$$[\alpha\beta,\,\gamma] = \frac{1}{2}\left(\frac{\partial a_{\beta\gamma}}{\partial u^\alpha} + \frac{\partial a_{\alpha\gamma}}{\partial u^\beta} - \frac{\partial a_{\alpha\beta}}{\partial u^\gamma}\right) \tag{2.88}$$

are Christoffel symbols of the first kind. After carrying out the operations indicated, we arrive at

$$
\begin{aligned}
\left\{{1 \atop 1\ \ 1}\right\}_a &= \frac{1}{A_0}\left\{\left[1+\frac{1}{\varrho^2\sin^2\vartheta}\left(\frac{\partial\varrho}{\partial\Lambda}\right)^2\right]\left[1+\frac{1}{\varrho}\frac{\partial^2\varrho}{\partial\vartheta^2}\right]\frac{1}{\varrho}\frac{\partial\varrho}{\partial\vartheta}\right.\\
&\quad \left. +\frac{1}{\varrho^3}\frac{\partial\varrho}{\partial\vartheta}\left(\frac{\partial\varrho}{\partial\Lambda}\right)^2\left[1-\frac{1}{r}\frac{\partial^2\varrho}{\partial\vartheta^2}\right]\right\},\\[2mm]
\left\{{1 \atop 1\ \ 2}\right\}_a &= \frac{1}{A_0}\left\{\left[1+\frac{1}{\varrho^2\sin^2\vartheta}\left(\frac{\partial\varrho}{\partial\Lambda}\right)^2\right]\right.\\
&\quad \times\left[\frac{1}{\varrho}\frac{\partial\varrho}{\partial\Lambda}+\frac{1}{\varrho^2}\frac{\partial\varrho}{\partial\vartheta}\frac{\partial^2\varrho}{\partial\vartheta\,\partial\Lambda}\right]-\frac{1}{\varrho^2}\frac{\partial\varrho}{\partial\vartheta}\frac{\partial\varrho}{\partial\Lambda}\\
&\quad \left. \times\left[\frac{1}{\varrho}\frac{\partial\varrho}{\partial\vartheta}+\cot\vartheta+\frac{1}{\varrho^2}\frac{\partial\varrho}{\partial\Lambda}\frac{\partial^2\varrho}{\partial\vartheta\,\partial\Lambda}\right]\right\},\\[2mm]
\left\{{1 \atop 2\ \ 2}\right\}_a &= \frac{-1}{A_0}\left\{\left[1+\frac{1}{\varrho^2\sin^2\vartheta}\left(\frac{\partial\varrho}{\partial\Lambda}\right)^2\right]\left[\frac{1}{\varrho}\frac{\partial\varrho}{\partial\vartheta}\sin^2\vartheta\right.\right.\\
&\quad \left. +\varrho^2\sin\vartheta\cos\vartheta-\frac{1}{\varrho^2}\frac{\partial\varrho}{\partial\vartheta}\frac{\partial^2\varrho}{\partial\varrho^2}\right]\\
&\quad \left. +\left[1+\frac{1}{\varrho\sin^2\vartheta}\frac{\partial^2\varrho}{\partial\Lambda^2}\right]\frac{1}{\varrho^2}\frac{\partial\varrho}{\partial\vartheta}\left(\frac{\partial\varrho}{\partial\Lambda}\right)^2\right\},\\[2mm]
\left\{{2 \atop 1\ \ 1}\right\}_a &= -\frac{1}{A_0}\frac{1}{\varrho\sin^2\vartheta}\frac{\partial\varrho}{\partial\Lambda}\\
&\quad \times\left\{1-\frac{1}{\varrho}\frac{\partial^2\varrho}{\partial\vartheta^2}+\frac{2}{\varrho^2}\left(\frac{\partial\varrho}{\partial\vartheta}\right)^2\right\},\\[2mm]
\left\{{2 \atop 1\ \ 2}\right\}_a &= \frac{1}{A_0}\left\{\left[1+\frac{1}{\varrho^2}\left(\frac{\partial\varrho}{\partial\vartheta}\right)^2\right]\left[\frac{1}{\varrho}\frac{\partial\varrho}{\partial\vartheta}+\cot\vartheta\right.\right.\\
&\quad \left. +\frac{1}{\varrho^2\sin^2\vartheta}\frac{\partial\varrho}{\partial\Lambda}\frac{\partial^2\varrho}{\partial\vartheta\,\partial\Lambda}\right]\\
&\quad \left. -\frac{1}{\varrho^2\sin^2\vartheta}\frac{\partial\varrho}{\partial\vartheta}\frac{\partial\varrho}{\partial\Lambda}\left[\frac{\partial\varrho}{\partial\Lambda}+\frac{\partial\varrho}{\partial\vartheta}\frac{\partial^2\varrho}{\partial\vartheta\,\partial\Lambda}\right]\right\},\\[2mm]
\left\{{2 \atop 2\ \ 2}\right\}_a &= \frac{1}{A_0}\left\{\left[1+\frac{1}{\varrho^2}\left(\frac{\partial\varrho}{\partial\vartheta}\right)^2\right]\right.\\
&\quad \times\left[1+\frac{1}{\varrho\sin^2\vartheta}\frac{\partial\varrho}{\partial\Lambda^2}\right]+\frac{1}{\varrho^2}\left(\frac{\partial\varrho}{\partial\vartheta}\right)^2\\
&\quad \left. \times\left[1+\cot\vartheta-\frac{1}{\varrho\sin^2\vartheta}\frac{\partial^2\varrho}{\partial\Lambda^2}\right]\right\}\frac{1}{\varrho}\frac{\partial\varrho}{\partial\Lambda}.
\end{aligned}
\tag{2.89}
$$

To be able to make use of the above relations, it is first necessary to express derivatives x_{τ}^1, i.e. $\partial\varrho/\partial\vartheta$ and $\partial\varrho/\partial\Lambda$, where ϱ is given by (2.62). This requires derivative $\partial Y_{jm}(\Omega)/\partial\vartheta$, for which formula (2.44) has already been introduced. For this derivative we easily get

$$\frac{\partial Y_{JM}(\Omega)}{\partial\Lambda} = i\,M\,Y_{JM}(\Omega)\,. \tag{2.90}$$

Using these relations, derivatives x_1^1 and x_2^1 can be expressed as follows:

$$x_1^1 = \frac{\partial\varrho}{\partial\vartheta} = \frac{a_1}{2}\sum_{JM} E_{JM}\{\sqrt{[(J-M)(J+M+1)]}\,\exp(-i\Lambda)\,Y_{JM+1}(\Omega)$$

$$-\sqrt{[(J+M)(J-M+1)]}\,\exp(i\Lambda)\,Y_{JM-1}(\Omega)\}, \tag{2.91}$$

$$x_2^1 = \frac{\partial\varrho}{\partial\Lambda} = ia_1\sum_{JM} M E_{JM}\,Y_{JM}(\Omega)\,. \tag{2.92}$$

The most important among the geometrical parameters of the geoid surface is the normal, which is the tangent to the line of force of the gravity field at the given point. From differential geometry (Sokolnikoff 1971) we know that the covariant component n_r of the unit normal is expressed as

$$n_r = \tfrac{1}{2}\,\varepsilon^{\alpha\beta}\,\varepsilon_{rst}\,x_\alpha^s\,x_\beta^t\,. \tag{2.93}$$

Tensors $\varepsilon^{\alpha\beta}$ and ε_{rst} are defined as

$$\varepsilon^{\alpha\beta} = e^{\alpha\beta}/\sqrt{A}\,, \quad \varepsilon_{rst} = \sqrt{(G)}e_{rst}\,. \tag{2.94}$$

System $e^{\alpha\beta}$ is defined as follows: $e^{12} = -e^{21} = 1$; $e^{11} = e^{22} = 0$. Similarly e_{rst} are unity if indices r, s, t are even permutations of numbers 1, 2, 3, and equal to -1 if the permutation is odd; $e_{rst} = 0$ if there are at least two identical numbers among indices r, s, t.

Explicitly the covariant components of the unit normal are

$$n_1 = \frac{1}{\sqrt{A_0}}\,, \quad n_2 = -\frac{1}{\sqrt{A_0}}\frac{\partial\varrho}{\partial\vartheta}\,, \quad n_3 = -\frac{1}{\sqrt{A_0}}\frac{\partial\varrho}{\partial\Lambda}\,, \tag{2.95}$$

and similarly the contravariant components

$$n^1 = n_1/g_{11} = n_1\,; \quad n^2 = n_2/g_{22} = -\frac{1}{\varrho^2}\frac{1}{\sqrt{A_0}}\frac{\partial\varrho}{\partial\vartheta}\,,$$

$$n^3 = n_3/g_{33} = -\frac{1}{\varrho^2\sqrt{(A_0)}\sin^2\vartheta}\frac{\partial\varrho}{\partial\Lambda}\,. \tag{2.96}$$

The projections of the normal into coordinate axes x^1, x^2, x^3 are the direction cosines of the normal:

$$\cos\alpha_\varrho = \frac{1}{\sqrt{A_0}}\,, \quad \cos\alpha_\vartheta = -\frac{1}{\varrho}\frac{1}{\sqrt{A_0}}\frac{\partial\varrho}{\partial\vartheta}\,, \quad \cos\alpha_\Lambda = -\frac{1}{\varrho}\frac{1}{\sqrt{A_0}\sin\vartheta}\frac{\partial\varrho}{\partial\Lambda}\,. \tag{2.97}$$

The direction cosines of the normal have already been expressed in (2.49) in terms of field intensity components. Comparing (2.97) and (2.51) yields

$$\frac{g_{\vartheta}}{g_{\varrho}} = -\frac{1}{\varrho}\frac{\partial\varrho}{\partial\vartheta}, \tag{2.98}$$

$$\frac{g_{\varLambda}}{g_{\varrho}} = -\frac{1}{\varrho\sin\vartheta}\frac{\partial\varrho}{\partial\varLambda}. \tag{2.99}$$

The rhs of Eqs. (2.98) and (2.99) are now expressed using Eqs. (2.91) and (2.92), i.e. in terms of harmonic coefficients E_{JM} of the geoid surface, whereas the lhs are determined by Eqs. (2.46)–(2.48), into which it is possible to substitute for ϱ from (2.62). The lhs are thus expressed as a combination of geopotential coefficients A_{JM} and harmonic coefficients E_{JM}, assuming, of course, that the geopotential coefficients are known.

The unit normal to the geoid contains terms representing the first derivatives of the geoid surface. We shall now turn our attention to the expressions related to the second derivatives. Since, from the point of view of physics, the detailed structure of the geoid reflects the dynamic processes in the mantle and lithosphere, we expect these expressions to help us in defining the region in which large changes occur in the direction of the normal; in geometric terms these are regions where the curvature changes rapidly. To preserve the physical and geometric meaning of the operations to be carried out, the derived quantities must be invariant to the coordinate system used; in other words they must reflect the inherent properties of the surface. Only such quantities have a physical and geometric meaning. In the case of the derivative this in fact means that we must adopt rules of differentiation such that the tensor character of the quantity does not change by differentiation, i.e. we must only use tensor derivatives.

We shall first consider the tensor derivative of the vector of unit normal n^r (Sokolnikoff 1971):

$$n^r_{/\alpha} = \frac{\partial n^r}{\partial u^{\alpha}} + \left\{\begin{matrix} r \\ m \quad n \end{matrix}\right\}_g n^m x^n_{\alpha}. \tag{2.100}$$

Unit normal n^r is a unit contravariant vector and thus satisfies the equation

$$g_{mn}n^m n^n = 1. \tag{2.101}$$

If tensor differentiation is applied to this equation, taking into account that the tensor derivative of the metric tensor is zero,

$$g_{mn}n^m n^n_{/\alpha} = 0. \tag{2.102}$$

This equation indicates that vector $n^r_{/\alpha}$ is perpendicular to normal n^r and, therefore, vector $n^r_{/\alpha}$ is tangential to the geoid.

Equation (2.100) can be modified to read

$$n^r_{/\alpha} = -a^{\beta\gamma}b_{\beta\alpha}x^r_{\gamma}, \tag{2.103}$$

which is Weingarten's formula. Tensor $b_{\alpha\beta}$ has the form

$$b_{\alpha\beta} = \tfrac{1}{2}\, \varepsilon^{\sigma\tau}\, \varepsilon_{mnp}\, x^{m}_{\alpha,\beta}\, x^{n}_{\sigma}\, x^{p}_{\tau}, \tag{2.104}$$

where $x^{m}_{\alpha,\beta}$ represents the covariant derivative of tensor x^{m}_{α} with respect to coordinate u^{β}. In geometry this derivative is expressed by the formula

$$x^{r}_{\alpha\beta} = \frac{\partial^2 x^r}{\partial u^{\alpha}\partial u^{\beta}} + \left\{ \begin{matrix} r \\ m \quad n \end{matrix} \right\}_{g} x^{m}_{\alpha} x^{n}_{\beta} - \left\{ \begin{matrix} \sigma \\ \alpha \quad \beta \end{matrix} \right\}_{\alpha} x^{r}_{\sigma}. \tag{2.105}$$

Index g indicates that the Christoffel symbol refers to metric tensor g_{mn}, and similarly index a indicates reference to metric symbol $a_{\alpha\beta}$. The Christoffel symbols referred to tensor g_{ij} are in explicit form:

$$\left\{ \begin{matrix} 1 \\ 2 \quad 2 \end{matrix} \right\}_{g} = -\varrho; \quad \left\{ \begin{matrix} 1 \\ 3 \quad 3 \end{matrix} \right\}_{g} = -\varrho \sin^2\vartheta; \quad \left\{ \begin{matrix} 2 \\ 1 \quad 2 \end{matrix} \right\}_{g} = \left\{ \begin{matrix} 2 \\ 2 \quad 1 \end{matrix} \right\}_{g} = \frac{1}{\varrho},$$

$$\left\{ \begin{matrix} 2 \\ 3 \quad 3 \end{matrix} \right\}_{g} = -\sin\vartheta\cos\vartheta; \quad \left\{ \begin{matrix} 3 \\ 1 \quad 3 \end{matrix} \right\}_{g} = \left\{ \begin{matrix} 3 \\ 3 \quad 1 \end{matrix} \right\}_{g} = \frac{1}{\varrho},$$

$$\left\{ \begin{matrix} 3 \\ 2 \quad 3 \end{matrix} \right\}_{g} = \left\{ \begin{matrix} 3 \\ 3 \quad 2 \end{matrix} \right\}_{g} = \cot\vartheta. \tag{2.106}$$

In view of (2.105) and using (2.89) and (2.106) covariant derivatives $x^{i}_{\alpha,\beta}$ can be expressed as

$$x^{1}_{1,1} = \frac{\partial^2 \varrho}{\partial\vartheta^2} - \varrho - \left\{ \begin{matrix} 1 \\ 1 \quad 1 \end{matrix} \right\}_{a} \frac{\partial\vartheta}{\partial\varrho} - \left\{ \begin{matrix} 2 \\ 1 \quad 1 \end{matrix} \right\}_{a} \frac{\partial\varrho}{\partial\Lambda},$$

$$x^{2}_{1,1} = \frac{2}{r}\frac{\partial\varrho}{\partial\vartheta} - \left\{ \begin{matrix} 1 \\ 1 \quad 1 \end{matrix} \right\}_{a}, \quad x^{3}_{1,1} = - \left\{ \begin{matrix} 2 \\ 1 \quad 1 \end{matrix} \right\}_{a},$$

$$x^{1}_{1,2} = \frac{\partial^2 \varrho}{\partial\vartheta\,\partial\varphi} - \left\{ \begin{matrix} 1 \\ 1 \quad 2 \end{matrix} \right\}_{a} \frac{\partial\varrho}{\partial\vartheta} - \left\{ \begin{matrix} 2 \\ 1 \quad 2 \end{matrix} \right\}_{a} \frac{\partial\varrho}{\partial\Lambda},$$

$$x^{2}_{1,2} = \frac{1}{\varrho}\frac{\partial\varrho}{\partial\Lambda} - \left\{ \begin{matrix} 1 \\ 1 \quad 2 \end{matrix} \right\}_{a},$$

$$x^{3}_{1,2} = \frac{1}{\varrho}\frac{\partial\varrho}{\partial\vartheta} + \cot\vartheta - \left\{ \begin{matrix} 2 \\ 1 \quad 2 \end{matrix} \right\}_{a},$$

$$x^{1}_{2,2} = \frac{\partial^2 \varrho}{\partial\Lambda^2} - \varrho \sin^2\vartheta - \left\{ \begin{matrix} 1 \\ 2 \quad 2 \end{matrix} \right\}_{a} \frac{\partial\varrho}{\partial\vartheta} - \left\{ \begin{matrix} 2 \\ 2 \quad 2 \end{matrix} \right\}_{a} \frac{\partial\varrho}{\partial\Lambda},$$

$$x^{2}_{2,2} = - \left\{ \begin{matrix} 1 \\ 2 \quad 2 \end{matrix} \right\}_{a} - \sin\vartheta\cos\vartheta,$$

$$x^{3}_{2,2} = \frac{2}{\varrho}\frac{\partial\varrho}{\partial\Lambda} - \left\{ \begin{matrix} 2 \\ 2 \quad 2 \end{matrix} \right\}_{a}. \tag{2.107}$$

Equation (2.104) yields the following expressions for the components of tensor $b_{\alpha\beta}$:

$$\sqrt{A_0}\, b_{11} = \frac{\partial^2 \varrho}{\partial \vartheta^2} - \varrho - \frac{2}{\varrho}\left(\frac{\partial \varrho}{\partial \vartheta}\right)^2 ,$$

$$\sqrt{A_0}\, b_{22} = \frac{\partial^2 \varrho}{\partial \Lambda^2} - \varrho \sin^2 \vartheta + \sin \vartheta \cos \vartheta \, \frac{\partial \varrho}{\partial \vartheta} - \frac{2}{\varrho}\left(\frac{\partial \varrho}{\partial \Lambda}\right)^2 ,$$

$$\sqrt{A_0}\, b_{12} = \frac{\partial^2 \varrho}{\partial \vartheta \, \partial \Lambda} - \frac{2}{\varrho}\frac{\partial \varrho}{\partial \vartheta}\frac{\partial \varrho}{\partial \Lambda} - \cot \vartheta \, \frac{\partial \varrho}{\partial \Lambda} . \qquad (2.108)$$

Tensors $a_{\alpha\beta}$ and $b_{\alpha\beta}$ now being available, we can use them to express the invariant surfaces of the geoid, the mean curvature of surface H and Gauss' total curvature K:

$$2H = a^{\alpha\beta} b_{\alpha\beta} , \qquad (2.109)$$

$$K = \frac{B}{A} = \frac{b_{11}b_{22} - b_{12}b_{21}}{a_{11}a_{22} - a_{12}a_{21}} . \qquad (2.110)$$

We can denote the tangent unit vector λ^α, i.e.

$$a_{\alpha\beta} \lambda^\alpha \lambda^\beta = 1 . \qquad (2.111)$$

Now let us consider a curve, which has tangent λ^α, on the geoid surface. Its normal curvature can be expressed also in terms of tensor $b_{\alpha\beta}$:

$$\varkappa = b_{\alpha\beta} \lambda^\alpha \lambda^\beta . \qquad (2.112)$$

Quadratic form (2.112) in directions λ^α or normal curvature can be altered using (2.111) to read

$$(b_{\alpha\beta} - \varkappa a_{\alpha\beta}) \lambda^\alpha \lambda^\beta = 0 . \qquad (2.113)$$

This means that different normal curvatures $\varkappa$ correspond to different directions of the tangent. It is easy to establish the conditions for directions $\lambda^\alpha_{(\gamma)}$ ($\gamma = 1, 2$) in which the normal curvature displays extreme values:

$$(b_{\alpha\beta} - \varkappa_{(\gamma)} a_{\alpha\beta}) \lambda^\alpha = 0 , \quad \gamma = 1, 2. \qquad (2.114)$$

[No summation over index (γ).] If the first equation of (2.114) is multiplied by $\lambda^\beta_{(2)}$ and the second by $\lambda^\beta_{(1)}$, and the latter is subtracted from the former, we get

$$(\varkappa_{(2)} - \varkappa_{(1)}) a_{\alpha\beta} \lambda^\alpha_{(2)} \lambda^{(\beta)}_{(1)} = 0 . \qquad (2.115)$$

Since $\varkappa_{(2)} \neq \varkappa_{(1)}$,

$$a_{\alpha\beta} \lambda^\alpha_{(2)} \lambda^\beta_{(1)} = 0 . \qquad (2.116)$$

Equation (2.116) indicates that directions $\lambda^\alpha_{(1)}$ and $\lambda^\alpha_{(2)}$, which are called the principal directions of curvature, are perpendicular to one another.

The condition for the extreme values of the normal curvature can be derived from the condition that the determinant of the homogeneous system of

Eqs. (2.114) for the principal directions of curvature $\lambda^{\alpha}_{(\gamma)}$ should be zero, i.e.

$$\det|b_{\alpha\beta} - \varkappa_{(\gamma)}a_{\alpha\beta}| = 0 , \tag{2.117}$$

which can be expressed as

$$\varkappa^2_{(\gamma)} - a^{\alpha\beta} b_{\alpha\beta} \varkappa_{(\gamma)} + B/A = 0 , \quad B = \det|b_{\alpha\beta}| . \tag{2.118}$$

In terms of curvatures H and K, from Eqs. (2.109) and (2.110), Eq. (2.118) will now read

$$\varkappa^2_{(\gamma)} - 2H\varkappa_{(\gamma)} + K = 0 . \tag{2.119}$$

For the extreme curvatures $\varkappa_{(1)}$ and $\varkappa_{(2)}$, therefore,

$$\varkappa_{(1)} + \varkappa_{(2)} = 2H , \quad \varkappa_{(1)}\varkappa_{(2)} = K . \tag{2.120}$$

Alternatively, the principal directions of curvature can be obtained by solving equation

$$h_{\alpha\beta} \lambda^{\alpha}_{(\gamma)} \lambda^{\beta}_{(\gamma)} = 0 , \tag{2.121}$$

in which tensor $h_{\alpha\beta}$ is defined in terms of $a_{\alpha\beta}$ and $b_{\alpha\beta}$:

$$h_{\alpha\beta} = e^{\gamma\delta} a_{\alpha\gamma} b_{\beta\delta} . \tag{2.122}$$

The differential equations of the lines on the geoid, whose tangents represent the principal directions of curvature, read

$$h_{\alpha\beta} du^{\alpha} du^{\beta} = 0 . \tag{2.123}$$

We have given the formulae for the fundamental geometric invariants of the geoid, i.e. for unit normal n^r, its tensor derivative $n^r_{,\alpha}$, Gauss' curvature K and mean curvature H, principal directions of curvature $\lambda^{\alpha}_{(1)}$ and $\lambda^{\alpha}_{(2)}$, and principal curvatures of the surface $\varkappa_{(1)}$ and $\varkappa_{(2)}$. Each of these quantities is independent of the coordinate system adopted and, consequently, represents the inherent properties of the surface.

The covariant derivatives of the normal can be expressed in two ways, either in terms of Christoffel symbols,

$$n^r_{,\alpha} = \frac{\partial n^r}{\partial u^{\alpha}} + \left\{ {r \atop m\ n} \right\}_g n^m x^n_{\alpha} , \tag{2.124}$$

or in terms of tensors $a^{\alpha\beta}$ and $b_{\alpha\beta}$,

$$n^r_{,\alpha} = - a^{\beta\gamma} b_{\beta\alpha} x^r_{\gamma} . \tag{2.125}$$

Both yield identical expressions:

$$n^1_{,1} = \frac{1}{\varrho\sqrt{A_0}} \frac{\partial\varrho}{\partial\vartheta} + \frac{\partial}{\partial\vartheta} \frac{1}{\sqrt{A_0}} ,$$

$$n^1_{,2} = \frac{1}{\varrho} \frac{1}{\sqrt{A_0}} \frac{\partial\varrho}{\partial\Lambda} + \frac{\partial}{\partial\Lambda} \frac{1}{\sqrt{A_0}} ,$$

$$n_{,1}^2 = -\frac{1}{\varrho^2\sqrt{A_0}}\frac{\partial^2\varrho}{\partial\vartheta^2} + \frac{1}{\varrho^3\sqrt{A_0}}\left(\frac{\partial\varrho}{\partial\vartheta}\right)^2 + \frac{1}{\varrho\sqrt{A_0}}$$
$$- \frac{1}{\varrho^2}\frac{\partial\varrho}{\partial\vartheta}\frac{\partial}{\partial\vartheta}\frac{1}{\sqrt{A_0}}\,,$$

$$n_{,2}^2 = -\frac{1}{\varrho^2\sqrt{A_0}}\frac{\partial^2\varrho}{\partial\vartheta\,\partial\Lambda} + \frac{1}{\varrho^3\sqrt{A_0}}\frac{\partial\varrho}{\partial\vartheta}\frac{\partial\varrho}{\partial\Lambda}$$
$$+ \frac{1}{\varrho^2\sqrt{A_0}}\cot\vartheta\frac{\partial\varrho}{\partial\Lambda} - \frac{1}{\varrho^2}\frac{\partial\varrho}{\partial\vartheta}\frac{\partial}{\partial\Lambda}\frac{1}{\sqrt{A_0}}\,,$$

$$n_{,1}^3 = -\frac{1}{\varrho^2\sqrt{A_0}\sin^2\vartheta}\frac{\partial^2\varrho}{\partial\vartheta\,\partial\Lambda} + \frac{1}{\varrho^3\sqrt{A_0}\sin^2\vartheta}\frac{\partial\varrho}{\partial\vartheta}\frac{\partial\varrho}{\partial\Lambda}$$
$$+ \frac{1}{\varrho^2\sqrt{A_0}\sin^2\vartheta}\cot\vartheta\frac{\partial\varrho}{\partial\Lambda}$$
$$- \frac{1}{\varrho^2\sin^2\vartheta}\frac{\partial\varrho}{\partial\Lambda}\frac{\partial}{\partial\vartheta}\frac{1}{\sqrt{A_0}}\,,$$

$$n_{,1}^3 = -\frac{1}{\varrho^2\sqrt{A_0}\sin^2\vartheta}\frac{\partial^2\varrho}{\partial\varphi^2} + \frac{1}{\varrho^3\sqrt{A_0}\sin^2\vartheta}\left(\frac{\partial\varrho}{\partial\Lambda}\right)^2$$
$$+ \frac{1}{\varrho\sqrt{A_0}} - \frac{1}{\varrho^2\sqrt{A_0}}\cot\vartheta\frac{\partial\varrho}{\partial\vartheta} - \frac{1}{\varrho^2\sin^2\vartheta}\frac{\partial\varrho}{\partial\Lambda}\frac{\partial}{\partial\Lambda}\frac{1}{\sqrt{A_0}}\,. \tag{2.126}$$

Direct substitution proves that

$$g_{ij}n^i n_{,\alpha}^j = 0\,. \tag{2.127}$$

Therefore, if $n_{,\alpha}^i$ is viewed as a three-dimensional vector, it must be perpendicular to the normal and lie in the tangent plane of the geoid.

The first and second derivatives of radius-vector ϱ with respect to angular coordinates ϑ and Λ occur in tensors $a_{\alpha\beta}$ and $b_{\alpha\beta}$, in the components of the normal and their tensor derivatives, and in the formulae for curvature. To represent these quantities on a global scale it is convenient to express them in terms of spherical harmonic series. This requires the relevant derivatives and their products to be expressed in terms of spherical harmonics $Y_{jm}(\Omega)$. Analytical tools are available for this purpose. The first derivatives are expressed in formulae (2.44) and (2.90) and their products in (2.80) and (2.81). As regards the second derivatives,

$$\frac{\partial^2 Y_{jm}(\Omega)}{\partial\vartheta^2} = \left\{-\left[j(j+1) - \frac{m^2}{\sin^2\vartheta}\right] + m\cot^2\vartheta\right\}Y_{jm}(\Omega)$$
$$+ \cot\vartheta\sqrt{[j(j+1) - m(m-1)]}\,e^{i\Lambda}Y_{jm-1}(\Omega)\,, \tag{2.128}$$

$$\frac{\partial^2 Y_{jm}(\Omega)}{\partial\varphi^2} = -m^2 Y_{jm}(\Omega)\,. \tag{2.129}$$

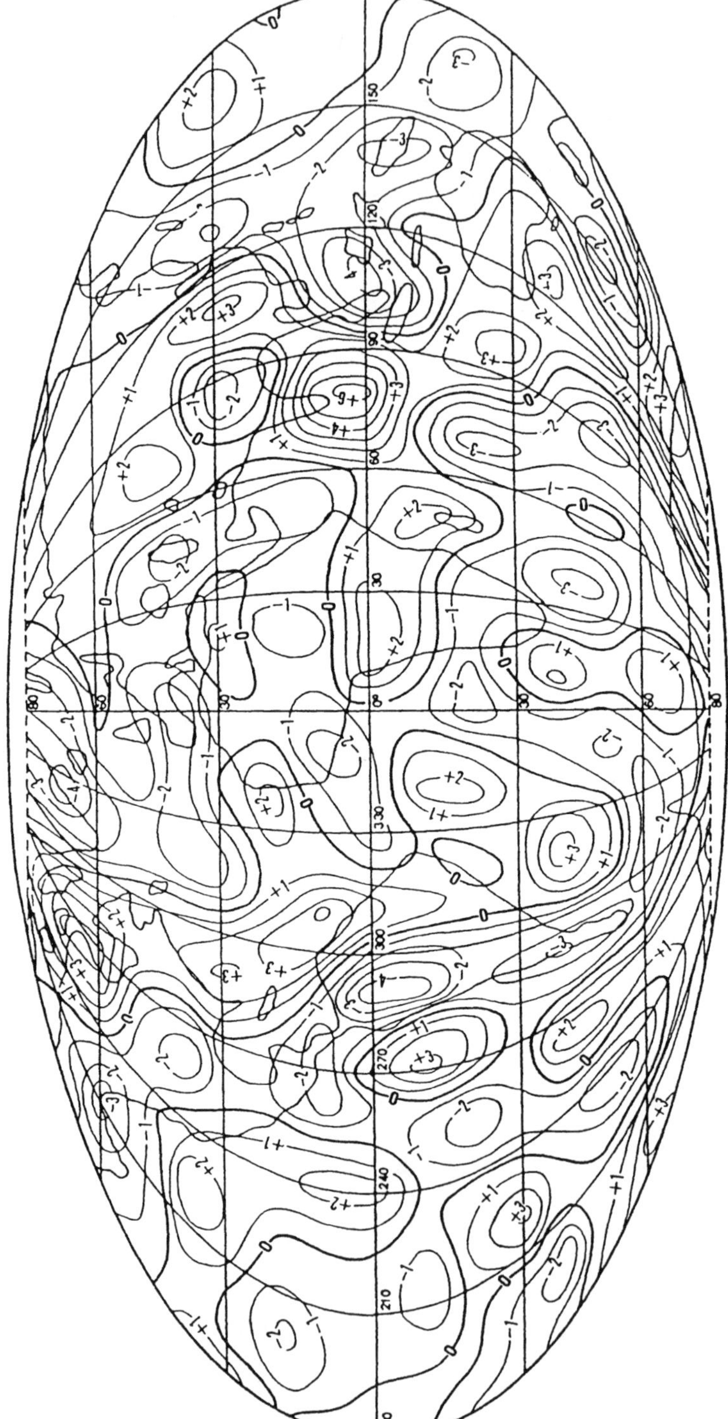

Fig. 2.2 Anomaly of radius of curvature of the equipotential surface of the geopotential (in 10^2 m) derived from the set of geopotential coefficients determined from the orbital dynamics of artificial satellites.

By using these relations and by repeated application of the theorem for the product of spherical harmonics (2.76) all the tensors and invariants mentioned can be expressed as series of spherical harmonics. The procedure is straightforward; however, the results cannot be produced in condensed form and will therefore be omitted.

The formulae derived apply to any surface which can be expressed in the form of (2.62) or (2.63). If this surface is taken to be the triaxial reference ellipsoid (2.140) the relevant curvature, for example, is referred to the ellipsoid. The difference between the radius of curvature of the geoid, $R_H = 2/H$, and the radius of curvature of the ellipsoid, $R_H^{(e)}$, defines anomaly ΔR_H:

$$\Delta R_H = R_H - R_H^{(e)} . \tag{2.130}$$

Figure 2.2 shows anomalies ΔR_H (in 10^2 m). The set of geopotential coefficients in Gaposchkin and Lambeck (1970) was used.

In the presatellite era there were no data on the global curvature of the geoid at all. This description of the shape properties of the equipotential surface should provide new ways of studying the structure of the Earth's mantle and crust.

2.5.3 The Earth's Triaxial Ellipsoid

If the surface of the geoid $W = W_0$ is to be represented by any other surface whose parameters of shape and dimension have to be derived, it is first necessary to formulate additional conditions which these parameters have to satisfy.

If the surface of the geoid (let us say Σ) was not flattened either at the poles or at the equator and, moreover, was not sufficiently close to a sphere, it would be natural to represent it by a sphere, and the only parameter to be determined would be its radius. If surface Σ was flattened at the poles but was sufficiently rotationally and equatorially symmetrical, it would evidently be best to represent it by a rotational ellipsoid. However, reality is different. As can be seen in Tables 2.3 and 2.4, the following zonal, tesseral and sectorial harmonic terms with significant amplitudes occur in expansion (2.63): (2, 0), (2, 2), (4, 0), (3, 0), (3, 1), (3, 3), (5, 5), (4, 2), (5, 3), (4, 4), (6, 4), (6, 6), (6, 2), (6, 0); the first of these terms ($A_n^{(0)}$, not included) are given in Table 2.5 in order of amplitude magnitude in the equatorial section ($\phi = 0$),

$$\alpha_{n,k} = R_0 \left(\frac{a_0}{R_0}\right)^n [(\bar{J}_n^{(k)})^2 + (\bar{S}_n^{(k)})^2]^{1/2} \bar{P}_n^{(k)}(0) . \tag{2.131}$$

Some of the tesseral and sectorial terms are shown schematically in Figs. 2.3–2.5. It would, therefore, be possible to replace surface Σ by a spheroid of the fourth degree, asymmetric with respect to the axis of rotation and to the plane of the equator. Retaining terms with amplitudes ≥ 5 m (numerical values

Table 2.5. Amplitudes of harmonic non-zonal terms in the expansion of the geoid's radius-vector by absolute magnitude

n	k	$\alpha_{n,k}$ (m)
2	2	34.7
3	3	21.3
3	1	− 21.2
5	5	10.4
4	2	− 8.0
5	3	− 5.7
4	4	5.2
6	4	− 5.0
6	6	3.9
6	2	3.8

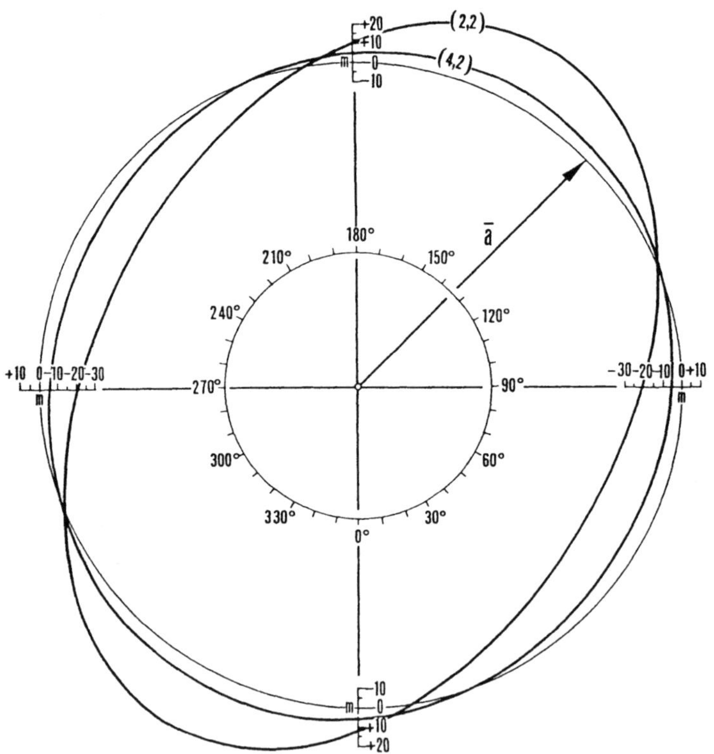

Fig. 2.3. Harmonic terms $n = 2$, $k = 2$ and $n = 4$, $k = 2$ plotted using the geopotential coefficients given in Table 1.3

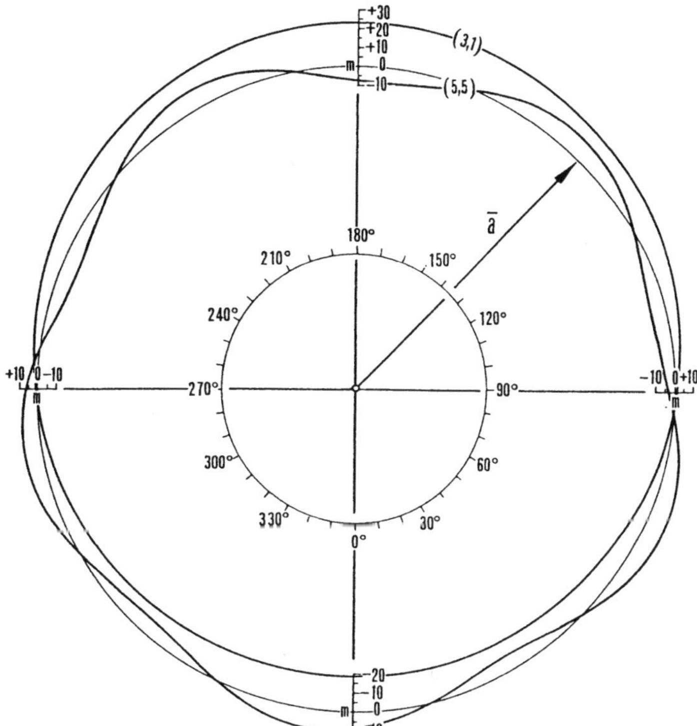

Fig. 2.4. Harmonic terms $n = 3$, $k = 1$ and $n = 5$, $k = 5$ plotted using the geopotential coefficients given in Table 1.3

in metres), its equation would read

$$\varrho_{\bar{n}=4} = 6\,370\,994.0 - 14\,266.7\,\mathrm{P}_2^{(0)}(\sin\phi)$$

$$+ (15.6\cos 2\varLambda - 8.9\sin 2\varLambda)\bar{\mathrm{P}}_2^{(2)}(\sin\phi)$$

$$+ 16.3\,\mathrm{P}_3^{(0)}(\sin\phi) + 13.0\,\bar{\mathrm{P}}_3^{(1)}(\sin\phi)\cos\varLambda$$

$$+ 5.7\,\bar{\mathrm{P}}_3^{(2)}(\sin\phi)\cos 2\varLambda + (4.6\cos 3\varLambda$$

$$+ 9.1\sin 3\varLambda)\bar{\mathrm{P}}_3^{(3)}(\sin\phi) + 19.8\,\mathrm{P}_4^{(0)}(\sin\phi)$$

$$+ 4.3\sin 2\varLambda\,\bar{\mathrm{P}}_4^{(2)}(\sin\phi) + 6.4\cos 3\varLambda\bar{\mathrm{P}}_4^{(3)}(\sin\phi)\,. \tag{2.132}$$

The amplitudes of all the other harmonic terms in the expansion of the radius-vector of the geoid surface, not included in (2.132), are smaller than 5 m.

However, in some dynamic problems the fundamental parameters of shape of the body must be introduced, namely polar (α) and equatorial (α_1) flattening. This, of course, requires the real body to be represented by a triaxial ellipsoid.

We shall assume that the centre of the Earth triaxial ellipsoid is in the mass centre of the Earth ($J_1^{(0)} = 0$, $J_1^{(1)} = 0$, $S_1^{(1)} = 0$). It is then defined uniquely by

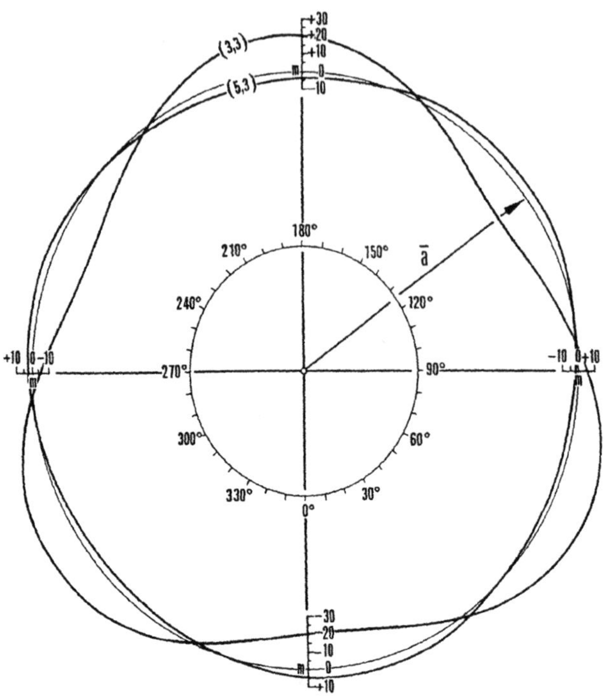

Fig. 2.5. Harmonic terms $n = 3$, $k = 3$ and $n = 5$, $k = 3$ plotted using the geopotential coefficients given in Table 1.3

four parameters, e.g. the semi-axes $a > b > c$ and the geocentric longitude Λ_a of the meridian in which a lies. Instead of semi-axes b and c it is better to use the eccentricities $e^2 = (a^2 - c^2)/a^2$, $e_1^2 = (a^2 - b^2)/a^2$, which are related to the flattenings as $e^2 = 2\alpha - \alpha^2$, $e_1^2 = 2\alpha_1 - \alpha_1^2$. The radius-vector ϱ_e of the triaxial ellipsoid as a function of the above parameters reads

$$\varrho_e = \varrho_e(a, e^2, e_1^2, \Lambda_a)$$
$$= \frac{a(1 - e^2)^{1/2}(1 - e_1^2)^{1/2}}{[1 - e^2 \cos^2\phi - e_1^2 \sin^2\phi - e_1^2(1 - e^2)\cos^2\phi \cos^2(\Lambda - \Lambda_a)]^{1/2}} \quad (2.133)$$

or, after expansion into a series of spherical harmonics retaining terms of the order of 10^{-10}, which is quite sufficient for further applications,

$$\varrho_e = a\left[1 + a_0^{(0)} + \sum_{m=1}^{4} \sum_{j=0}^{2m} (a_{2m}^{(j)} \cos j\Lambda + b_{2m}^{(j)} \sin j\Lambda) P_{2m}^{(j)}(\sin\phi)\right], \quad (2.134)$$

$$a_0^{(0)} = -\tfrac{1}{6}e^2 - \tfrac{11}{120}e^4 - \tfrac{1}{6}e_1^2 - \tfrac{103}{1680}e^6 + \tfrac{1}{20}e^2 e_1^2 - \tfrac{1823}{40\,320}e^8$$

$$- \tfrac{11}{120}e_1^4 + \tfrac{13}{560}e^4 e_1^2 + \tfrac{13}{560}e^2 e_1^4 + \tfrac{139}{10\,080}e^6 e_1^2$$

$$= -\tfrac{1}{3}\alpha - \tfrac{1}{5}\alpha^2 - \tfrac{13}{105}\alpha^3 - \tfrac{5}{63}\alpha^4 - \tfrac{1}{3}\alpha_1 - \tfrac{1}{5}\alpha_1^2$$

$$+ \tfrac{1}{5}\alpha\alpha_1 + \tfrac{3}{35}\alpha^2\alpha_1\,,$$

$$a_2^{(0)} = -\tfrac{1}{3}e^2 - \tfrac{5}{42}e^4 + \tfrac{1}{6}e_1^2 - \tfrac{3}{56}e^6 + \tfrac{1}{28}e^2 e_1^2 - \tfrac{293}{11\,088}e^8 + \tfrac{5}{84}e_1^4$$

$$- \tfrac{1}{112}e^4 e_1^2 + \tfrac{1}{28}e^2 e_1^4 - \tfrac{313}{22\,176}e^6 e_1^2$$

$$= -\tfrac{2}{3}\alpha - \tfrac{1}{7}\alpha^2 + \tfrac{1}{21}\alpha^3 + \tfrac{10}{99}\alpha^4 + \tfrac{1}{3}\alpha_1 + \tfrac{1}{14}\alpha_1^2$$

$$+ \tfrac{1}{7}\alpha\alpha_1 - \tfrac{1}{7}\alpha^2\alpha_1\,,$$

$$a_4^{(0)} = \tfrac{3}{35}e^4 + \tfrac{57}{770}e^6 - \tfrac{3}{35}e^2 e_1^2 + \tfrac{2271}{40\,040}e^8 + \tfrac{9}{280}e_1^4 - \tfrac{18}{385}e^4 e_1^2$$

$$- \tfrac{213}{6160}e^2 e_1^4 - \tfrac{837}{40\,040}e^6 e_1^2$$

$$= \tfrac{12}{35}\alpha^2 + \tfrac{96}{385}\alpha^3 + \tfrac{15}{143}\alpha^4 + \tfrac{9}{70}\alpha_1^2 - \tfrac{12}{35}\alpha\alpha_1 + \tfrac{6}{35}\alpha^2\alpha_1\,,$$

$$a_6^{(0)} = -\tfrac{5}{231}e^6 - \tfrac{41}{1386}e^8 + \tfrac{5}{154}e^4 e_1^2 - \tfrac{15}{616}e^2 e_1^4 + \tfrac{89}{2772}e^6 e_1^2$$

$$= -\tfrac{40}{231}\alpha^3 - \tfrac{148}{693}\alpha^4 + \tfrac{20}{77}\alpha^2\alpha_1\,,$$

$$a_8^{(0)} = \tfrac{7}{1287}e^8 - \tfrac{14}{1287}e^6 e_1^2 = \tfrac{112}{1287}\alpha^4\,,$$

$$\left.\begin{matrix} a_2^{(2)} \\ b_2^{(2)} \end{matrix}\right\} = \left(\tfrac{1}{12}e_1^2 - \tfrac{1}{56}e^2 e_1^2 + \tfrac{5}{168}e_1^4 - \tfrac{1}{96}e^4 e_1^2 - \tfrac{1}{336}e^2 e_1^4 - \tfrac{307}{44\,352}e^6 e_1^2\right){\cos 2\Lambda_a \atop \sin 2\Lambda_a}$$

$$= \left(\tfrac{1}{6}\alpha_1 + \tfrac{1}{28}\alpha_1^2 - \tfrac{1}{14}\alpha\alpha_1 - \tfrac{1}{28}\alpha^2\alpha_1\right){\cos 2\Lambda_a \atop \sin 2\Lambda_a}\,,$$

$$\left.\begin{matrix} a_4^{(2)} \\ b_4^{(2)} \end{matrix}\right\} = \left(-\tfrac{1}{140}e^2 e_1^2 + \tfrac{1}{280}e_1^4 - \tfrac{1}{440}e^4 e_1^2 - \tfrac{19}{6160}e^2 e_1^4 - \tfrac{109}{160\,160}e^6 e_1^2\right){\cos 2\Lambda_a \atop \sin 2\Lambda_a}$$

$$= \left(\tfrac{1}{70}\alpha_1^2 - \tfrac{1}{35}\alpha\alpha_1 - \tfrac{3}{770}\alpha^2\alpha_1\right){\cos 2\Lambda_a \atop \sin 2\Lambda_a}\,,$$

$$\left.\begin{matrix} a_4^{(4)} \\ b_4^{(4)} \end{matrix}\right\} = \left(\tfrac{1}{2240}e_1^4 - \tfrac{1}{9856}e^2 e_1^4\right){\cos 4\Lambda_a \atop \sin 4\Lambda_a} = \tfrac{1}{5600}\alpha_1^2 {\cos 4\Lambda_a \atop \sin 4\Lambda_a}\,,$$

$$\left.\begin{matrix} a_6^{(2)} \\ b_6^{(2)} \end{matrix}\right\} = \left(\tfrac{1}{924}e^4 e_1^2 - \tfrac{1}{924}e^2 e_1^4 + \tfrac{5}{5544}e^6 e_1^2\right) = \tfrac{2}{231}\alpha^2\alpha_1 {\cos 2\Lambda_a \atop \sin 2\Lambda_a}\,,$$

$$\left.\begin{matrix} a_6^{(4)} \\ b_6^{(4)} \end{matrix}\right\} = -\tfrac{1}{44\,352}e^2 e_1^4 {\cos 4\Lambda_a \atop \sin 4\Lambda_a}\,,$$

$$\left.\begin{matrix} a_8^{(2)} \\ b_8^{(2)} \end{matrix}\right\} = -\tfrac{1}{5148}e^6 e_1^2 {\cos 2\Lambda_a \atop \sin 2\Lambda_a} \tag{2.135}$$

(terms with α^4, α_1^2, $\alpha^2\alpha_1^2$ and smaller have been omitted).

To determine four parameters $p_1 = a$, $p_2 = e^2$, $p_3 = e_1^2$, $p_4 = \Lambda_a$, four conditions should be imposed in advance. For example,

$$\int_\Sigma [\varrho - \varrho_e(a, e^2, e_1^2, \Lambda_a)] \frac{\partial}{\partial p_i} \varrho_e(a, e^2, e_1^2, \Lambda_a) \, d\Sigma = 0; \quad i = 1, \ldots, 4,$$

which follows from the fundamental condition

$$\int_\Sigma [\varrho - \varrho_e(a, e^2, e_1^2, \Lambda_a)]^2 \, d\Sigma = \min., \tag{2.136}$$

minimizing the integral of the square of the difference of the radius-vectors of both surfaces.

Apart from (2.136) also

$$\int_\Sigma (W_0 - W_e)^2 \, d\Sigma = \min. \tag{2.137}$$

(W_e is the gravity potential on the surface of the ellipsoid being derived), or

$$\int_\Sigma (g - g_e)^2 \, d\Sigma = \min. \tag{2.138}$$

(g is the acceleration of gravity on surface Σ and g_e is the acceleration of gravity on the ellipsoid), or

$$\int_\Sigma (K - K_e)^2 \, d\Sigma = \min. \tag{2.139}$$

(K, K_e represent Gauss' curvature of surface Σ and of the ellipsoid, respectively), etc.

In integrals (2.136)–(2.139) Σ is the given surface of the geoid and $d\Sigma$ is its element to which Eq. (2.75) applies.

Under condition (2.136) and with a value of the geopotential scale factor $R_0 = 6\,363\,672.5 \pm 0.3$ m, the set of geopotential coefficients of several geopotential models yielded the parameters of the triaxial Earth ellipsoid given in Table 2.6. Parameters $\bar{a}$ and $\bar{\alpha}$ represent the semimajor axis and the flattening of the corresponding rotational Earth ellipsoid ($\alpha_1 = 0$).

Table 2.6. Parameters of the Earth's triaxial and rotational ellipsoid according to various geopotential models

Model	a (m)	$1/\alpha$	$1/\alpha_1$	Λ_a	$\bar{a}$ (m)	$1/\bar{\alpha}$
GEM-T2 (1990)	6 378 171.36	297.774	91 470	$-14.94°$	6 378 136.50	298.2575
GEM-T1 (1988)	6 378 171.36	297.774	91 476	$-14.94°$	6 378 136.49	298.2577
GRIM 4 S1 (1991)	6 378 171.36	297.774	91 457	$-14.95°$	6 378 136.49	298.2578

The most recent parameters (rounded-off values) are

$$a = 6\,378\,171.4 \text{ m}, \quad 1/\alpha = 297.774, \quad 1/\alpha_1 = 91\,500,$$

$$\Lambda_a = -14.9°, \quad \bar{a} = 6\,378\,136.5 \text{ m}, \quad 1/\bar{\alpha} = 298.258. \tag{2.140}$$

The relation between $a, \bar{a}$ and $\alpha, \bar{\alpha}$ is a function of equatorial flattening α_1; evidently

$$\bar{a} = \frac{1}{2\pi} \int_0^{2\pi} (\varrho_e)_{\Phi=0} \, d\Lambda = a(1 - \tfrac{1}{2}\alpha_1),$$

$$\bar{\alpha} = \frac{1}{2\pi} \int_0^{2\pi} \frac{(\varrho_e)_{\Phi=0,\Lambda} - a(1-e^2)^{1/2}}{(\varrho_e)_{\Phi=0,\Lambda}} \, d\Lambda = \alpha - \tfrac{1}{2}\alpha_1 + \tfrac{1}{2}\alpha\alpha_1.$$

The fact that triaxiality is a significant feature is illustrated by Fig. 2.6, which shows schematically the differences between the radius-vectors of the surfaces of the geoid and rotational Earth ellipsoid in equatorial section, and by Fig. 2.7, which illustrates the differences d between the diameters of the surfaces of the geoid and rotational Earth ellipsoid; the extremes of function d are just 90° apart.

An alternative expansion of radius-vector ϱ_e of ellipsoid (2.133) into a series of spherical harmonics with complex spherical harmonic coefficients E_{jm} is

$$\varrho_e = a(1 + t_e), \tag{2.141}$$

where a is the semimajor axis and t_e is expressed retaining terms of up to order

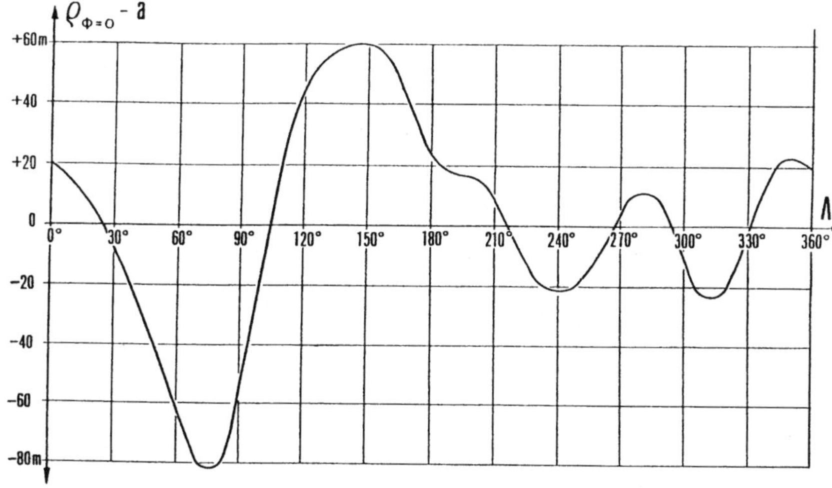

Fig. 2.6. Difference between the radius-vector of the geoid surface, $\rho_\phi = 0$, and of the Earth's ellipsoid of rotation $\bar{a}$ in the equatorial section (in metres)

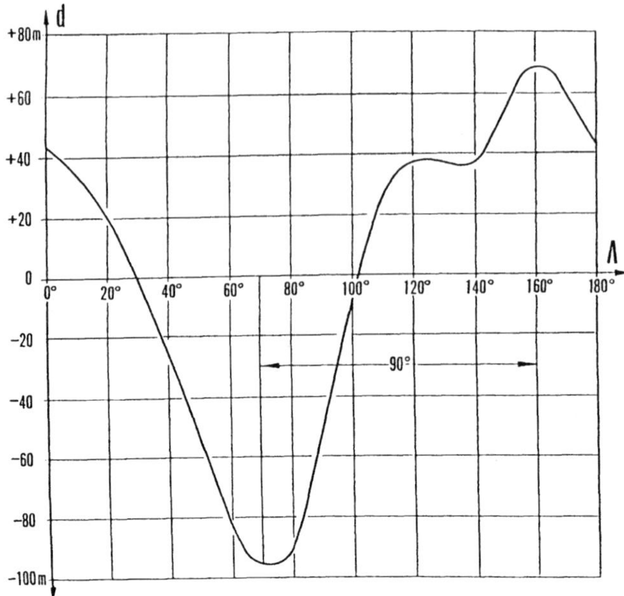

Fig. 2.7. Differences d between the diameters of the geoid surface and of the Earth's ellipsoid of rotation (in metres)

10^{-8} as follows:

$$t_e = L_{00}Y_{00}(\Omega) + L_{20}Y_{20}(\Omega) + L_{40}Y_{40}(\Omega) + L_{60}(\Omega)$$
$$+ L_{22}Y_{22}(\Omega) + L_{2-2}Y_{2-2}(\Omega) + L_{42}Y_{42}(\Omega) + L_{4-2}Y_{4-2}(\Omega). \quad (2.142)$$

Coefficients L are expressed in terms of flattenings α and α_1 as

$$L_{00} = -2\sqrt{(\pi)}\left(\frac{1}{3}\alpha + \frac{1}{5}\alpha^2 + \frac{1}{3}\alpha_1 - \frac{1}{5}\alpha\alpha_1 + \frac{13}{7\cdot15}\alpha^3\right),$$

$$L_{20} = -\frac{2\sqrt{\pi}}{\sqrt{5}}\left(\frac{2}{3}\alpha + \frac{1}{7}\alpha^2 - \frac{1}{3}\alpha_1 - \frac{1}{7}\alpha\alpha_1 - \frac{1}{3\cdot7}\alpha^3\right),$$

$$L_{40} = \frac{8\sqrt{\pi}}{5\cdot7}\left(\alpha^2 - \alpha\alpha_1 + \frac{8}{11}\alpha^3\right), \quad L_{60} = -\frac{80\sqrt{\pi}}{3\cdot7\cdot11\cdot\sqrt{13}}\alpha^3,$$

$$L_{2\pm2} = \sqrt{\left(\frac{2\pi}{15}\right)}\left(\alpha_1 - \frac{3}{7}\alpha\alpha_1\right)\exp(\mp2i\Lambda_a),$$

$$L_{4\mp2} = -\frac{2}{7}\sqrt{\left(\frac{2\pi}{5}\right)}\alpha\alpha_1\exp(\mp2i\Lambda_a). \quad (2.143)$$

Coefficients L clearly contain terms of various orders of flattening α. For some purposes it is convenient to group terms of the same order; t_e from (2.142) can then be expressed as

$$t_e = t_1 + t_2 + t_3 \,, \tag{2.144}$$

where t_i is $O(\alpha^i)$. Regrouping now yields

$$t_1 = -\frac{2\sqrt{\pi}}{3}\,\alpha\left[Y_{00}(\Omega) + \frac{2}{\sqrt{5}}\,Y_{20}(\Omega) \right],$$

$$t_2 = -\frac{2\sqrt{\pi}}{5}\,\alpha^2\left[Y_{00}(\Omega) + \frac{\sqrt{5}}{7}\,Y_{20}(\Omega) - \frac{4}{7}\,Y_{40}(\Omega) \right]$$

$$\quad -\frac{2\sqrt{\pi}}{3}\,\alpha_1\left[Y_{00}(\Omega) - \frac{1}{\sqrt{5}}\,Y_{20}(\Omega) \right]$$

$$\quad +\sqrt{\left(\frac{2\pi}{15}\right)}\,\alpha_1\,Y_{2\pm2}(\Omega)\exp(\mp 2i\Lambda_a)\,,$$

$$t_3 = \frac{2\sqrt{\pi}}{5}\,\alpha\alpha_1\left[Y_{40}(\Omega) + \frac{\sqrt{5}}{7}\,Y_{20}(\Omega) - \frac{4}{7}(\Omega) \right]$$

$$\quad +\frac{2\sqrt{\pi}}{7}\,\alpha^3\left[-\frac{13}{15}\,Y_{00}(\Omega) + \frac{\sqrt{5}}{15}\,Y_{20}(\Omega) + \frac{32}{5\cdot11}\,Y_{40}(\Omega) \right.$$

$$\quad \left. -\frac{40}{3\cdot11\sqrt{13}}\,Y_{60}(\Omega) \right]$$

$$\quad -\frac{1}{7}\sqrt{\left(\frac{2\pi}{5}\right)}\,\alpha\alpha_1\left[\sqrt{(3)}\,Y_{2\pm2}(\Omega) + 2Y_{4\pm2}(\Omega)\right]\exp(\mp 2i\Lambda_a)\,. \tag{2.145}$$

In Eqs. (2.145) the expressions in which two signs occur should be considered as the sums of two terms.

The differences between radius-vectors of the geoid ϱ (2.63) and (2.62), and of the ellipsoid ϱ_e (2.134) and (2.141), respectively, referred to as anomalies of the geoid radius-vector, reflect the deviations of the shape of the geoid relative to the ideal triaxial ellipsoid, and can be used for mass interpretations of the body.

Figure 2.8 shows the differences $\varrho - \varrho_e$ (in metres), with the triaxial Earth ellipsoid being defined by parameters $a = 6\,378\,173$ m (mean equatorial semi-axis $\bar{a} = 6\,378\,139$ m), $\alpha = 1:297.786$ (mean polar flattening $\bar{\alpha} = 1:298.258$), $\alpha_1 = 1:94\,000$, $\Lambda_a = 14.8°$W. The region of the maximum anomaly lies south of India, where the geoid displays a depression of over 100 m, and the horizontal gradients of anomalies $\varrho - \varrho_e$ are considerable (~ 200 m/5000 km).

For the sake of illustration, Fig. 2.9 shows the geoid section in the equatorial plane, Fig. 2.10 shows it in the plane of meridian $\Lambda = \Lambda_a = 14.8°$W, and Fig. 2.11 shows it in the plane of the meridian perpendicular to the preceding meridian.

For mass interpretations it is convenient to use deflections of the vertical from the ellipsoid normals whose components in the meridian and prime

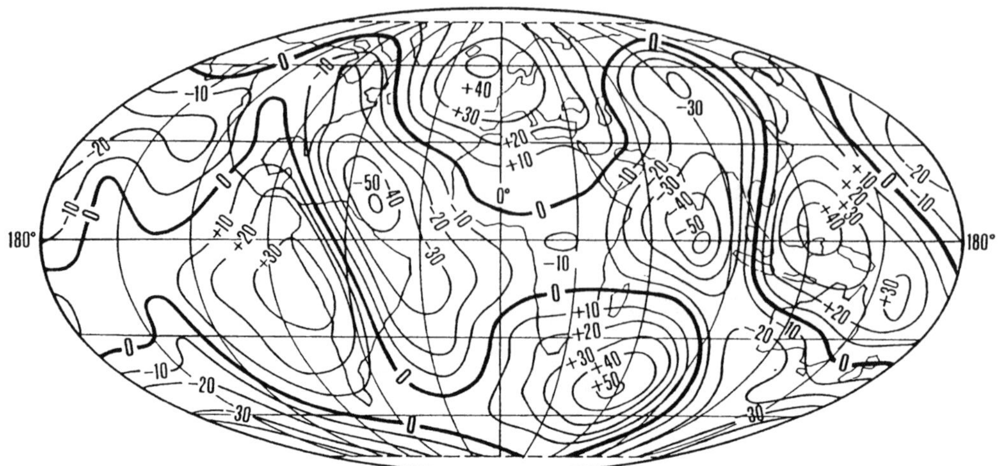

Fig. 2.8. Global pattern of differences between the radius-vectors of the geoid and Earth's triaxial ellipsoid derived from satellite data (in metres)

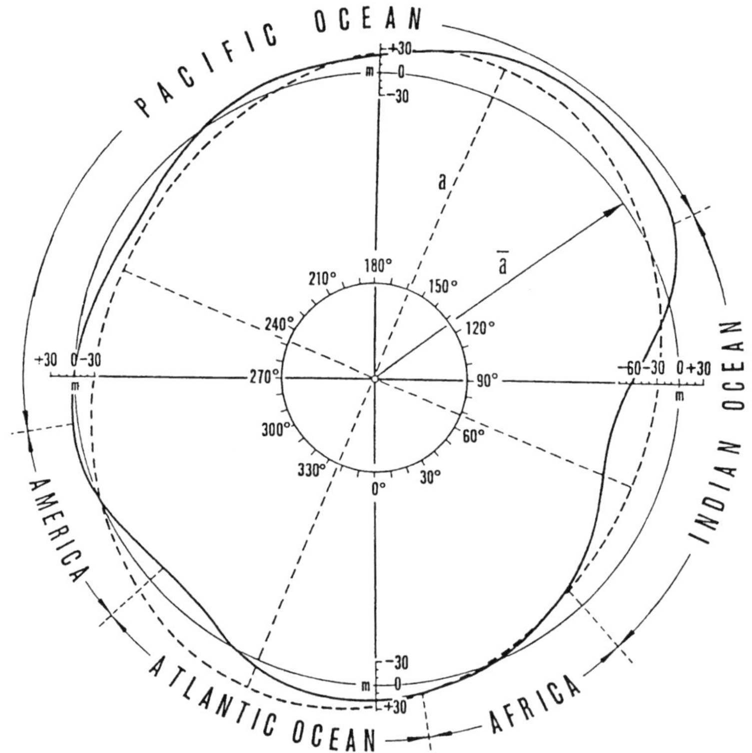

Fig. 2.9. Equatorial plane section of the geoid

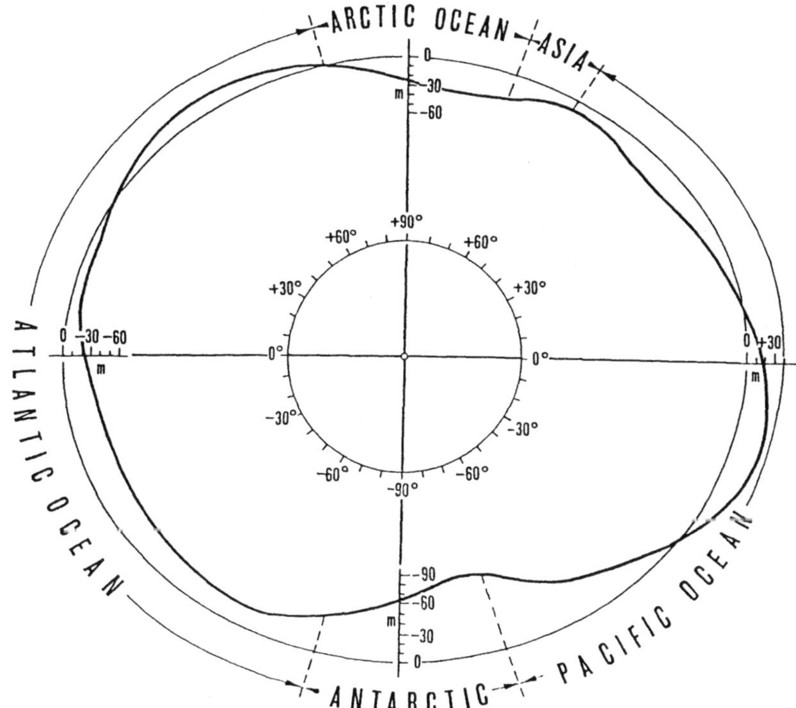

Fig. 2.10. Geoid section of meridional plane $\Lambda = \Lambda_a = -14.8°$ containing the largest axis of the Earth's triaxial ellipsoid

vertical are

$$\xi = \cos\alpha_\vartheta - \cos\alpha_\vartheta|_e \,; \quad \eta = \cos\alpha_A - \cos\alpha_A|_e \,, \tag{2.146}$$

where $\cos\alpha_\vartheta$, $\cos\alpha_A$ are projections of the unit geoid normal on to the coordinate axes given by (2.97), quantities with index e being referred to the Earth ellipsoid.

Figures 2.12 and 2.13 show the components of deflections $\xi = \text{const}$ and $\eta = \text{const}$ (in seconds of arc) on the rotational ellipsoid with parameters

$$\bar{a} = 6\,378\,160 \text{ m}\,, \quad \bar{\alpha} = 1:298.247\,. \tag{2.147}$$

Figure 2.14 is a schematic illustration of the total deflections of the vertical in a $10° \times 10°$ grid in the same system. The set of satellite constants from Gaposchkin and Lambeck (1970) was used. In the presatellite era these data were available only in parts of some continents and they were practically lacking in the regions of oceans and seas. The principal global anomalies in the Indian and Pacific oceans were unknown at the time.

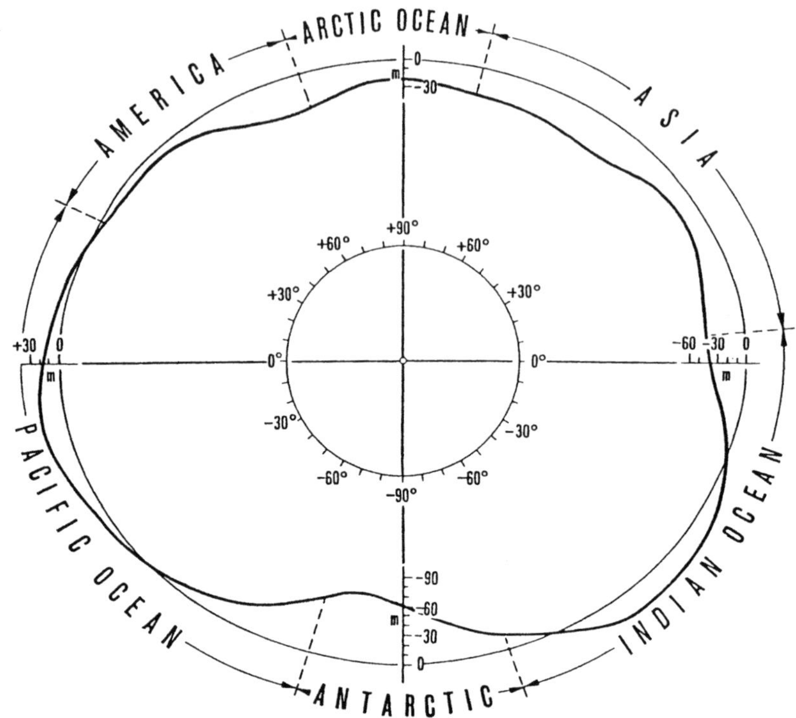

Fig. 2.11. Geoid section of the meridional plane containing the minor axis of the equatorial ellipse (plane perpendicular to meridian $\Lambda = \Lambda_a$; Fig. 2.10)

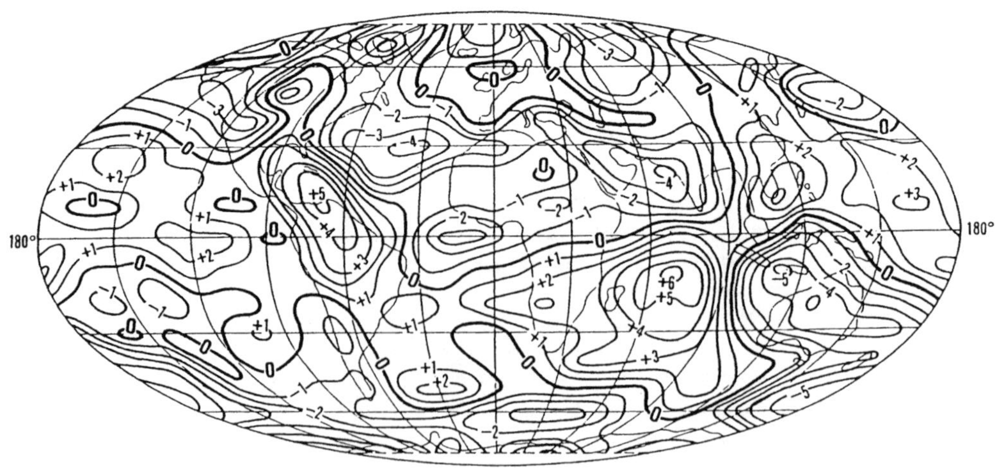

Fig. 2.12. Components of deflections of the vertical ξ in the meridional plane derived from satellite data (in seconds of arc)

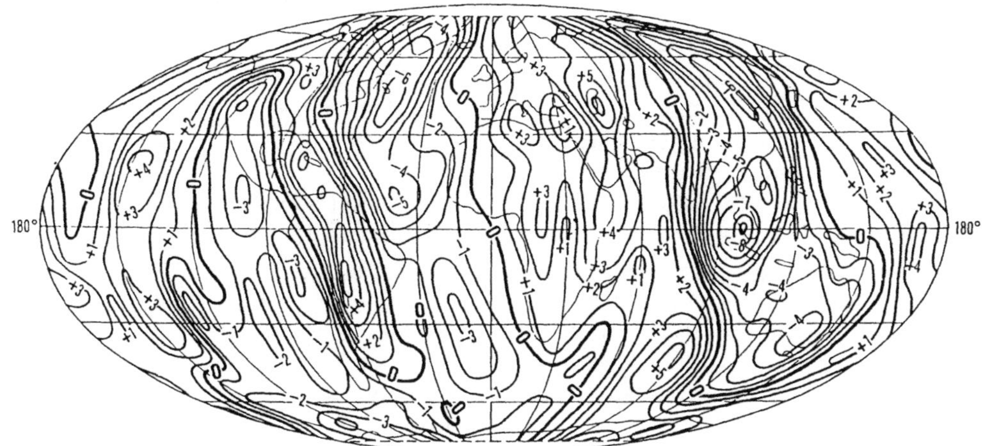

Fig. 2.13. Components of deflections of the vertical η in the plane of the prime vertical derived from satellite data (in seconds of arc)

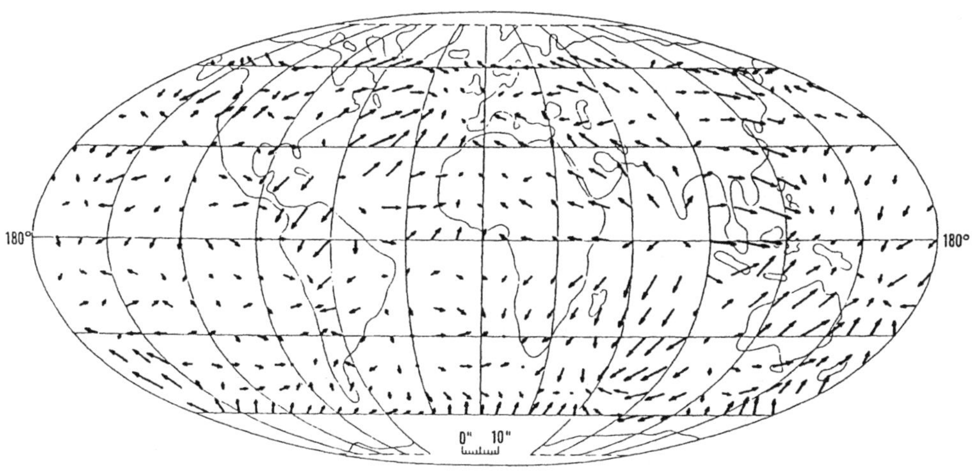

Fig. 2.14. Pattern of deflections of the vertical derived from satellite data

2.5.4 Determination of the Coefficients in the Harmonic Development of the Geoid's Radius-Vector and of the Geopotential Scale Factor R_0

In Section 2.5.2 we derived Eqs. (2.98) and (2.99), which relate coefficients E_{jm} and $A_n^{(k)}$, $B_n^{(k)}$ of the expansion of the radius-vector of equipotential surface (2.62) and (2.63), respectively, with geopotential coefficients A_{jm} and $J_n^{(k)}$, $S_n^{(k)}$ in the expansion of gravity potential $W_0 = GM/R_0$. For a given value of $W_0(R_0)$ and set of geopotential coefficients $A_{jm}(J_n^{(k)}, S_n^{(k)})$ these formulae define coefficients $E_{jm}(A_n^{(k)}, B_n^{(k)})$. Although the expansions of gravitational potential (2.3) and

(2.7) and expansions of radius-vector (2.62) and (2.63) formally look alike, there is a principal difference between expansion coefficients A_{jm} and E_{jm}. The external gravitational potential is a harmonic function of the coordinates and, consequently, the dependence on coordinate ϱ is expressed in terms of the powers of ϱ^{-j-1}. This implies that the geopotential coefficients are universal for the external gravitational field as a whole and are indeed constant, provided that time variations of the field are not admitted; otherwise they are, in general, slowly varying functions of time. As regards the expansion of radius-vector ϱ of the equipotential surface, the situation is quite different. The radius-vector is not a harmonic function and expansion (2.62) is only a formal expression of the fact that every continuous, piecewise smooth function of angular coordinates ϑ, Λ, finite everywhere, can be developed into a series of spherical harmonics $Y_{jm}(\vartheta, \Lambda)$. From a formal point of view, therefore, the coefficients of the radius-vector expansion are spectral coefficients of a Fourier series, based on orthonormal spherical harmonics $Y_{jm}(\vartheta, \Lambda)$, valid only for the equipotential surface in question. Consequently, coefficients E_{jm} (and $A_n^{(k)}$, $B_n^{(k)}$) are not universal for the set of external equipotential surfaces as a whole, but are functions of scale factor R_0, or of the parameters of the geometric surface a_1, α, α_1, approximating the equipotential in question.

Quantities W_0, GM, (R_0), which are the fundamental quantities defining the dimension of the Earth uniquely, could in fact be determined only indirectly and with low accuracy in the presatellite era.

With the aid of artificial satellites, the geocentric gravitational constant has already been determined with an accuracy of $\pm 5 \times 10^{-9}$ (see Sect. 1.7) and $R_0(W_0)$ with an accuracy of $\pm 5 \times 10^{-8}$. There are three methods of determining $R_0(W_0)$:

1. From the geocentric positions of satellite stations $M(x_j; \varrho, \phi, \Lambda)$ determined by methods of satellite geodesy, their sea level heights H determined by levelling (Fig. 2.15), and from geopotential coefficients $J_n^{(k)}$ and $S_n^{(k)}$ (Sect. 1.6.2). Since angle ε between geocentric radius-vector ϱ of the satellite station and the normal to the geoid is small (always $< 12'$), coordinates $(\varrho_0, \phi_0, \Lambda_0)$ of point M_0 on the geoid, corresponding to satellite station M on the Earth's surface, can be computed with sufficient accuracy (of at least ± 0.1 m in factor R_0 being determined) from

$$\varrho_0 = \varrho - H \sec \varepsilon = \left[\sum_{j=1}^{3} x_j^2 \right]^{1/2} - H \, ,$$

$$\phi_0 = \phi = \arctan \frac{x_3}{(x_1^2 + x_2^2)^{1/2}} \, , \quad \Lambda_0 = \Lambda = \arctan \frac{x_2}{x_1} \, , \tag{2.148}$$

and factor R_0, in view of (2.63), can then be expressed as

$$R_0 = \varrho_0 \left\{ 1 + \sum_{n=2}^{\bar{n}} \sum_{k=0}^{n} \left(\frac{a_0}{\varrho_0} \right)^n (J_n^{(k)} \cos k\Lambda_0 + S_n^{(k)} \sin k\Lambda_0) \right.$$

$$\left. \times P_n^{(k)}(\sin \phi_0) + \frac{1}{3} q \left(\frac{a_0}{\varrho_0} \right)^{-3} [1 - P_2^{(0)}(\sin \phi_0)] \right\}^{-1} \, . \tag{2.149}$$

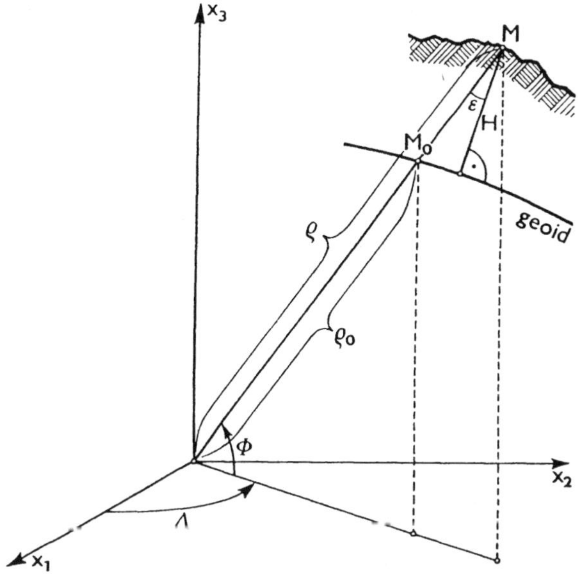

Fig. 2.15. Geocentric position of the satellite station and of the geoid

This method is not, however, theoretically quite exact because the set of coefficients $J_n^{(k)}$, $S_n^{(k)}$ describes the external gravitational field, i.e. it is valid at point M but not at point M_0 on the geoid within terrestrial masses. The inaccuracy is the higher the larger the sea level heights of the satellite stations involved. An exact solution would require the effect of the external masses in the space between the geoid and Earth to be considered, i.e. regularization of the Earth. However, the results that this method yields differ practically but little from the results of the second method, which does not have this drawback.

2. By satellite altimetry, i.e. measuring the distance h_S of satellite (altimeter) S (Fig. 2.16), whose position $(\varrho_S, \phi_S, \Lambda_S)$ is known, from the level of oceans and seas. Since altitude h_S may amount to several hundred kilometres, the angle between ϱ_S and normal SS_0 to equipotential surface $W = W_S$, passing through the altimeter, cannot be neglected; S_0' is the subsatellite point on geoid surface $W = W_0$. For practical purposes points S_0 and S_0' on the geoid need not be distinguished, i.e. angle ε_S need not be considered. The equation of equipotential surface $W = W_S$ reads, in analogy to (2.63),

$$\varrho_{W=W_S} = R_S\left[1 + A_0^{(0)} + \sum_{n=2}^{\bar{n}} \sum_{k=0}^{n} (A_n^{(k)} \cos k\Lambda + B_n^{(k)} \sin k\Lambda)P_n^{(k)}(\sin \phi)\right], \tag{2.150}$$

where

$$R_S = GM/W_S, \tag{2.151}$$

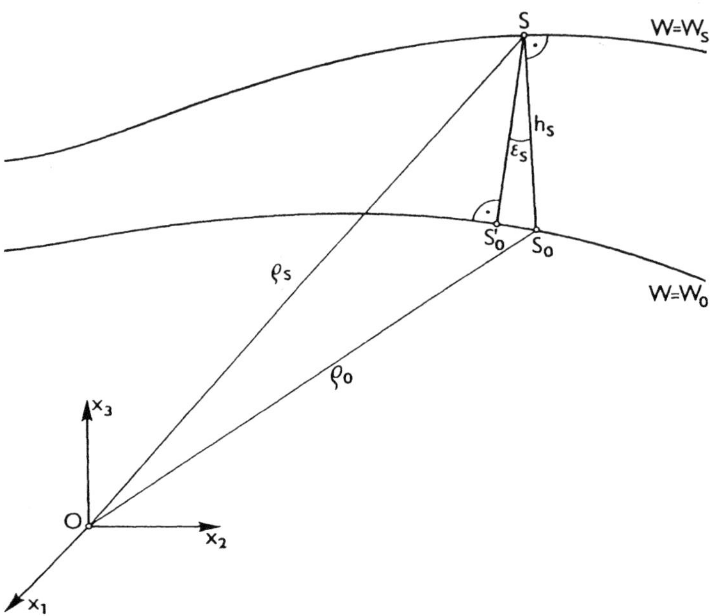

Fig. 2.16. Satellite altimetry

and W_S is the value of the geopotential at point $S(\varrho_S, \phi_S, \Lambda_S)$ calculated from

$$W_S = \frac{GM}{\varrho_S}\left\{ 1 + \sum_{n=2}^{\bar{n}} \sum_{k=0}^{n} \left(\frac{a_0}{\varrho_S}\right)^n (J_n^{(k)}\cos k\Lambda_S + S_n^{(k)}\sin k\Lambda_S) \right.$$

$$\left. \times \mathrm{P}_n^{(k)}(\sin \Phi_S) + \frac{1}{3}q\left(\frac{a_0}{\varrho_S}\right)^{-3}[1 - \mathrm{P}_2^{(0)}(\sin \phi_S)] \right\}. \tag{2.152}$$

The coordinates $x_{j,0}$ of point S_0 are

$$x_{1,0} = \varrho_S \cos\phi_S \cos\Lambda_S - \left[h_S \cos\phi_S \cos\Lambda_S \right.$$

$$\left. + \mu\left(\frac{\partial\varrho}{\partial\phi}\right)_S \sin\phi_S \cos\Lambda_S + \mu\sec\phi_S\left(\frac{\partial\varrho}{\partial\Lambda}\right)_S \sin\Lambda_S \right] T^{-1/2},$$

$$x_{2,0} = \varrho_S \cos\phi_S \sin\Lambda_S - \left[h_S \cos\phi_S \sin\Lambda_S + \mu\left(\frac{\partial\varrho}{\partial\phi}\right)_S \sin\phi_S \sin\Lambda_S \right.$$

$$\left. - \mu\sec\phi_S\left(\frac{\partial\varrho}{\partial\Lambda}\right)_S \cos\Lambda_S \right] T^{-1/2},$$

$$x_{3,0} = \varrho_S \sin\phi_S - \left[h_S \sin\phi_S - \mu\left(\frac{\partial\varrho}{\partial\phi}\right)_S \cos\phi_S \right] T^{-1/2},$$

$$\mu = \frac{h_S}{\varrho_S}, \tag{2.153}$$

where

$$T = 1 + \frac{1}{\varrho_S^2}\left(\frac{\partial\varrho}{\partial\phi}\right)_S^2 + \frac{1}{\varrho_S^2\cos^2\phi_S}\left(\frac{\partial\varrho}{\partial\Lambda}\right)_S^2,$$

$$\left(\frac{\partial\varrho}{\partial\phi}\right)_S = R_S \sum_{n=2}^{\bar{n}} \sum_{k=0}^{n} (A_n^{(k)}\cos k\Lambda_S + B_n^{(k)}\sin k\Lambda_S)$$

$$\times [P_n^{(k+1)}(\sin\phi_S) - k\tan\phi_S P_n^{(k)}(\sin\phi_S)],$$

$$\left(\frac{\partial\varrho}{\partial\Lambda}\right)_S = R_S \sum_{n=2}^{\bar{n}} \sum_{k=0}^{n} k(B_n^{(k)}\cos k\Lambda_S - A_n^{(k)}\sin k\Lambda_S)P_n^{(k)}(\sin\phi_S). \tag{2.154}$$

The required coordinates ϱ_0, ϕ_0, Λ_0 of point S_0 on the geoid can be obtained from (2.153) and, after substituting them in (2.149), factor R_0 can be calculated. The effect of small angle ε_S has been neglected; however, it can easily be taken into account. The advantage of this method is that no regularization is required because there are no terrestrial masses, assuming that we neglect the mass of the atmosphere, between equipotential surfaces $W = W_S$ and $W = W_0$ in the areas of oceans and seas. The accuracy of the method is, of course, limited by the errors incurred in determining the position of altimeter, S, i.e. the satellite orbital elements, and by Eq. (2.150) being approximate.

3. By making use of the astro-geodetic heights of the geoid, ζ_a, above the geodetic reference ellipsoid $E(a, e^2)$. It is assumed that coordinates $\Delta x_j^{(0)}$ of centre O_E of the ellipsoid (Fig. 2.17) relative to the Earth's centre of mass are

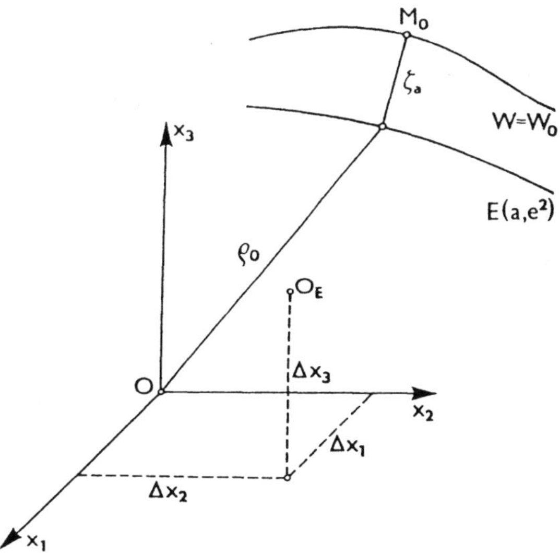

Fig. 2.17. Geocentric position of the geoid and of the ellipsoid of reference

known. The coordinates of point M_0 on the geoid then read

$$\varrho_0 = \left[\sum_{j=1}^{3} (\Delta x_j^{(0)} + X_j)^2 \right]^{1/2} ,$$

$$\phi_0 = \arctan\left\{ (\Delta x_3^{(0)} + X_3) \left[\sum_{j=1}^{2} (\Delta x_j^{(0)} + X_j)^2 \right]^{-1/2} \right\} ,$$

$$\Lambda_0 = \arctan[(\Delta x_2^{(0)} + X_2)(\Delta x_1^{(0)} + X_1)^{-1}] , \qquad (2.155)$$

$$X_1 = (N_0 + \zeta_a)\cos B_0 \cos L_0 , \quad X_2 = (N_0 + \zeta_a)\cos B_0 \sin L_0 ,$$

$$X_3 = [N_0(1 - e^2) + \zeta_a]\sin B_0 , \quad N_0 = a(1 - e^2 \sin^2 B_0)^{-1/2} ; \qquad (2.156)$$

a and e^2 are the semimajor axis and square of eccentricity of the geodetic reference ellipsoid, and B_0 and L_0 are the geodetic coordinates of point M_0. In (2.155) it has been assumed that axes x_j are parallel with geodetic axes X_j. Should this not be so, then corrections may be introduced; however, the system is practically never tilted more than 1.5″ and the effect may be neglected. After substituting (2.155) into (2.149) we again calculate factor R_0.

Contemporary studies indicate that the present most probable value of

$$R_0 = (6\,363\,672.5 \pm 0.3)\,\mathrm{m} . \qquad (2.157)$$

This is the radius of a sphere of spherically symmetric mass on whose surface the value of the gravitational potential is equal to the value of the gravity potential on the surface of geoid W_0. Given the geocentric gravitational constant (1.152), the latter comes out as

$$W_0 = 62\,636\,856.5 \pm 3.0\,\mathrm{m^2\,s^{-2}} . \qquad (2.158)$$

2.5.5 Power Series of the Geoid's Radius-Vector

Not only does the first power of the geoid's radius vector $\varrho(\vartheta, \Lambda)$ occur in the formulae for the gravity potential, but generally its p-th power, where p is an integer, also occurs there. Consequently, it is necessary to determine not only the spherical coefficients in expansion (2.62) but also those for the p-th power in general; this may formally be expressed as

$$\varrho^p(\Omega) = a_1^p \sum_{J=0}^{J} \sum_{M=-J}^{J} E_{JM}^{(p)} Y_{JM}(\Omega) . \qquad (2.159)$$

Spherical coefficients $E_{jM}^{(p)}$ will be determined with accuracy $O(\alpha^4)$, while terms $O(\alpha^5)$ and those of higher orders will be neglected. This accuracy is sufficient because it exceeds that of measuring the geoid heights. The largest term in expansion (2.62) is $E_{00} Y_{00}(\Omega) \sim 1$, and term $O(\alpha)$ is coefficient E_{20}. The other terms are at least $O(\alpha^3)$. These largest terms will be considered in deriving coefficients $E_{jM}^{(p)}$ of expansion (2.159) separately. We shall introduce quantity

$$F_{00} = E_{00} Y_{00}(\Omega) = E_{00}/2\sqrt{\pi}\,, \tag{2.160}$$

which is very close to 1. According to the binomial theorem

$$
\begin{aligned}
\varrho^p(\Omega) = a_1^p &\left[F_{00} + \sum_{\substack{NM \\ N \geqslant 1}} E_{NM} Y_{NM} \right]^p = a_1^p \left\{ F_{00}^p + \binom{p}{1} F_{00}^{p-1} \sum_{\substack{NM \\ N \geqslant 1}} E_{NM} Y_{NM} \right. \\
&+ F_{00}^{p-2} \binom{p}{2} \left[E_{20} Y_{20} + \sum_{\substack{NM \\ \neq (2,0)}} E_{NM} Y_{NM} \right] \\
&\times \left[E_{20} Y_{20} + \sum_{\substack{KL \\ \neq (2,0)}} E_{KL} Y_{KL} \right] \\
&+ \binom{p}{3} F_{00}^{p-3} \left[E_{20} Y_{20} + \sum_{\substack{NM \\ \neq (2,0)}} E_{NM} Y_{NM} \right] \\
&\times \left[E_{20} Y_{20} + \sum_{\substack{KL \\ \neq (2,0)}} E_{KL} Y_{KL} \right] \\
&\times \left[E_{20} Y_{20} + \sum_{\substack{RS \\ \neq (2,0)}} E_{RS} Y_{RS} \right] + \binom{p}{4} F_{00}^{p-4} E_{20}^4 Y_{20}^4 \right\}. \tag{2.161}
\end{aligned}
$$

We have omitted the arguments of spherical harmonics $Y_{NM}(\Omega)$ in Eqs. (2.161) to save space. By omitting terms $O(\alpha^2)$ and those of higher orders in expansion (2.161) we arrive at

$$
\begin{aligned}
\varrho^p(\Omega) = a_1^p &\left\{ F_{00}^p + p F_{00}^{p-1} \sum_{\substack{NM \\ N \geqslant 1}} E_{NM} Y_{NM} + \binom{p}{2} F_{00}^{p-2} \right. \\
&\times \sum_{\substack{NMKL \\ N \geqslant 2, K \geqslant 2}} E_{NM} E_{KL} Y_{NM} Y_{KL} \\
&+ \binom{p}{3} F_{00}^{p-3} \left[E_{20}^3 Y_{20}^3 + 3 E_{20}^2 Y_{20}^2 \sum_{\substack{NM \\ \neq (2,0)}} E_{NM} Y_{NM} \right] \\
&+ \binom{p}{4} F_{00}^{p-4} E_{20}^4 Y_{20}^4 \right\}. \tag{2.162}
\end{aligned}
$$

We shall express the product of the spherical harmonics in terms of the Clebsch–Gordan expansion (2.76) and (2.77). In particular,

$$Y_{20}^2(\Omega) = \frac{1}{2\sqrt{\pi}} \left\{ Y_{00}(\Omega) + \frac{2\sqrt{5}}{7} Y_{20}(\Omega) + \frac{2 \cdot 3}{7} Y_{40}(\Omega) \right\},$$

$$Y_{20}^3(\Omega) = \frac{\sqrt{5}}{28\sqrt{\pi}} \left\{ 2Y_{00}(\Omega) + 3\sqrt{(5)}Y_{20}(\Omega) + \frac{2^2 3^2}{11} Y_{40}(\Omega) \right.$$

$$\left. + \frac{2 \cdot 3^2 \cdot 5}{11\sqrt{13}} Y_{60}(\Omega) \right\},$$

$$Y_{20}^4(\Omega) = \frac{5}{56\sqrt{\pi}} \left\{ 3Y_{00}(\Omega) + \frac{2^2 \cdot 5\sqrt{5}}{11} Y_{20}(\Omega) + \frac{2^2 3^2 \cdot 17}{11 \cdot 13} Y_{40}(\Omega) \right.$$

$$\left. + \frac{2^3 3^2}{11\sqrt{13}} Y_{60}(\Omega) + \frac{2^3 3^2 \cdot 7}{11 \cdot 13\sqrt{17}} Y_{80}(\Omega) \right\}. \tag{2.163}$$

By substituting (2.76) and (2.163) into (2.161) and comparing the coefficients with the spherical harmonics of expansions (2.161) and (2.159) we arrive at the following expressions for coefficients $E_{NM}^{(p)}$:

$$E_{JM}^{(p)} = E_{00}^{p-1}\left[\left[E_{00}\delta_J^0 + pE_{JM}(1 - \delta_J^0) + \binom{p}{2}\frac{1}{E_{00}\sqrt{(2J+1)}} \right.\right.$$

$$\times \sum_{\substack{NMKL \\ N \geq 2, K \geq 2}} E_{NM}E_{KL}\sqrt{[(2N+1)(2K+1)]}C_{N0\,K0}^{J0}C_{NMKL}^{JM}$$

$$+ \binom{p}{3}\frac{2E_{20}^2}{E_{00}^2}\left\{ \frac{\sqrt{5}}{14} E_{20}\left(2\delta_J^0 + 3\sqrt{(5)}\delta_J^2 + \frac{2^2 3^2}{11}\delta_J^4 + \frac{2 \cdot 3^2 \cdot 5}{11\sqrt{13}}\delta_J^6\right)\right.$$

$$+ 3\left[\frac{1}{2} E_{JM}(1 - \delta_J^0)(1 - \delta_J^1)(1 - \delta_J^2)\delta_M^0 + \frac{5}{7}\frac{1}{\sqrt{(2J+1)}}\sum_{\substack{NM \\ \neq(2,0)}} E_{NM} \right.$$

$$\times \sqrt{(2N+1)}C_{N020}^{J0}C_{NM20}^{JM} + \frac{3^2}{7}\frac{1}{\sqrt{(2J+1)}}\sum_{\substack{NM \\ \neq(2,0)}} E_{NM}\sqrt{(2N+1)}$$

$$\left.\left. \times C_{40N0}^{J0}(C_{40NM}^{JM}\right]\right\} + \binom{p}{4}\frac{E_{20}^4}{E_{00}^3}\frac{5}{7}\left[3\delta_J^0 + \frac{2^5 5\sqrt{5}}{11}\delta_J^2 + \frac{2^2 3^2 \cdot 17}{11 \cdot 13}\delta_J^4 \right.$$

$$\left.\left.\left. + \frac{2^3 3^3}{11\sqrt{13}}\delta_J^6 + \frac{2^3 3^2 \cdot 7}{11 \cdot 13\sqrt{17}}\delta_J^8 \right]\right]\right]. \tag{2.164}$$

In particular, the largest coefficient

$$E_{00}^{(p)} = \frac{E_{00}^p}{(2\sqrt{\pi})^{p-1}} \left\{ 1 + \binom{p}{2}\frac{1}{E_{00}^2}\sum_{\substack{NM \\ N \geq 2}} |E_{NM}|^2 \right.$$

$$\left. + \binom{p}{3}\frac{2}{7}\frac{E_{20}^2}{E_{00}^3}(\sqrt{(5)}E_{20} + 9E_{40}) + \binom{p}{4}\frac{3 \cdot 5}{7}\frac{E_{20}^4}{E_{00}^4} \right\}. \tag{2.165}$$

2.6 True Gravity Anomalies

The magnitude of acceleration of gravity in outer space can be expressed in terms of the first derivatives of the gravity potential, (2.46)–(2.48):

$$g = (g_\varrho^2 + g_\vartheta^2 + g_\Lambda^2)^{1/2} \, . \tag{2.166}$$

If (ϱ, ϕ, Λ) is a gravity point on the Earth's surface at which gravity acceleration g' was determined by gravimetric measurement in the universal Potsdam system, the correction of the Potsdam gravity standard δg_p can be derived from a sufficient number of differences $g(\varrho, \phi, \Lambda) - g'(\varrho, \phi, \Lambda)$, i.e.

$$\delta g_p = \frac{1}{m} \sum_{i=1}^{m} [g_i(\varrho, \phi, \Lambda) - g_i'(\varrho, \phi, \Lambda)] \, ; \tag{2.167}$$

the correction amounts to $\sim -14.0 \times 10^{-5} \, \mathrm{m \, s^{-2}}$. The satellite method can be used to improve the accuracy of correction δg_r, independently of surface gravimetric measurements, if relative data observed worldwide are used. However, the problem is now only of historical significance because there are instruments which measure absolute gravity, without reference to the initial point required in earlier relative gravity measurements.

By substituting the geoid's radius-vector (2.63) into Eq. (2.166) we obtain the gravity pattern on the fundamental equipotential surface $W = W_0$ as

$$g_{W=W_0} = \frac{GM}{R_0^2} \left[1 + g_0^{(0)} + \sum_{n=2}^{\bar{n}} \sum_{k=0}^{n} (g_n^{(k)} \cos k\Lambda + h_n^{(k)} \sin k\Lambda) P_n^{(k)}(\sin \phi) \right].$$

$$\tag{2.168}$$

The numerical values of the coefficients up to $\bar{n} = 12$ are given in Tables 2.7 and 2.8. The geopotential coefficients (Lerch et al. 1978) of the GEM 10B model, geocentric gravitational constant (1.152) and factor (2.157), and the angular

Table 2.7. Coefficients of the zonal terms in the expansion of gravity on the geoid (not normalized)

n	$g_n^{(0)} \, (10^{-9})$	$\dfrac{GM}{R^2} g_n^{(0)} \, (10^{-5} \, \mathrm{m \, s^{-2}})$
2	3 516 979	3461.71
3	5 108	5.02
4	− 13 121	− 12.91
5	915	0.90
6	− 2 709	− 2.66
7	2 215	2.17
8	1 481	1.45
9	957	0.94
10	2 223	2.18

Table 2.8. Coefficients of the tesseral and sectorial terms in the expansion of gravity on the geoid (fully normalized)

n	k	$\bar{g}_n^{(k)}$ (10^{-9})	$\frac{GM}{R_0^2}\bar{g}_n^{(k)}$ $(10^{-5}\,\mathrm{m\,s}^{-2})$	$\bar{h}_n^{(k)}$ (10^{-9})	$\frac{GM}{R_0^2}\bar{h}_n^{(k)}$ $(10^{-5}\,\mathrm{m\,s}^{-2})$	n	k	$\bar{g}_n^{(k)}$ (10^{-9})	$\frac{GM}{R_0^2}\bar{g}_n^{(k)}$ $(10^{-5}\,\mathrm{m\,s}^{-2})$	$\bar{h}_n^{(k)}$ (10^{-9})	$\frac{GM}{R_0^2}\bar{h}_n^{(k)}$ $(10^{-5}\,\mathrm{m\,s}^{-2})$
2	1	1	0.00	0	0.00	9	4	− 160	− 0.15	89	0.08
	2	2443	2.40	− 1403	− 1.38		5	− 151	− 0.14	− 431	− 0.42
							6	450	0.44	1775	1.74
3	1	4089	5.02	510	0.50		7	− 894	− 0.88	− 734	− 0.72
	2	1800	4.02	− 1251	− 1.23		8	1571	1.54	− 65	− 0.06
	3	1436	1.77	2856	2.81		9	− 400	− 0.39	790	0.77
4	1	− 1614	− 1.58	− 1432	− 1.41	10	1	731	0.71	− 1192	− 1.17
	2	1065	1.04	2027	1.99		2	− 815	− 0.80	− 279	− 0.27
	3	2999	2.95	− 620	− 0.61		3	− 91	− 0.08	− 1449	− 1.42
	4	− 585	− 0.57	929	0.91		4	850	− 0.83	− 652	− 0.64
5	1	− 220	− 0.21	− 388	− 0.38		5	− 512	− 0.50	− 391	− 0.38
	2	2639	− 2.59	− 1315	− 1.29		6	− 366	− 0.36	− 718	− 0.70
	3	− 1867	− 1.83	− 874	− 0.86		7	77	0.07	54	0.05
	4	− 1219	− 1.20	187	0.18		8	376	0.37	− 755	− 0.74
	5	636	0.62	− 2689	− 2.64		9	1163	1.14	− 426	− 0.41
6	1	− 390	− 0.38	145	0.14		10	936	0.92	− 275	− 0.27
	2	245	0.24	− 1822	− 1.79	11	1	60	0.05	− 162	− 0.15
	3	308	0.30	22	0.02		2	270	0.26	− 916	− 0.90
	4	− 495	− 0.48	− 2351	− 2.31		3	− 423	− 0.41	− 1382	− 1.36
	5	− 1338	− 1.31	− 2732	− 2.68		4	− 470	− 0.46	− 640	− 0.63
	6	26	0.02	− 1181	− 1.16		5	438	0.43	588	0.57

n	m				
7	1	1670	1.64	597	0.58
	2	1984	1.95	639	0.62
	3	1504	1.48	−1300	−1.28
	4	−1700	−1.67	−764	−0.75
	5	37	0.03	162	0.16
	6	−2211	−2.17	846	0.83
	7	−28	−0.02	136	0.13
8	1	197	0.19	362	0.35
	2	513	0.50	509	0.50
	3	−125	−0.12	−665	−0.65
	4	−1762	−1.73	539	0.53
	5	−139	−0.13	606	0.59
	6	−479	−0.47	2226	2.19
	7	474	0.46	514	0.50
	8	−893	−0.87	152	0.15
9	1	1243	1.22	113	0.11
	2	194	0.19	−227	−0.22
	3	−1252	−1.23	−716	−0.70

n	m				
	6	−102	−0.10	356	0.35
	7	40	0.03	−962	−0.94
	8	−36	−0.03	239	0.23
	9	−288	0.28	364	0.35
	10	−48	−0.04	−138	−0.13
	11	496	0.48	−602	−0.59
12	1	−638	−0.62	−559	−0.55
	2	53	0.05	261	0.25
	3	533	0.52	193	0.19
	4	−819	−0.80	−9	−0.00
	5	479	0.47	127	0.12
	6	98	0.09	437	0.43
	7	−202	−0.19	411	0.40
	8	−312	−0.30	240	0.23
	9	438	0.43	167	0.16
	10	−19	−0.01	330	0.32
	11	156	0.15	−35	−0.03
	12	−10	−0.01	−116	−0.11

velocity of rotation,

$$\omega = 7.292\,115 \times 10^{-5}\,\text{rad s}^{-1}\,; \quad q = 0.003\,461\,392\,, \tag{2.169}$$

were used to compute them. Figure 2.18 shows g along the Earth's equator as derived from satellite data (solid curve S). The results of gravimetric measurements made on the Earth's surface, which are but sporadic (parts of curve T), are shown for comparison. The latter do not reflect the large anomaly in the region of $\Lambda = 60° - \Lambda = 90°$E at all. The displacement of both curves is due to the known error in the Potsdam gravity standard. In Fig. 2.19 g along the equator is shown together with the geoid's radius-vector to illustrate the correlation of the two quantities in the region of their extremes.

For $\bar{n} = 4$ we obtain the gravity acceleration on the fourth-degree spheroid retaining the dominating harmonic terms with amplitudes $\geqq 2 \times 10^{-5}\,\text{m s}^{-2}$ $(GM/R_0^2 = 984\,287.2 \times 10^{-5}\,\text{m s}^{-2})$:

$$
\begin{aligned}
g_{\bar{n}=4} = {}& 979\,767.6 + 3461.7\mathrm{P}_2^{(0)}(\sin\phi) \\
&+ 2.4\bar{\mathrm{P}}_2^{(2)}(\sin\phi)\cos 2\Lambda + 5.0\mathrm{P}_3^{(0)}(\sin\phi) \\
&+ 4.0\bar{\mathrm{P}}_3^{(1)}(\sin\phi)\cos\Lambda + 1.8\bar{\mathrm{P}}_3^{(2)}(\sin\phi)\cos 2\Lambda \\
&+ 2.8\bar{\mathrm{P}}_3^{(3)}(\sin\phi)\sin 3\Lambda - 12.9\mathrm{P}_4^{(0)}(\sin\phi) \\
&+ 3.0\bar{\mathrm{P}}_4^{(3)}(\sin\phi)\cos 3\Lambda\,.
\end{aligned}
\tag{2.170}
$$

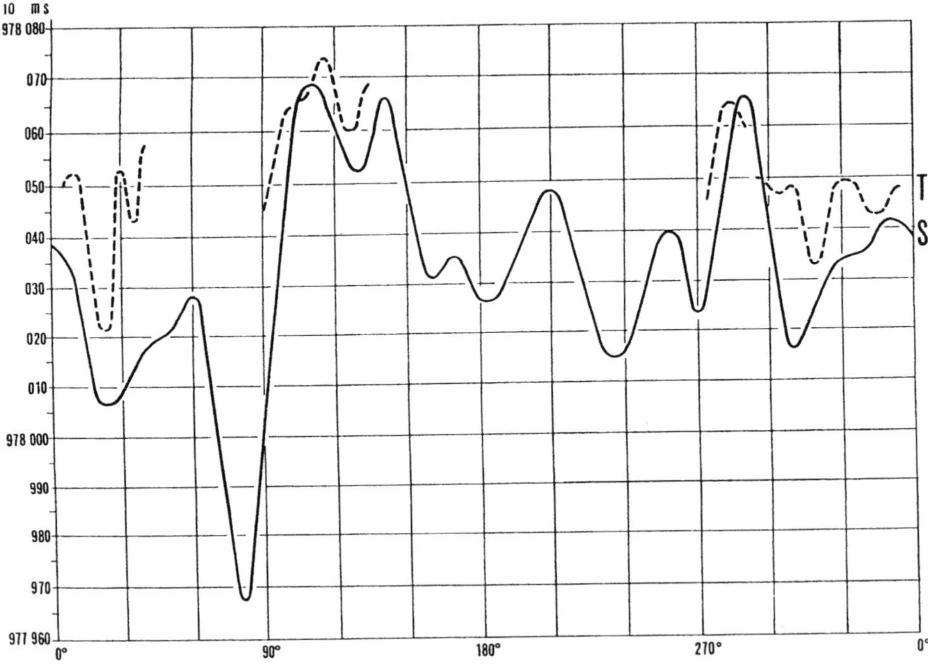

Fig. 2.18. Gravity acceleration g (in 10^{-5} m s^{-2}) in the geoid's meridional section derived from satellite data (*bold curve S*) and from measurements by gravity meters (*dashed curve T*)

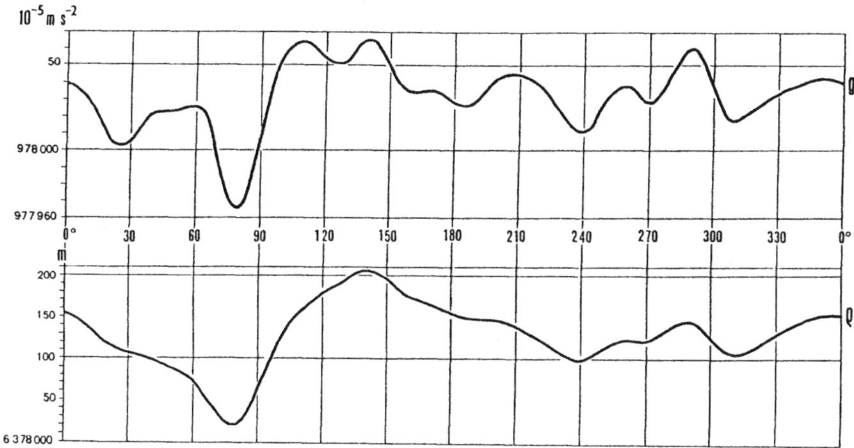

Fig. 2.19. Comparison of the acceleration of gravity g (in 10^{-5} m s^{-2}) with the geoid radius-vector ρ (in m) in the geoid's equatorial section

The gravity along the geodetic equator (in units of 10^{-5} m s^{-2}) is obtained by putting $\phi = 0$:

$$g_{\bar{n}=4}(\phi = 0) = 978\,032.0 - 6.5\cos\Lambda + 4.7\cos 2\Lambda + 5.9\sin 3\Lambda . \qquad (2.171)$$

If we restrict ourselves to $\bar{n} = 2$, $k = 0$ in (2.170), we arrive at $g_{n=2}$ on Clairaut's spheroid.

The differences $(g - g_{n=2})$ or $(g - g_{n=4})$ are referred to as true gravity anomalies (gravity disturbances), reflecting a finer field structure the higher the degree of the spheroid introduced. These differ from the so-called mixed gravity anomalies used in classical gravimetry.

2.7 Structure of the Gravitational Field over the Northern and Southern Hemispheres

The set of geopotential coefficients describes the gravitational field as a whole uniformly; however, at the same time, it reflects its peculiarities in anomalous regions the more truly, the higher the degree $\bar{n}$ of the harmonic terms retained. The three basic shape parameters, α, α_1, Λ_a, defining the shape and orientation of the best-fitting triaxial ellipsoid, characterize the general shape of the geoid, but these global parameters are not typical of some of the strongly anomalous regions.

For example, let us consider the parallel sections of surface $W = W_0$, i.e. with $\phi = $ const, and substitute an ellipse for each such section. We find that these ellipses generally differ in shape and orientation from the ellipses corresponding to the analogous sections of the best-fitting triaxial Earth ellipsoid.

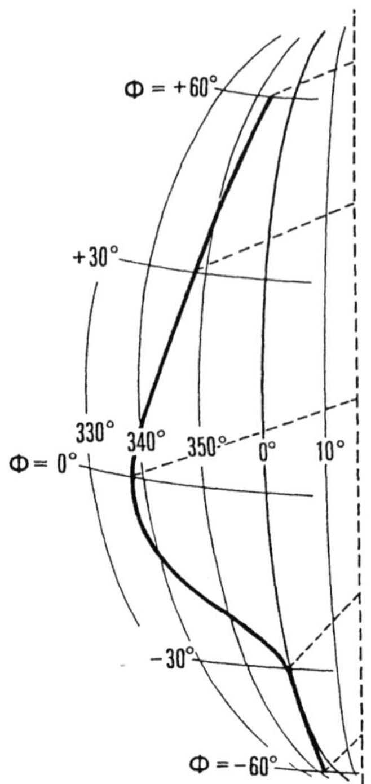

Fig. 2.20. Longitudes of meridians $\bar{\Lambda}_a$ containing the major axes of the ellipses approximating geoid equatorial sections

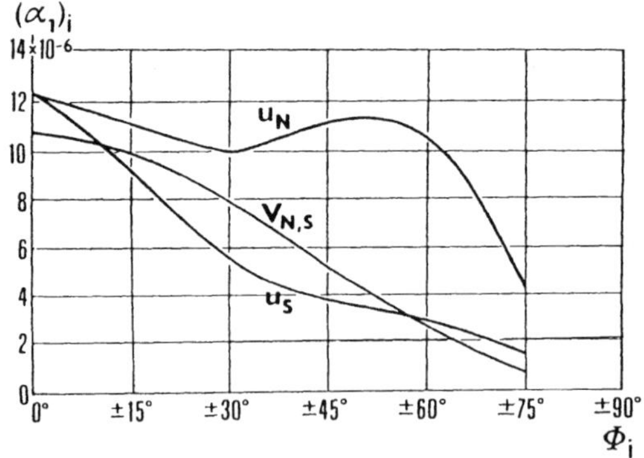

Fig. 2.21. Flattening $(\alpha_1)_i$ of parallel sections (u_N in the Northern Hemisphere and u_S in the Southern Hemisphere) and of the Earth's triaxial ellipsoid ($v_{N,S}$)

Let us denote ϕ_i the geocentric latitude of the i-th section, and $(e_1^2)_i$ the square of the eccentricity of the analogous section (of the same latitude ϕ_i) of the surface of the triaxial Earth ellipsoid, parameters $e^2(\alpha)$, $e_1^2(\alpha_1)$; it then holds that

$$(e_1^2)_i = \frac{e_1^2(1-e^2)\cos^2\phi_i}{1-e^2\cos^2\phi_i - e_1^2\sin^2\phi_i}. \tag{2.172}$$

Table 2.9. Mean values of the geoid's radius-vector in plane sections of the geoid symmetrical relative to the equator, and their differences

Southern Hemisphere			Northern Hemisphere			Difference (N − S)	Difference (N − S)
ϕ_i	$\bar{\rho}_i$ (m)	$\bar{\rho}_i \sin \phi_i$ (m)	ϕ_i	$\bar{\rho}_i$ (m)	$\bar{\rho}_i \sin \phi_i$ (m)	$\bar{\rho}_i$ (m)	$\bar{\rho}_i \sin_i \phi_i$ (m)
15°	6 376 693.8	1 650 409.8	− 15°	6 376 705.7	1 650 412.9	− 11.9	− 3.1
30°	6 372 767.1	3 186 383.5	− 30°	6 372 780.0	3 186 390.0	− 12.9	− 6.5
45°	6 367 420.0	4 502 445.8	− 45°	6 367 425.9	4 502 450.0	− 5.9	− 4.2
60°	6 362 085.6	5 509 727.7	− 60°	6 363 077.7	5 509 720.9	+ 7.9	+ 6.8
75°	6 358 189.7	6 141 539.7	− 75°	6 358 161.5	6 141 512.4	+ 28.2	+ 27.2
90°	6 356 774.9	6 356 774.9	− 90°	6 356 732.0	6 356 732.0	+ 42.9	+ 42.9

The lengths of the semimajor axes of all these ellipses

$$(\Lambda_a)_i = \Lambda_a , \tag{2.173}$$

i.e. always equal to the length of the meridian in which the semimajor axis a of the triaxial ellipsoid is located. Let us impose the condition

$$\int_0^{2\pi} (\varrho_{\phi=\phi_i} \cos \phi_i - r_{\phi=\phi_i})^2 \, d\phi = \min ; \tag{2.174}$$

$\varrho_{\phi=\phi_i}$ is the radius-vector of the geoid section created by plane, $\phi = \phi_i$, $r_{\phi=\phi_i}$ is the radius-vector of the best-fitting ellipse, parameters $\bar{a}_i, (\bar{e}_1)_i, (\bar{\Lambda}_a)_i$,

$$r_{\phi=\phi_i} = \bar{a}_i \left\{ \frac{1 - (\bar{e}_1^2)_i}{1 - (\bar{e}_1^2)_i \cos^2 [\Lambda - (\bar{\Lambda}_a)_i]} \right\}^{1/2} . \tag{2.175}$$

The solution yields parameters $(\bar{e}_1)_i, (\bar{\Lambda}_a)_i$, which differ in general from $(e_1)_i, (\Lambda_a)_i$. The values of $(\bar{\Lambda}_a)_i$ are shown schematically in Fig. 2.20 and those of flattening $(\bar{\alpha}_1)_i, [(\bar{e}_1)_i^2 = 2(\bar{\alpha}_1)_i - (\bar{\alpha}_1)_i^2]$ are shown in Fig. 2.21 (u_N, u_S are the curves for the Northern and Southern hemispheres, respectively; $v_{N,S}$ represents the flattening of the section of the Earth's triaxial ellipsoid).

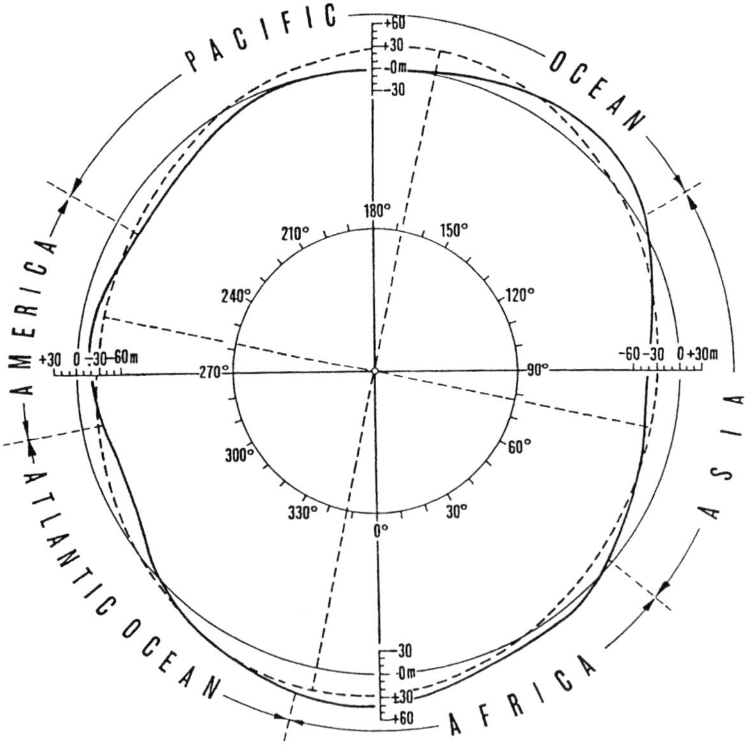

Fig. 2.22. Plane section $\phi = +30°$ of the geoid

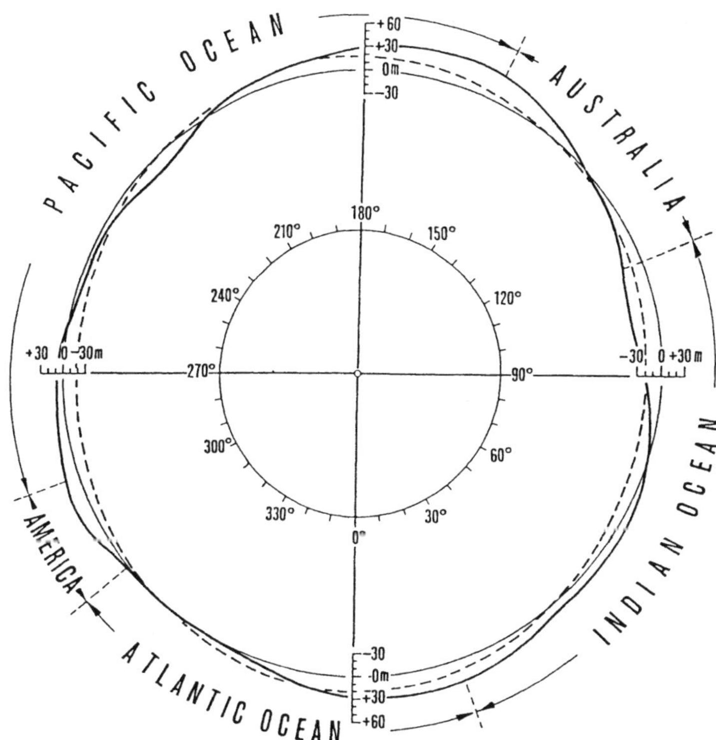

Fig. 2.23. Plane section $\phi = -30°$ of the geoid

The mean values of the radius-vector of the geoid surface in the Northern $(\bar{\varrho}_N)$ and Southern $(\bar{\varrho}_S)$ hemispheres also differ:

$$\bar{\varrho}_N = 6\,370\,987.7 \text{ m}; \quad \bar{\varrho}_S = 6\,370\,991.8 \text{ m}, \tag{2.176}$$

$\bar{\varrho} = 6\,370\,989.8$ m being the mean value for the Earth as a whole.

Table 2.9 gives the mean values $\bar{\varrho}_i$ of the geoid's radius-vector in the said sections $\phi = \phi_i$, symmetrical relative to the equator:

$$\varrho_i = \frac{1}{2\pi} \int_0^{2\pi} \varrho_i \, d\Lambda = R_0 \left[1 + A_0^{(0)} + \sum_{n=2}^{\bar{n}} A_n^{(0)} P_n^{(0)}(\sin \phi_i) \right]. \tag{2.177}$$

The sections are illustrated in Figs. 2.22–2.25 for $\phi_i = +30°, -30°, +60°$ and $-60°$. Table 2.10 contains the parameters of the rotational ellipsoid best fitting the geoid surface under condition (2.136) separately for the Northern Hemisphere, the Southern Hemisphere and the Earth as a whole. Table 2.11 gives the main values of gravity acceleration in $10^{-5}\,\text{m s}^{-2}$ (mgal):

$$\bar{g}_i = \frac{1}{2\pi} \int_0^{2\pi} g_i \, d\Lambda = \frac{GM}{R_0^2} \left[1 + g_0^{(0)} + \sum_{n=2}^{\bar{n}} g_n^{(0)} P_n^{(0)}(\sin \phi_i) \right]. \tag{2.178}$$

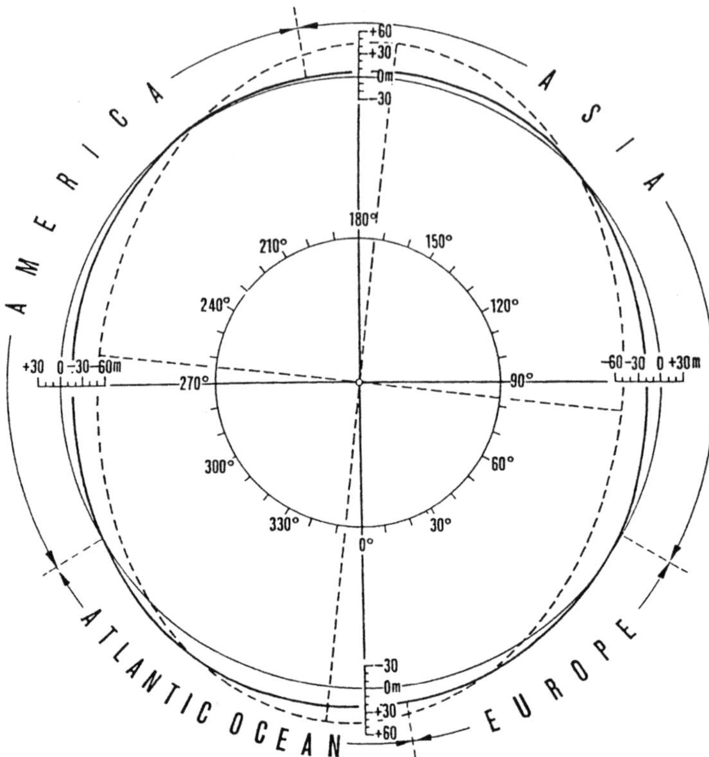

Fig. 2.24. Plane section $\phi = +60°$ of the geoid

Figures 2.26–2.28 show the differences in sections $\phi = \phi_i$ for $\phi_i = \pm 30°$, $\pm 45°$ and $\pm 60°$.

One can thus conclude that the mass structure of the Earth is considerably asymmetric, not only relative to the axis of rotation but also relative to the equatorial plane. These asymmetries affect some of the geodynamic phenomena connected with perturbations due to 'third bodies', namely, the Moon.

The geoid surface above oceans can be mapped by direct satellite altimetry. No substantial differences, compared with the geoid determined gravimetrically, were observed. Fig. 2.29 presents the surface of the oceanic geoid as derived from the SEASAT satellite data (Marsh and Martin 1982).

2.8 Theory of the Order of Flattening

In the preceding sections we have discussed the external gravity field regardless of its sources. The distribution of the field sources, i.e. the density distribution within the Earth, is of little importance from some geodetic points of view;

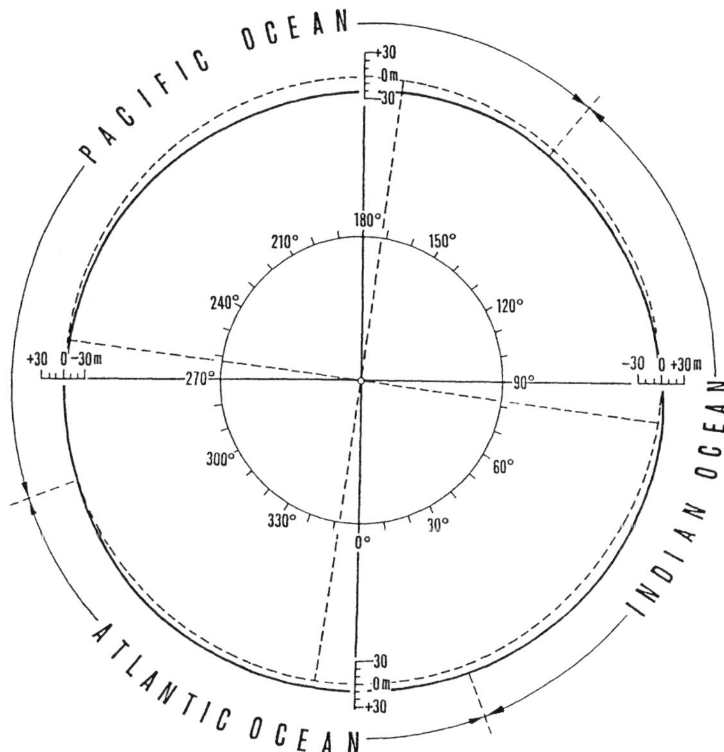

Fig. 2.25. Plane section $\phi = -60°$ of the geoid

Table 2.10. Parameters of the rotational ellipsoid best fitting the surface of the geoid in the Northern and Southern Hemispheres, and for the Earth as a whole

Region	a (m)	e^2	$1/\alpha$
Northern Hemisphere	6 378 131.4	0.006 689 510	298.475
Southern Hemisphere	6 378 145.8	0.006 699 214	298.042
Earth as a whole	6 378 138.6	0.006 694 362	298.257

however, in geophysics, i.e. with regard to the internal dynamics of the Earth, it is of primary concern. Consequently, in this section and in the following sections we shall be dealing with the relations between the external (and also the internal) gravity field and the distribution of density within the Earth. We shall attempt to present a theory which will be valid without any a priori restrictive assumptions related to field-source distribution. In this section alone, however, we shall make an exception and assume hydrostatic distribution of masses within the Earth for the following reason. The hypothesis of hydrostatic distribution of masses

Table 2.11. Mean values of the acceleration of gravity in plane sections of the geoid symmetrical relative to the equator, and their differences

Northern Hemisphere		Southern Hemisphere		Difference
ϕ_i	$\bar{g}_i$ $(10^{-5}\,\mathrm{m\,s^{-2}})$	ϕ_i	$\bar{g}_i$ $(10^{-5}\,\mathrm{m\,s^{-2}})$	$(\mathrm{N}-\mathrm{S})$ $(10^{-5}\,\mathrm{m\,s^{-2}})$
15°	978 386.7	− 15°	978 390.8	− 4.1
30°	979 343.3	− 30°	979 346.1	− 2.8
45°	980 644.0	− 45°	980 645.8	− 1.7
60°	981 937.9	− 60°	981 937.2	+ 0.7
75°	982 881.4	− 75°	982 873.3	+ 8.1
90°	983 227.2	− 90°	983 204.1	+ 23.0

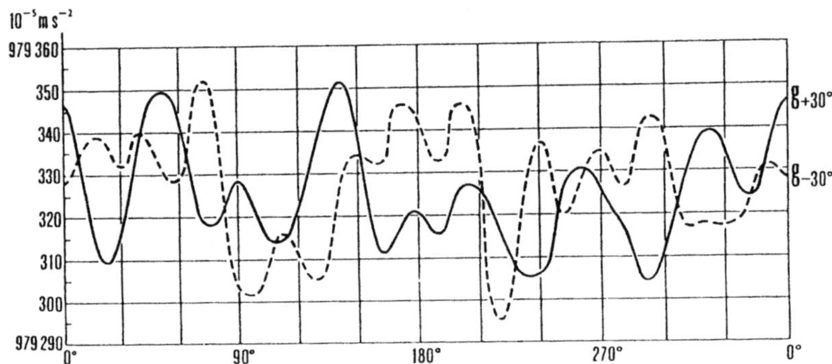

Fig. 2.26. Acceleration of gravity along parallels $\phi = \pm 30°$

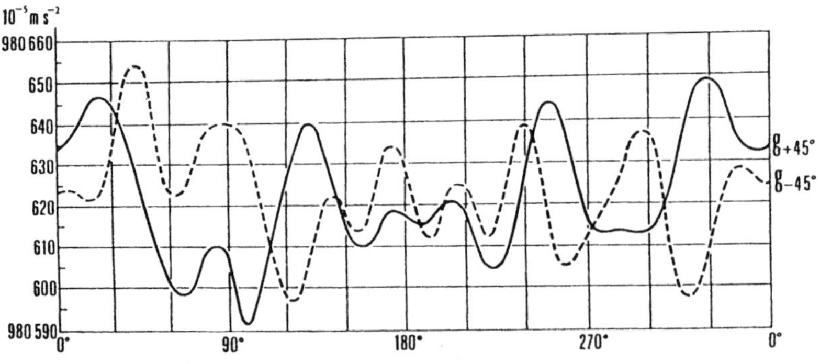

Fig. 2.27. Acceleration of gravity along parallels $\phi = \pm 45°$

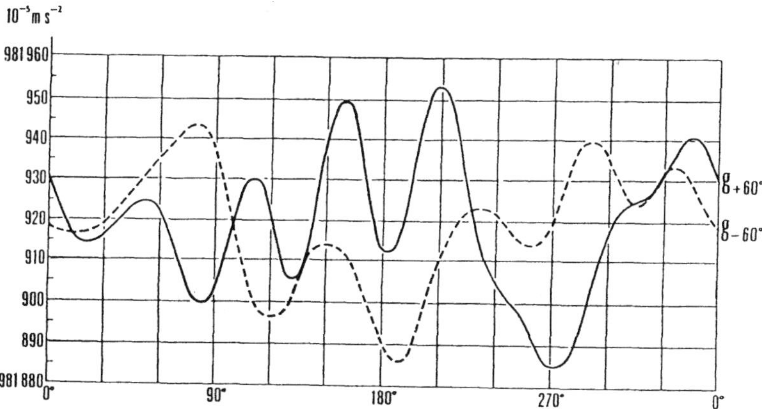

Fig. 2.28. Acceleration of gravity along parallels $\phi = \pm 60°$

within the Earth was the dominant hypothesis in geodesy and geophysics, prevailing even long after 1958 when the observation of perturbations of satellite orbits indubitably proved the existence of terms in the geopotential expansion which were 'forbidden' with regard to the hypothesis of hydrostatic mass distribution within the Earth. However, even in the presatellite era there existed opinions which cast doubt on the validity of the hypothesis of hydrostatic mass distribution within the Earth, and which pointed out the existence of dynamic processes, the result of which is the upsetting of hydrostatic equilibrium, in the Earth's crust, mantle and core. Although the hypothesis of hydrostatic equilibrium evidently does not hold exactly within the Earth, it has certainly retained its significance as a limiting theory of the 'dead Earth', free of all internal forces with the exception of gravitation and the centrifugal forces. It is therefore pointless to try to achieve the same high accuracy as in the preceding sections and a relative accuracy of the order of the Earth's flattening is quite sufficient. This approach will enable us to reduce the general theory to the classical case of Clairaut's theory and, in some cases, to obtain a better insight into the relations of the general theory.

2.8.1 Clairaut's Theory of the External Field

In the $O(\alpha)$ approximation the gravitational potential is represented only by the two largest geopotential coefficients, A_{00} and A_{20}, and the geoid surface is also approximated by an expansion whose parameters are the two coefficients E_{00} and E_{20}. This simplified surface, approximating the geoid, is referred to as Clairaut's spheroid. In the above approximation the formula for the gravity

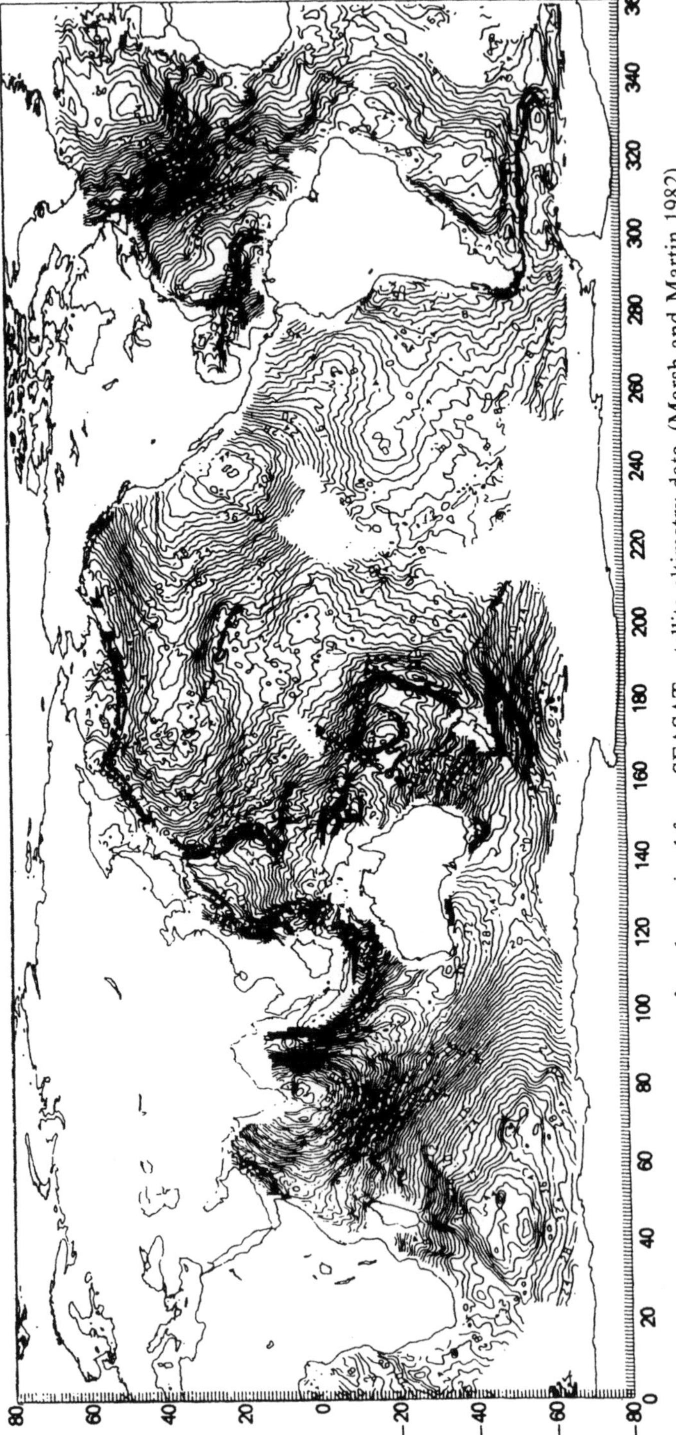

Fig. 2.29. Mean ocean surface determined from SEASAT satellite altimetry data. (Marsh and Martin 1982)

potential will reduce to

$$W(\varrho, \vartheta) = \frac{GM}{\varrho}\left\{A_{00}Y_{00}(\Omega) + \left(\frac{a_0}{\varrho}\right)^2 A_{20}Y_{20}(\Omega)\right.$$

$$\left. + q\left(\frac{\varrho}{a_0}\right)^3 \frac{2\sqrt{\pi}}{3}\left[Y_{00}(\Omega) - \frac{1}{\sqrt{5}}Y_{20}(\Omega)\right]\right\}, \tag{2.179}$$

where $A_{00} = 2\sqrt{\pi}$. The equation of equipotential surface (2.62) will read

$$\varrho(\Omega) = a_1[E_{00}Y_{00}(\Omega) + E_{20}Y_{20}(\Omega)]. \tag{2.180}$$

If spherical harmonics $Y_{00}(\Omega)$ and $Y_{20}(\Omega)$ are expressed using (2.14), (2.180) will become

$$\varrho(\vartheta) = a_1[1 - \alpha\cos^2\vartheta], \tag{2.181}$$

in which we have put

$$E_{20} = -\frac{4\sqrt{\pi}}{3\sqrt{5}}\alpha, \tag{2.182}$$

and

$$E_{00} = 2\sqrt{(\pi)}\left(1 - \frac{\alpha}{3}\right) = 2\sqrt{\pi} + \frac{\sqrt{5}}{2}E_{20}. \tag{2.183}$$

Equation (2.182) defines coefficient E_{20} in terms of flattening α, and E_{00} can be expressed in terms of coefficient E_{20}. We shall now derive the relation between potential coefficient A_{20} and spheroid coefficient E_{20}. Eqs. (2.98) and (2.99) represent the relations between potential and geometric coefficients in general; in this particular case we shall retain terms $O(\alpha)$. Firstly, $g_\vartheta = O(\alpha)$ and $\frac{1}{a_1}\frac{\partial\varrho}{\partial\vartheta} = O(\alpha)$, and secondly, we shall put $g_r = \frac{GM}{\varrho^2}$ and $\varrho = a_1$ in Eq. (2.98), so that in the $O(\alpha)$ approximation

$$g_\vartheta = -\frac{GM}{a_1^2}\frac{\partial\varrho}{a_1\partial\vartheta}. \tag{2.184}$$

We shall express g_ϑ using Eqs. (2.47) and (2.50):

$$g_\vartheta = -\frac{GM}{a_0^2}[\exp(-i\Lambda)Y_{21}(\Omega) - \exp(i\Lambda)Y_{2-1}(\Omega)]$$

$$\times\left[-\sqrt{\left(\frac{3}{2}\right)}A_{20} + \sqrt{\left(\frac{2\pi}{3\cdot5}\right)}q\right] \tag{2.185}$$

and, in view of (2.91),

$$\frac{\partial\varrho}{\partial\vartheta} = a_1\sqrt{(\tfrac{3}{2})}E_{20}[\exp(-i\Lambda)Y_{21}(\Omega) - \exp(i\Lambda)Y_{2-1}(\Omega)]. \tag{2.186}$$

By substituting into (2.184) we can now express the coefficient of surface expansion E_{20} in terms of zonal potential coefficient A_{20}:

$$E_{20} = A_{20} - \tfrac{2}{3}q\sqrt{\frac{\pi}{5}}. \tag{2.187}$$

If we express E_{20} with regard to (2.182) in terms of flattening α and coefficient A_{20} in terms of zonal parameter $J_2 = - J_2^{(0)}$,

$$A_{20} = - J_2 \frac{2\sqrt{\pi}}{\sqrt{5}}, \tag{2.188}$$

and alter Eq. (2.187) to read

$$\alpha = \tfrac{3}{2}J_2 + \tfrac{1}{2}q, \tag{2.189}$$

we arrive at the familiar relation between geometric parameter α and parameter J_2.

If we only retain the $O(\alpha)$ terms, expressed in terms of A_{20}, in the formula for the radial component of the gravity acceleration, it is easy to express it in its familiar form

$$g_r = g_{re}(1 + \beta\cos^2\vartheta), \tag{2.190}$$

where g_{re} is the radial component of the gravity acceleration at the equator,

$$g_{re} = \frac{GM}{a_1^2}(1 - q + \tfrac{3}{2}J_2) = \frac{GM}{a_1^2}(1 + \alpha - \tfrac{3}{2}q); \tag{2.191}$$

the radial component at the pole

$$g_{rP} = \frac{GM}{a_1^2}(1 + q). \tag{2.192}$$

Coefficient β then comes out as

$$\beta = \frac{g_{rP} - g_{ra}}{g_{re}} = \tfrac{5}{2}q - \alpha. \tag{2.193}$$

2.8.2 Internal Gravitational Field of the Hydrostatic Earth. Clairaut's Differential Equation

Observations indicate that the Earth behaves like an elastic body with respect to short-period forces such as seismic forces. With respect to long-period forces, however, it behaves like a viscous body. We can thus adopt a viscoelastic body as an adequate model. The equilibrium conditions are expressed by the equation

$$\mathbf{V}\cdot\tau + \mathbf{g}\sigma = 0, \tag{2.194}$$

where $\mathbf{g}$ represents the internal gravitational force acting on a unit volume.

Quantity τ is the stress tensor which, for a viscous body, reads

$$\tau = -pI + \zeta[\nabla\mathbf{v} + (\nabla\mathbf{v})^T]\,, \tag{2.195}$$

p being hydrostatic pressure, I identity tensor and $\mathbf{v}$ the velocity of motion of the substance. The viscous behaviour is characterized by coefficient ζ. For the body as a whole to be in equilibrium it is necessary to add the equation of continuity $\partial\sigma/\partial t + \nabla\cdot(\sigma\mathbf{v}) = 0$ and other constraints to Eqs. (2.194) and (2.195). Differential equation (2.194) together with (2.195) and the appropriate boundary conditions defines the relation between density distribution σ and stress τ, which, apart from hydrostatic pressure p, also contains components describing the effect of friction of the flowing viscous substance. The solution of this equation is of fundamental importance for describing mantle dynamics and lithosphere tectonics, and is thus in the limelight of interest, e.g. Baumgardner (1988), Čadek (1989), Čadek and Matyska (1991). To be able to solve these equations it is necessary to know the density distribution in the mantle as well as the 3-D viscosity pattern. This condition is the principal difficulty because only very limited data are available on the distribution of the viscosity coefficient within the Earth. Knowledge of current motions of lithospheric plates (Ricard and Vigny 1989, Ricard et al. 1989) is used to formulate the boundary conditions. The discussion of the problems related to solving Eq. (2.194) would exceed the scope of this book and therefore the reader is referred to the work of, for example, Sabadini et al. (1991) for details.

We shall now assume that the pressure and density distribution of the Earth obeys the laws of hydrostatics. Observations of artificial satellite orbits and the geopotential coefficients derived from them have definitely proved that this assumption is not realistic and, consequently, that the Earth is not a hydrostatic body. Thus, equilibrium within the Earth is described by Eqs. (2.194) and (2.195). Deviations from the hydrostatic density distribution within the Earth are due to non-hydrostatic stresses and non-zero velocities of substance motion. In this sense the assumption of a hydrostatic Earth is only justified as a theoretical limiting model of the Earth with no internal dynamics. As we shall prove in Section 2.11, the ratio of the density deviations to the mean density is a quantity of the order of the Earth's flattening. This proves that it is pointless to consider an accuracy higher than that of quantity $O(\alpha)$ in the hydrostatic theory.

If $\mathbf{v} = 0$ and if the internal forces are only gravity forces, $\nabla\cdot\tau = -\nabla p$, and Eq. (2.194) reduces to

$$\nabla p = \sigma\nabla W = \sigma\mathbf{g}\,. \tag{2.196}$$

Let us consider a surface of constant density σ within the Earth. Equation (2.196) is then easy to integrate:

$$p + \sigma W = \text{const}\,. \tag{2.197}$$

This equation indicates that surfaces of equal pressure are also surfaces of equal gravity potential. Therefore, under the assumptions given, surfaces of equal density, pressure and gravity potential coincide. Moreover, if we apply operation curl to Eq. (2.196), we shall find that $\mathbf{V}p \times \mathbf{V}W = 0$, i.e. the iso-surfaces of the pressure gradient and gravity are parallel.

Denoting the largest semi-axis of the surface of constant density, considered to be a spheroid, a_1,

$$\varrho(r) = a_1 r [1 - f_2(r) Y_{20}(\Omega)] \, . \tag{2.198}$$

The equal-density surfaces thus form a system of spheroids dependent on parameter r, $0 \leqslant r \leqslant 1$. If we take the form of $Y_{20}(\Omega)$ per (2.14) into account, we find that (2.198) is a system of spheroids with semi-axes $a_1 r [1 + \tfrac{3}{4}\sqrt{(5/\pi)} f_2(r)]$ and flattening

$$\alpha(r) = -\frac{3}{4}\sqrt{\left(\frac{5}{\pi}\right)} f_2(r) \, , \tag{2.199}$$

which depends on the reduced radius-vector r. In view of (2.199) we shall consider quantity $f_2(r)$ to be a quantity of the order of the flattening.

We will consider the gravitational effect of a layer of constant density, whose boundary is defined by radius-vector $\varrho(r)$, per (2.198), and whose thickness is $\mathrm{d}\varrho(r)$. The potential of this layer is

$$\mathrm{d}V_e(P) = G\sigma(r) \iint \frac{\varrho'^2(r)\,\mathrm{d}\varrho'(r)}{d(P, P')} \,\mathrm{d}\Omega' \, , \tag{2.200}$$

where $d(P, P')$ is the distance between potential point $P(\varrho, \Omega)$ and point $P'(\varrho', \Omega')$ located in the layer.

We shall first assume that potential point $P(\varrho, \Omega)$ is outside the layer. $d(P, P')$ can then be expressed by a series of Legendre polynomials,

$$1/d(P, P') = \frac{1}{\varrho(r)} \sum_{j=0}^{\infty} \left(\frac{\varrho'(r)}{\varrho}\right)^j \mathrm{P}_j(\cos \Psi), \quad \varrho \geqslant \varrho', \tag{2.201}$$

Ψ being the angle between the radius-vectors to points P and P'. Applying the addition theorem for Legendre polynomials [analogy of (1.69)] yields

$$\mathrm{P}_j(\cos \Psi) = \frac{4\pi}{2j + 1} \sum_{m=-j}^{j} Y^*_{jm}(\Omega') Y_{jm}(\Omega) \, . \tag{2.202}$$

The reciprocal distance (2.201) becomes

$$\frac{1}{d(P, P')} = \frac{4\pi}{\varrho} \sum_{j=0}^{\infty} \left(\frac{\varrho'}{\varrho}\right)^j \frac{1}{2j + 1} \sum_{m=-j}^{j} Y^*_{jm}(\Omega') Y_{jm}(\Omega) \, . \tag{2.203}$$

In view of (2.198),

$$\mathrm{d}\varrho'(r) = a_1 \{1 - [f_2(r) + r f'_2(r)] Y_{20}(\Omega')\} \,\mathrm{d}(r) \, . \tag{2.204}$$

Given accuracy $O(\alpha)$ we arrive at

$$\varrho^{j+2}(r)\,\mathrm{d}\varrho(r) \sim a_1^{j+3} r^{j+2} [1 - \{(j + 3) f_2(r) + r f'_2(r)\} Y_{20}(\Omega)]\,\mathrm{d}r \, . \tag{2.205}$$

By substituting into (2.200) and taking into account the orthogonality of spherical harmonics $Y_{jm}(\Omega)$, we obtain the following expression for the external potential:

$$dV_e(P) \sim \frac{4\pi a_1^3 G\sigma(r)}{\varrho}\, dr'\left\{r'^2 - \tfrac{1}{5}(a_1/\varrho)^2 \frac{d}{dr}\left[r'^5 f_2(r')\right]\right\}, \tag{2.206}$$

$$\varrho \geqslant \varrho'(r).$$

At the internal point the reciprocal distance reads

$$\frac{1}{d(P, P')} = \frac{4\pi}{\varrho'(r)}\sum_{j=0}^{\infty}\left[\frac{\varrho}{\varrho'(r)}\right]^j \frac{1}{2j+1}\sum_{m=-j}^{j} Y_{jm}^*(\Omega')Y_{jm}(\Omega). \tag{2.207}$$

Quite analogously the expression for the gravitational potential at the internal point of the layer is

$$dV_i(P) \sim 4\pi a_i^2\, G\sigma(r)\,dr\left[r - \frac{\varrho^2}{a_1^2}\frac{1}{5}f'_2(r)Y_{20}(\Omega)\right]. \tag{2.208}$$

We shall now deal with the gravitational field of an inhomogeneous spheroid composed of the layers described. The gravitational potential at internal point $P(x, \Omega)$, x being the reduced radius-vector, reads

$$V(P) = \int_0^x dV_e + \int_x^1 dV_i. \tag{2.209}$$

After substituting into (2.209) for dV_e and dV_i from (2.206) and (2.208), respectively, we get

$$V(P) = \frac{4\pi a_1^3 G}{\varrho}\int_0^x r^2\sigma(r)\,dr - \frac{4\pi a_1^3 G}{5\varrho}\left(\frac{a_1}{\varrho}\right)^2 Y_{20}(\Omega)$$

$$\times \int_0^x \sigma(r)\frac{d}{dr}\left[r^5 f_2(r)\right]dr + 4\pi a_1^2 G\int_x^1 r\sigma(r)\,dr$$

$$- \frac{4\pi a_1^2 G}{5}\left(\frac{\varrho}{a_1}\right)^2 Y_{20}(\Omega)\int_x^1 \sigma(r)f'_2(r)\,dr. \tag{2.210}$$

The gravity potential $W(P)$ at the internal point is now obtained by adding the potential of the centrifugal forces, i.e.

$$W(P) = V(P) + \tfrac{1}{2}\omega^2\varrho^2\sin^2\vartheta = V(P) + \tfrac{1}{2}\omega^2\varrho^2\left[\frac{2}{3} - \frac{4}{3}\sqrt{\left(\frac{\pi}{5}\right)}Y_{20}(\Omega)\right]. \tag{2.211}$$

The square of the angular velocity, ω^2, can be expressed in terms of Helmert's parameter q, using (2.17), and mean density $\bar{\sigma} = M/(\tfrac{4}{3}\pi a_1^3)$:

$$\omega^2 = \tfrac{4}{3}\pi G q\bar{\sigma}. \tag{2.212}$$

The equipotential surface on which potential point P is located, can be expressed as

$$\varrho = a_1 x [1 - f_2(x) Y_{20}(\Omega)] , \tag{2.213}$$

which means that the potential is constant, i.e. $W(P) = W_P$, for all points P of surface (2.213). Since the equipotential surface in a hydrostatic Earth is also a surface of constant density, we are able to introduce mean density σ_x of the spheroid bounded by surface (2.213):

$$\sigma_x = 4\pi \int_0^{\varrho x} \sigma(\varrho')\varrho'^2 \, \mathrm{d}\varrho' / (\tfrac{4}{3}\pi a_1^3 x^2) = \frac{3}{x^3} \int_0^x \sigma(r) r^2 \, \mathrm{d}r . \tag{2.214}$$

It is easy to prove that

$$\frac{\mathrm{d}\sigma_x}{\mathrm{d}x} = -\frac{3}{x} [\sigma_x - \sigma(x)] . \tag{2.215}$$

Substituting into (2.211) for ϱ from (2.213) and retaining terms of the order of α only, we arrive at

$$\frac{W_P}{4\pi a_1^2 G} = \left\{ \sigma_x \frac{x^2}{3} + \int_x^1 r\sigma(r) \, \mathrm{d}r \right\}$$

$$+ Y_{20}(\Omega) \left\{ \frac{\sigma_x}{3} x^2 f_2(x) - \frac{1}{5x^3} \int_0^x \sigma(r) \frac{\mathrm{d}}{\mathrm{d}r} [r^5 f_2(r)] \, \mathrm{d}r \right.$$

$$\left. - \frac{x^2}{5} \int_x^1 \sigma(r) \frac{\mathrm{d}f_2(r)}{\mathrm{d}r} \, \mathrm{d}r - \frac{2}{9} \sqrt{\left(\frac{\pi}{5}\right)} q x^2 \bar{\sigma} \right\} . \tag{2.216}$$

Since W_P is a constant and Eq. (2.216) expresses identity, the coefficient with $Y_{20}(\Omega)$ must be equal to zero, i.e.

$$\frac{\sigma_x}{3} x^2 f_2(x) - \frac{1}{5x^3} \int_0^x \sigma(r) \frac{\mathrm{d}}{\mathrm{d}r} [r^5 f_2(r)] \, \mathrm{d}r$$

$$- \frac{x^2}{5} \int_x^1 \sigma(r) \frac{\mathrm{d}f_2(r)}{\mathrm{d}r} \, \mathrm{d}r - \frac{2}{9} \sqrt{\left(\frac{\pi}{5}\right)} q x^2 \bar{\sigma} = 0 . \tag{2.217}$$

Multiplying Eq. (2.217) by $9x^3 \sqrt{5}/(4\sqrt{\pi})$, introducing flattening $\alpha(x)$ instead of $f_2(x)$ from (2.199) yields

$$\alpha(x)\sigma_x x^5 - \frac{3}{5} \int_0^x \sigma(r) \frac{\mathrm{d}}{\mathrm{d}r} [r^5 \alpha(r)] \, \mathrm{d}r - \frac{3x^5}{5} \int_x^1 \sigma(r) \frac{\mathrm{d}\alpha(r)}{\mathrm{d}r} \, \mathrm{d}r - \tfrac{1}{2} q x^5 \bar{\sigma} = 0 . \tag{2.218}$$

Equation (2.218) is known as Clairaut's differential equation. It defines the flattening of the equipotential surfaces of a hydrostatic body with depth. The change in the flattening depends on the density distribution, $\sigma(x)$, within the body.

Other modifications of Clairaut's differential equation follow. By differentiating (2.218) with respect to x we get

$$\frac{d\alpha}{dx} x\sigma_x - 3 \int_x^1 \sigma(r) \frac{d\alpha(r)}{dr} dr + 2\alpha\sigma_x - \tfrac{5}{2}q\bar{\sigma} = 0.$$
(2.219)

Further differentiation yields

$$\frac{d^2\alpha}{dx^2} x\sigma_x + 6\sigma(x) \frac{d\alpha}{dx} - 6\frac{\alpha}{x} [\sigma_x - \sigma(x)] = 0.$$
(2.220)

By reverting to the unreduced radius-vector, $\varrho = xa_1$, (2.220) can be expressed as follows:

$$\frac{d^2\alpha}{d\varrho^2} \sigma\varrho + 6\sigma \frac{d\alpha}{d\varrho} - 6\frac{\alpha}{\varrho} (\sigma_\varrho - \sigma) = 0.$$
(2.221)

Radau (1885) introduced an important transformation:

$$\eta = \frac{\varrho}{\alpha} \frac{d\alpha}{d\varrho}.$$
(2.222)

Parameter η is referred to as Radau's parameter, and it is denoted η_a for $\varrho = a$. This transformation can be used to modify Clairaut's differential equation to an approximate form (Jeffreys 1959):

$$\frac{d}{d\varrho} [\sigma\varrho^5 \sqrt{(1 + \eta)}] = 5\sigma\varrho^4.$$
(2.223)

Based on the solution of (2.223) the moment of inertia C of the hydrostatic Earth can be expressed with an accuracy of 10^{-3} in terms of η_a as follows:

$$\frac{C}{Ma_1^2} = \tfrac{2}{3}[1 - \tfrac{2}{5}\sqrt{(1 + \eta_a)}],$$

$$\frac{C - \tfrac{1}{2}(A + B)}{Ma_1^2} = \tfrac{2}{3}(\alpha_a - \tfrac{1}{2}q).$$
(2.224)

James and Kopal (1963) derived Clairaut's differential equation more accurately, up to the third power of the Earth's flattening, and they also derived and discussed its solution. Alessandriny (1989) produced its solution for a hydrostatic Earth.

2.9 Internal Sources of the Gravitational Field

The internal sources of the gravitational field are given by the density distribution within the Earth and by the shape of the Earth's surface. In this section we shall deal with the relations between the density distribution and the external gravitational field; as the first step we shall derive the relations between the density distribution and geopotential coefficients.

Gravitational potential $V(P)$ at external point P can be expressed as

$$V(P) = G \int K(P, P')\sigma(P')\,d\tau' , \tag{2.225}$$

where G is the gravitational constant, $\sigma(P')$ is the density at internal point P', and $d\tau'$ is the volume element. Integral kernel $K(P, P')$ is equal to the reciprocal distance $d(P, P')$ between point $P'(\varrho', \vartheta', \Lambda')$ and potential point $P(\varrho, \vartheta, \Lambda)$,

$$K(P, P') = d^{-1}(P, P') = \frac{1}{\varrho} \sum_{j=0}^{\infty} \left(\frac{\varrho'}{\varrho}\right)^{j} P_{j}(\cos \gamma) , \tag{2.226}$$

where γ is the angle between directions (ϑ, Λ) and (ϑ', Λ'). The integral kernel is separated using the addition theorem for Legendre polynomials (2.202). The expansion of gravitational potential $V(P)$ into spherical harmonics is assumed in the form of (2.3). By substituting (2.226) into (2.225) and comparing the coefficients of spherical harmonics $Y_{jm}(\vartheta, \Lambda)$ we arrive at

$$A_{jm} = \frac{1}{Ma^{j}} \frac{4\pi}{2j + 1} \int \varrho'^{j}\sigma(P')Y_{jm}^{*}(\Omega')\,d\tau' ,$$

$$j = 0, \ldots ; \quad m = -j + 1, \ldots , j . \tag{2.227}$$

Equation (2.227) expresses the relation between the density distribution within the Earth, σ, and complex geopotential coefficients A_{jm}.

2.9.1 Physical Interpretation of the Geopotential Coefficients. Tensor of Inertia

The physical interpretation of the geopotential coefficients follows from Eq. (2.227). Coefficients A_{jm} can be interpreted either as spectral coefficients of the density distribution within the Earth, or as moments of order j relative to the origin of coordinates. With this in mind, we shall discuss the special cases of $j = 1$ and $j = 2$.

Let us denote the coordinates of the centre of mass (x_0, y_0, z_0):

$$x_0 = \frac{1}{M} \int x\sigma\,d\tau = \frac{1}{M} \int \varrho \sin \vartheta \cos \Lambda \sigma\,d\tau ;$$

$$y_0 = \frac{1}{M} \int y\sigma\,d\tau = \frac{1}{M} \int \varrho \sin \vartheta \sin \Lambda \sigma\,d\tau ;$$

$$z_0 = \frac{1}{M} \int z\sigma\,d\tau = \frac{1}{M} \int \varrho \cos \vartheta \sigma\,d\tau . \tag{2.228}$$

From Eqs. (2.14) the coordinates of the centre of mass can be expressed in terms of spherical harmonics $Y_{jm}(\vartheta, \Lambda)$ and then, using (2.227), in terms of coefficients A_{jm}:

$$x_0 = \sqrt{\left(\frac{2\pi}{3}\right)} \frac{1}{M} \int \varrho [Y_{1-1}(\Omega) - Y_{11}(\Omega)] \sigma \, d\tau$$

$$= -\frac{a}{2} \sqrt{\left(\frac{3}{2\pi}\right)} (A_{11} - A_{1-1}),$$

$$y_0 = -\frac{i}{M} \sqrt{\left(\frac{2\pi}{3}\right)} \int \varrho [Y_{11}(\Omega) + Y_{1-1}(\Omega)] \sigma \, d\tau = \frac{ia}{2} \sqrt{(\tfrac{3}{2})}(A_{11} + A_{1-1}),$$

$$z_0 = \frac{2\sqrt{\pi}}{M\sqrt{3}} \int \varrho Y_{10}(\Omega) \sigma \, d\tau = \frac{a}{2} \sqrt{\left(\frac{3}{\pi}\right)} A_{10} . \tag{2.229}$$

The coordinates of the centre of mass are expressed in terms of constants $A_{11}, A_{1-1}, A_{10}(J_1^{(1)}, S_1^{(1)}, J_1^{(0)})$. If the centre of mass is located at the origin of coordinates, then $A_{1m} = 0$, where $m = -1, 0, 1$.

If, in the first approximation, the Earth is assumed to be a rigid body rotating with angular velocity $\omega\,(\omega_1, \omega_2, \omega_3)$, the kinetic energy of its rotation is described by the formula

$$T_{rot} = \tfrac{1}{2} I_{ik} \omega_i \omega_k , \tag{2.230}$$

where I_{ik} are the components of the tensor of inertia

$$I_{ik} = \int_M \left[\sum_{j=1}^{3} (\delta_{ik} x_j^2 - x_i x_k) \right] dm , \quad I_{ik} = I_{ki} \tag{2.231}$$

which, resolved into components, read

$$I_{11} = A = \int_M (x_2^2 + x_3^2) dm ; \quad I_{22} = B = \int_M (x_1^2 + x_3^2) dm ;$$

$$I_{33} = C = \int_M (x_1^2 + x_2^2) dm ; \quad I_{23} = -D = -\int_M x_2 x_3 \, dm ;$$

$$I_{13} = -E = -\int_M x_1 x_3 \, dm ; \quad I_{12} = -F = -\int_M x_1 x_2 \, dm . \tag{2.232}$$

We shall now switch from Cartesian coordinates to spherical coordinates $(\varrho, \vartheta, \Lambda)$:

$$x_1 = \varrho \sin \vartheta \cos \Lambda , \quad x_2 = \varrho \sin \vartheta \sin \Lambda , \quad x_3 = \varrho \cos \vartheta . \tag{2.233}$$

Taking into account formulae (2.14), which define the spherical harmonics up to the second degree, we are easily able to derive the following relations for the components of the tensor of inertia:

$$I_{33} = C = \frac{4\sqrt{\pi}}{3} \int \varrho^2 \left[Y_{00}(\Omega) - \frac{1}{\sqrt{5}} Y_{20}(\Omega) \right] \sigma \, d\tau \,,$$

$$C - \frac{A+B}{2} = -2 \sqrt{\left(\frac{\pi}{5}\right)} \int \varrho^2 Y_{20}(\Omega) \sigma \, d\tau \,,$$

$$B - A = 2 \sqrt{\left(\frac{2\pi}{3 \cdot 5}\right)} \int \varrho^2 [Y_{22}(\Omega) + Y_{2-2}(\Omega)] \sigma \, d\tau \,,$$

$$I_{23} = D = -i \sqrt{\left(\frac{2\pi}{3 \cdot 5}\right)} \int \varrho^2 [Y_{21}(\Omega) + Y_{2-1}(\Omega)] \sigma \, d\tau \,,$$

$$I_{13} = E = \sqrt{\left(\frac{2\pi}{3 \cdot 5}\right)} \int \varrho^2 [Y_{21}(\Omega) - Y_{2-1}(\Omega)] \sigma \, d\tau \,,$$

$$I_{12} = F = i \sqrt{\left(\frac{2\pi}{3 \cdot 5}\right)} \int \varrho^2 [Y_{22}(\Omega) - Y_{2-2}(\Omega)] \sigma \, d\tau \,. \tag{2.234}$$

The integrals occurring in Eqs. (2.234) can be expressed in terms of geopotential coefficients A_{jm} (2.227):

$$\left(C - \frac{A+B}{2}\right) \Big/ (Ma^2) = J_2 = -\frac{1}{2} \sqrt{\left(\frac{5}{\pi}\right)} A_{20} \,,$$

$$A_{22} = \sqrt{\left(\frac{3\pi}{2 \cdot 5}\right)} [(B-A) + 2iF]/(Ma^2) \,,$$

$$A_{2-2} = \sqrt{\left(\frac{3\pi}{2 \cdot 5}\right)} [(B-A) - 2iF]/(Ma^2) \,,$$

$$A_{21} = \sqrt{\left(\frac{2 \cdot 3 \cdot \pi}{5}\right)} [E - iD]/(Ma^2) \,,$$

$$A_{2-1} = -\sqrt{\left(\frac{2 \cdot 3 \cdot \pi}{5}\right)} [E + iD]/(Ma^2) \,. \tag{2.235}$$

Using transformation relations (2.8), formulae (2.235) can be expressed in terms of real geopotential coefficients:

$$J_2^{(2)} = \frac{B-A}{4Ma^2} \,; \quad S_2^{(2)} = \frac{F}{2Ma^2} \,;$$

$$J_2^{(1)} = \frac{E}{Ma^2} \,; \quad S_2^{(1)} = \frac{D}{Ma^2} \,. \tag{2.236}$$

The geopotential coefficients of degree $j = 1$ thus define the position of the Earth's centre of mass, while the coefficients of degree $j = 2$ define the differences between the principal moments of inertia and the positions of the axes of the central ellipsoid of inertia. We shall not give the developed formulae of the geopotential coefficients of degree $j > 2$; the odd ones are due to the mass asymmetry of the body relative to the equatorial plane; all coefficients of order $m \neq 0$ (i.e. tesseral and sectorial) express mass asymmetry relative to the smallest axis of the ellipsoid of inertia. This axis is close to the instantaneous axis of rotation and would be identical with the former if it were not for free nutation.

2.9.2 Transformation of the Coordinate System into the Principal Axes of the Earth's Inertia Tensor

The ellipsoid of inertia can be coordinated with the tensor of inertia:

$$I_{ij}x_i x_j = 1 \; . \tag{2.237}$$

The problem of transforming the tensor of inertia to the diagonal form can be reduced to the problem of finding a coordinate system in which the quadric (2.237) is expressed in canonic form, i.e. products of inertia $I_{ij} = 0$, $i \neq j$. A system of this type is known to be orthogonal and represents a natural coordinate system. Its position can be determined using the method known from geometry with the aid of eigenvalues and matrices of eigenvectors of the moment of inertia. However, this procedure is not convenient in this particular case because the elements of the tensor of inertia cannot be measured directly, and only the differences between the moments of inertia can be determined indirectly. In this case we have only five relations (2.235), expressing combinations of the elements of the tensor of inertia in terms of geopotential coefficients, for six independent elements of the tensor of inertia. We shall therefore adopt a different method of determining the positions of the principal axes of the ellipsoid of inertia relative to the existing coordinate system in which the geopotential coefficients have been determined.

For products of inertia E, D, F to be zero, it is sufficient, in view of (2.235), for

$$A_{21} = A_{2-1} = 0 \; ; \quad \mathrm{Im}\, A_{22} = 0 \; , \tag{2.238}$$

or

$$J_2^{(1)} = S_2^{(1)} = S_2^{(2)} = 0 \; .$$

We shall now rotate the coordinate system in which geopotential coefficients A'_{jm} have been determined into a new system with constants A_{jm} by means of Euler angles $\bar{\alpha}$, $\bar{\beta}$, $\bar{\gamma}$. The transformation from the geopotential coefficients in the primed system to the unprimed system is expressed by Eqs. (2.25) for $j = 2$ which, with a view to (2.29), we shall record as follows:

$$A_{2m} = \sum_{M = -2}^{2} A'_{2M} d^2_{mM}(\bar{\beta}) \exp[-\mathrm{i}(m\bar{\alpha} + M\bar{\gamma})] \; , \quad m = -2, -1, 0, 1, 2 \; . \tag{2.239}$$

Matrix $d^2_{mM}(\bar{\beta})$ is given in Table 2.2.

From (2.238) and (2.239) we can derive three trigonometric equations for the unknown Euler angles $(\bar{\alpha}, \bar{\beta}, \bar{\gamma})$:

$$0 = \sum_{M=-2}^{2} A'_{2M} d^2_{\pm 1 M}(\bar{\beta}) \exp[-i(\pm \bar{\alpha} + M\bar{\gamma})], \tag{2.240}$$

$$0 = \text{Im} \sum_{M=-2}^{2} d^2_{2M}(\bar{\beta}) A'_{2M} \exp[-i(2\bar{\alpha} + M\bar{\gamma})]. \tag{2.241}$$

Denoting $\bar{\vartheta}, \bar{\Lambda}$ the angles defining the direction of the new axis x_3, and $\bar{\omega}$ the angle through which it is necessary to rotate the old axis x'_1 about the new axis x_3 for the former to reach the position of the new axis x_1, we find that

$$\tan \bar{\vartheta} = \tan \frac{\bar{\beta}}{2} \bigg/ \sin \frac{\bar{\alpha} + \bar{\gamma}}{2}; \quad \bar{\Lambda} = \frac{\pi}{2} + \frac{\bar{\alpha} - \bar{\gamma}}{2};$$

$$\cos \frac{\bar{\omega}}{2} = \cos \frac{\bar{\beta}}{2} \cos \frac{\bar{\alpha} + \bar{\gamma}}{2}. \tag{2.242}$$

In other words, by solving (2.240) and (2.241) we obtain the Euler angles $\bar{\alpha}, \bar{\beta}, \bar{\gamma}$ that transform the coordinate system into the new system in which the tensor of inertia has the diagonal (canonic) form. Angles $\bar{\vartheta}, \bar{\Lambda}$ define the direction of the new axis x_3; $\varrho, \bar{\vartheta}, \bar{\Lambda}$ $(0 \leqslant \varrho \leqslant \infty)$ are points located on the new axis x_3. By rotating the old axis x'_1 (or x'_2) about the new axis x_3 through angle $\bar{\omega}$, the old axes will achieve their new positions x_1 (or x_2, respectively).

Let us discuss a simpler and particular case in which $\bar{\beta} = 0$, i.e. to transform the tensor of inertia into diagonal form it is sufficient to rotate the coordinate system about axis $x_3 = x'_3$. Note that, in view of Table 2.2, in this particular case

$$d^2_{mM}(0) = \delta_{mM} \tag{2.243}$$

and without loss of generality we can put $\bar{\gamma} = 0$. Equations (2.240) then read

$$0 = A'_{2 \pm 1} \exp[-i(\pm \bar{\alpha})]. \tag{2.244}$$

This implies that the axis of the existing system x'_3 is identical with the axis of the system x_3, in which the quadric of inertia is expressed in canonic form only if

$$A'_{2 \pm 1} = 0, \quad \text{and} \quad J'^{(1)}_2 = S'^{(1)}_2 = 0, \tag{2.245}$$

respectively.

Assume that the given system of geopotential coefficients satisfies condition (2.245); Eqs. (2.240) are then satisfied identically. Equation (2.241) serves to determine the angle of rotation $\bar{\alpha}$. $\text{Im}[A'_{22} \exp(-2i\bar{\alpha})] = 0$, i.e.

$$\bar{\alpha}_{22} = \tfrac{1}{2} \arctan \left(\frac{S'^{(2)}_2}{J'^{(2)}_2} \right). \tag{2.246}$$

With regard to the position of axis x'_1 (or x'_2) relative to axis x_1 (or x_2, respectively) it has been rotated through an angle of $14.85°$ to the west; this corresponds very well with the value of parameter Λ_a in Table 2.6. We can thus draw the conclusion that the largest semi-axis of the Earth's triaxial ellipsoid, derived under condition (2.136), is very close to the appropriate axis of the Earth's ellipsoid of inertia. As regards axis x_3, the satellite data, i.e. geopotential coefficients $J_2^{(1)}$ and $S_2^{(1)}$, are able to define it.

Moreover, no methods exist for direct observation of the principal moments A, B, C relative to the axis of the ellipsoid of inertia. Dynamic satellite methods are able to determine geopotential coefficients $J_2^{(0)}$, $J_2^{(2)}$, $S_2^{(2)}$, or rather only the differences,

$$B - A = 4J_{2,2}Ma^2, \tfrac{1}{2}(A + B) - C = J_2^{(0)}Ma^2 \,,$$

$$C - B = -(J_2^{(0)} + 2J_{2,2})Ma^2, C - A = -(J_2^{(0)} - 2J_{2,2})Ma^2 \,, \qquad (2.247)$$

$$J_{2,2} = [(J_2^{(2)})^2 + (S_2^{(2)})^2]^{1/2} \,,$$

assuming, of course, that the total mass of the Earth is known with sufficient accuracy. However, compared with (2.247) the mass is known only very approximately, with an accuracy corresponding to the contemporaneous accuracy of Newton's gravitational constant. Indeed, including the mass of the atmosphere, which amounts to about $10^{-6}\,M$,

$$M = \frac{GM}{G} = \frac{(398\,600.441 \pm 0.001)\,10^9\,\mathrm{m}^3\,\mathrm{s}^{-2}}{(6672.59 \pm 0.30)\,10^{-14}\,\mathrm{m}^3\,\mathrm{s}^{-2}\,\mathrm{kg}^{-1}} = (5.9737 \pm 0.0003)\,10^{24}\,\mathrm{kg} \,,$$

$$(2.248)$$

since, with regard to relative errors, practically

$$\frac{\delta M}{M} = \frac{\delta G}{G} = 5 \times 10^{-5} \,. \qquad (2.249)$$

Using this value of the Earth's mass and geopotential coefficients

$$J_2^{(0)} = (1082.6269 \pm 0.0006) \times 10^{-6}; \quad J_{2,2} = (1.815 \pm 0.001) \times 10^{-6}$$

[which corresponds to the fully normalized value $\bar{J}_{2,2} = (2.812 \pm 0.001) \times 10^{-6}$], determined with the optional factor of length $a_0 = 6378\,136.5$ m, we obtain, after rotating reference axes x_1, x_2 to coincide with the equatorial axes of the ellipsoid of inertia,

$$C - A = (2.6397 \pm 0.0002) \times 10^{35}\,\mathrm{kg\,m}^2 \,,$$

$$C - B = (2.6221 \pm 0.0002) \times 10^{35}\,\mathrm{kg\,m}^2 \,,$$

$$B - A = (1.764 \pm 0.001) \times 10^{33}\,\mathrm{kg\,m}^2 \,, \qquad (2.250)$$

taking

$$Ma^2 = (2.4301 \pm 0.0001) \times 10^{38}\,\mathrm{kg\,m}^2 \,. \qquad (2.251)$$

Ratio $\dfrac{C - \tfrac{1}{2}(A + B)}{C} = H$ (the coefficient in the precession constant) can be

determined from observing the lunisolar precession of the vector of the angular momentum, or of the vector of the Earth's rotation (see Sect. 3.9). Its most recent value (Kinoshita and Souchay 1990) is

$$H = 0.003\,273\,956\,7 \pm 2 \times 10^{-9} = 1/(305.4408 \pm 0.0002)\,. \qquad (2.252)$$

Consequently,

$$A = (8.0095 \pm 0.0004) \times 10^{37}\ \mathrm{kg\,m^2}\,,$$
$$B = (8.0097 \pm 0.0004) \times 10^{37}\ \mathrm{kg\,m^2}\,,$$
$$C = (8.0359 \pm 0.0004) \times 10^{37}\ \mathrm{kg\,m^2}\,, \qquad (2.253)$$

or

$$A/Ma^2 = 0.329\,592\,1 \pm 0.000\,000\,3\,,$$
$$B/Ma^2 = 0.329\,599\,4 \pm 0.000\,000\,3\,,$$
$$C/Ma^2 = 0.330\,678\,4 \pm 0.000\,000\,3\,. \qquad (2.254)$$

The accuracy in determining the principal moments of inertia (2.253) is, therefore, still relatively low. The accuracy of the differences of the moments (2.250) is higher; however, even it is affected by the error in the Earth's mass. The accuracy of the relative values

$$\frac{B - A}{Ma^2} = 4J_{2.2} = (7.262 \pm 0.004) \times 10^{-6}\,,$$

$$\frac{C - A}{Ma^2} = -\,(J_2^{(0)} - 2J_{2.2}) = (1086.258 \pm 0.002) \times 10^{-6}\,,$$

$$\frac{C - B}{Ma^2} = -\,(J_2^{(0)} + 2J_{2.2}) = (1078.996 \pm 0.002) \times 10^{-6}\,. \qquad (2.255)$$

is of the order of errors of the geopotential coefficients.

2.10 Density Models of the Earth

The significance of density lies in the fact that it is an interdisciplinary parameter. Density represents the distribution of sources of the Earth's gravitational field, and also occurs in the equations of motion describing the Earth as an elastic body:

$$\tau_{ij,j} + \sigma F_i = \sigma \frac{\partial^2 u_i}{\partial t^2}\,, \quad i,j = 1, 2, 3, \qquad (2.256)$$

where τ_{ij} is the tensor of elastic stresses and F_i and u_i are components of the body force and displacement, respectively. The natural oscillations of the Earth, which represent the motion in which gravitational body forces are reflected in a non-negligible way, obey Eqs. (2.256). The laws of propagation of seismic waves within the Earth are derived from the equation of equilibrium (2.256), the effect of the body forces being neglected. The Earth's tidal motions are also affected by the density distribution. Furthermore, density plays an important part in convective motions within the Earth's core and mantle. It is one of the thermodynamic parameters of state which describe the thermodynamic states of substances in terms of the equation of state $p = p(\sigma, T)$, in which pressure p is a function of density σ and temperature T. Since density links mechanical and thermodynamic processes taking place within the Earth and, moreover, the Earth's basic data, such as total mass and moments of inertia, depend on it, considerable attention was already being devoted to density models in the eighteenth century, beginning with Legendre. At the time the density laws were being related to the hydrostatic theory of the Earth. This theory was expected to impose substantial constraints on constructing the density law via derived parameters such as Radau's parameter η and flattening α. However, contrary to expectations, it was found that a broad class of density models yielded quantities predicted by the theory of the Earth's hydrostatic equilibrium. In 1958 observations of variations in the orbital elements of artificial satellites proved beyond doubt that the Earth was not in hydrostatic equilibrium.

2.10.1 Mean Spherically Symmetrical Models of the Earth

Mean density models are related to a spherically symmetrical Earth. The first model, proposed by Clairaut, was $\sigma(x) = 1/x^m$, but this proved to be unrealistic because it yielded infinite density at the Earth's centre. Legendre adopted the following model in which the density was a function of the reduced radius-vector x:

$$\sigma(x) = \sigma_0 \sin Ax/(Ax) \,. \tag{2.257}$$

This enabled Clairaut's differential equation to be integrated. Laplace proved that Legendre's density law yielded the following relation for compressibility:

$$dp = c\sigma \, d\sigma \,, \tag{2.258}$$

i.e. the density variation, $d\sigma$, due to increment of pressure dp is proportional to density σ. Laplace produced a density model which depended on two parameters and other models were formulated by Roche (1854), Thomson and Tait (1879), Dziewonski et al. (1975), Press (1968, 1970), Wang (1972), Lambeck (1976), Dziewonski (1984), etc.

The current mean density models are not based exclusively on gravimetric data but combine the results of observing the Earth's free oscillations with dispersion of long-period surface waves. They also take into account the astro-geodetic data relating to the Earth's mass and moment of inertia. We shall discuss here some of the more recent models.

Bullen (1975) proposed Earth models designated A and B. In constructing model A, the Earth was divided into zones A–G, as shown in Table 2.12. In zones B, D and E Bullen assumed that the Williamson–Adams equation would apply:

$$\frac{\mathrm{d}\sigma}{\mathrm{d}\varrho} = -\frac{Gm\sigma}{\varrho^2\phi}. \tag{2.259}$$

Here σ is density, m is the mass of a sphere with radius ρ and $\phi = v_p^2 - \frac{4}{3}v_s^2 = k/\sigma$ is the so-called seismic ratio which can be determined if the distribution of velocities of longitudinal, v_p, and shear, v_s, seismic waves is known [k is the elastic modulus of incompressibility, $1/k = (1/\sigma)(\mathrm{d}\sigma/\mathrm{d}p)$]. Equation (2.259) holds for a homogeneous body with an adiabatic temperature gradient. The seismic ratio ϕ was determined from Jeffreys' models of seismic wave velocities (Jeffreys 1939; Jeffreys and Bullen 1967).

In integrating the Williamson–Adams equation in zone B, the density value at the upper boundary, $\sigma = 3.32\ \mathrm{g\,cm}^{-3}$, was taken as the boundary value. Zone C was proved to be inhomogeneous and, consequently, the Williamson–Adams equation could not be applied there. The density in zone C was approximated by a power function in five variables. The details of the properties of model A can be found in Bullen (1975). The results are summarized in Table 2.13. The value of the moment of inertia was taken to be $C = 0.3335\ Ma_1^2$. Bullen's model indicates an interesting feature: the modulus of incompressibility k and pressure p are nearly continuous as they cross the boundary between the core and mantle. The values of $\mathrm{d}k/\mathrm{d}p$ are 3.0 and 2.8 at the upper and lower core-mantle boundaries, respectively, although density σ and seismic wave velocities v_P and v_S display considerable discontinuities there. This led Bullen to formulate the so called $k - p$ hypothesis which assumes the dependence of the modulus of incompressibility k on pressure p to be a smooth continuous function in the lower mantle and core. The $k - p$ hypothesis was used in constructing model B,

Table 2.12. Division of the Earth into Bullen's zones. (Bullen 1975)

Zone	Region	Depth interval (km)	Seismic wave velocity gradient
A	Earth's crust	0–33	Irregular
B	Upper mantle	33–410	Normal
C		410–1000	Larger than normal
D	Lower mantle	1000–2900	Normal (constant in zone 2700–2900 km)
E	Outer core	2900–4980	Normal, P waves only
F	Transition zone	4980–5120	Negative, P waves only
G	Inner core	5120–6370	Very small, non-negative

Table 2.13. Bullen's model A. (Bullen 1975) σ, density; p, pressure; μ, shear modulus; v_P (v_S), velocity of P (S) waves; k, modulus of incompressibility

Zone	Depth (km)	σ (g/cm^3)	p (10^{11} N/m^2)	k (10^{11}N/m^2)	μ (10^{-1}N/m^2)	g (m/s^2)	v_P (km/s)	v_S (km/s)
....	0		0			9.822		
A		(2.84)		(0.65)	(0.35)		(6.30)	(3.55)
....	33	3.32	0.004	1.15	0.63	9.846	7.75	4.35
B	413	3.64	0.141	1.73	0.90	9.960	8.97	4.96
C	984	4.55	0.379	3.49	1.83	9.966	11.42	6.35
...	2000	5.11	0.870	5.10	2.46	10.010	12.79	6.92
D	2898	5.56	1.360	6.39	2.97	10.730	13.64	7.30
....	2898	9.98	1.360	6.55	0	10.730	8.10	0
E	4000	11.42	2.470	10.33	0	7.870	9.51	0
....	4980	12.17	3.200	13.26	0	4.780	10.44	0
F	5120	12.25	3.280			4.310	9.40	
....	5120	12.25					11.16	
G	6371	12.51	3.610			0	11.31	

which does not differ much from model A. The exception is zone G of the inner core which, in contrast to the situation in model A, is considered to be solid in model B.

The next model we shall mention is the model of density distribution and seismic wave velocities, developed by Press (1968) and derived by the Monte Carlo method. The travel-time curves of longitudinal and transverse seismic waves, mass and the Earth's moment of inertia ($M = 5.976 \times 10^{24}$ kg, $C = 0.3308\, Ma_1^2$) were used as input data. Model parameters v_P and v_S were selected randomly and tested by comparison with the travel-time curves of S and P waves, with the data on the total mass and moment of inertia of the Earth and, finally, with the observed values of periods of the Earth's free oscillations.

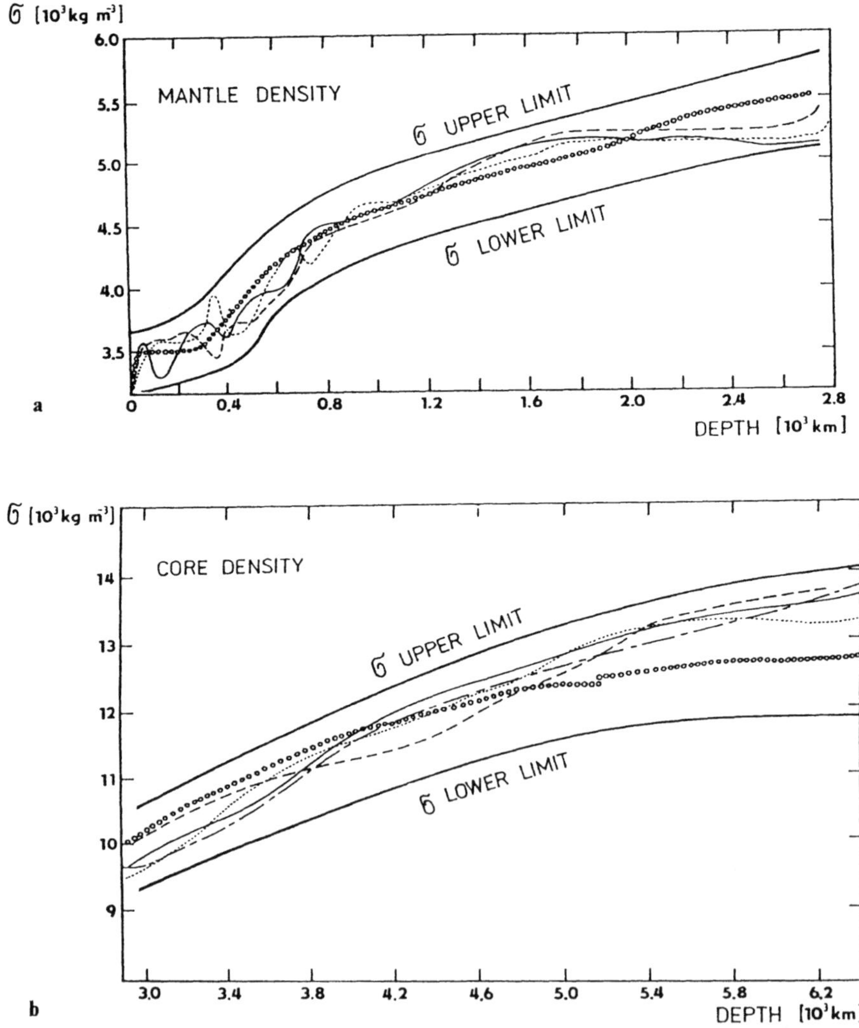

Fig. 2.30. Density model of **a** Earth's mantle and **b** Earth's core. (After Press 1968)

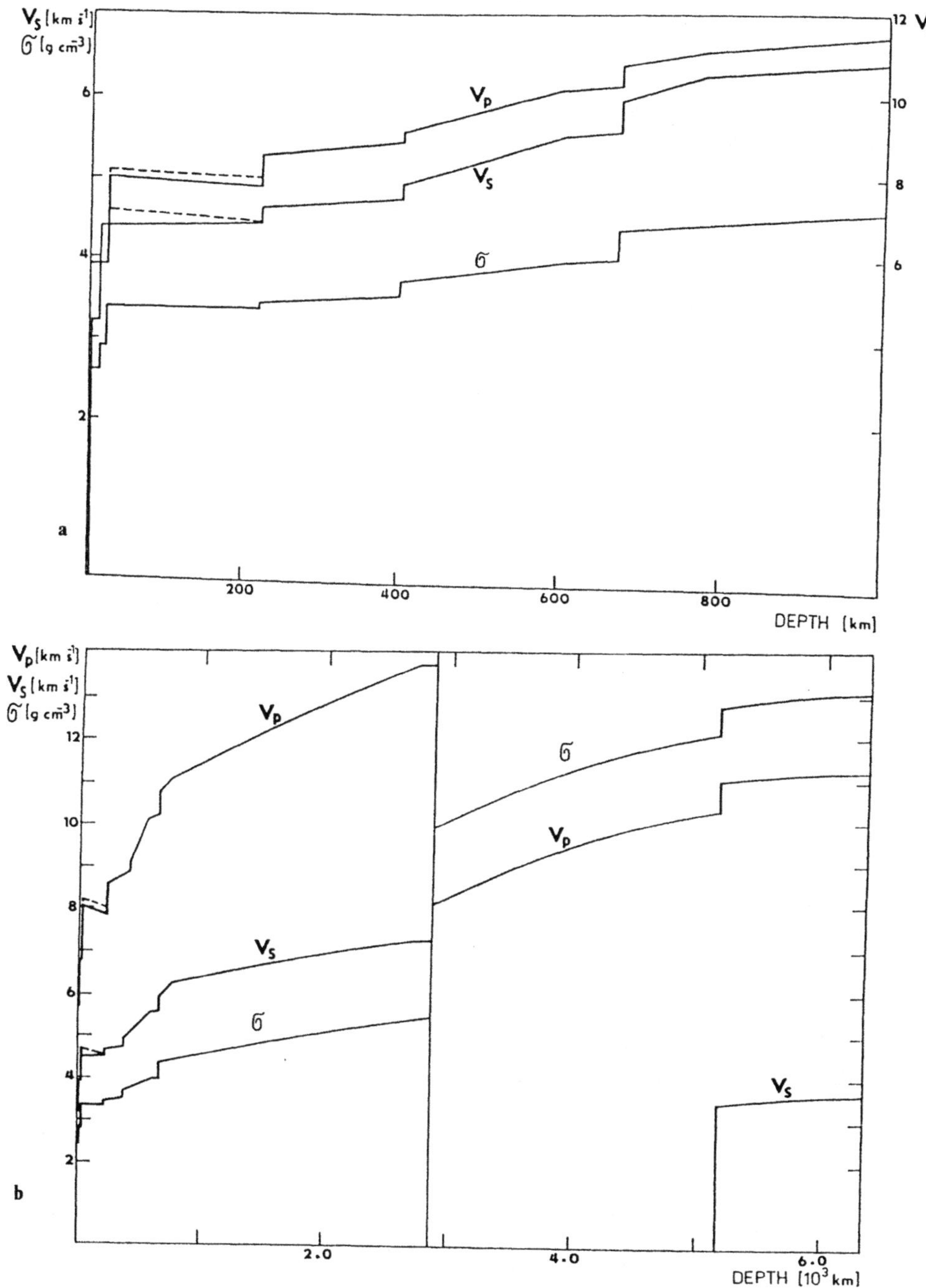

Fig. 2.31. a PREM Earth model, upper mantle (Dziewonski and Anderson 1981). Velocity of longitudinal seismic waves, v_P, velocity of shear seismic waves v_S, density σ. The *dashed curve* represents the horizontal velocity component and the *solid curve* the vertical velocity component. **b** PREM Earth model for the Earth as a whole

A total of 5×10^6 models were generated, of which six satisfied the conditions imposed: three satisfactory models for velocity v_S and density σ are shown in Fig. 2.30a and b.

The Preliminary Reference Earth Model (PREM) (Dziewonski and Anderson 1981) is founded on the most comprehensive set of Earth data. This is a mean model for a radially symmetrical Earth whose figure is assumed to be spherical. In the same way as Bullen's and Press' models it is based on astro-geodetic data: $M = 5.974 \times 10^{24}$ kg, moment of inertia $C = 0.3308\,M\bar{a}^2$ (the same as in Bullen's models). Nine hundred observations of frequencies of free oscillations and surface waves represent the input data of this model. The most extensive set of input data is founded on observations of arrival times of seismic body waves. Roughly 2 million observations of p-wave arrivals and 250 000 observations of S-wave arrivals from 26 000 foci, whose depths did not exceed 100 km, were used. It was found that an isotropic model would not satisfy the long-period branches of surface-wave dispersion curves, on the one hand, and of global data, on the other hand. An anisotropic layer of the transverse isotropy type was therefore introduced; the vertical and horizontal velocities in this differ by 2–4%. This eliminated the discrepancy between the dispersion of Love and Rayleigh waves. The introduction of the anisotropic wave also overcame the difficulties of satisfying long- and short-period (period less than 200 s) toroidal and spheroidal models with a single model.

Figure 2.31a and b illustrates seismic wave velocities v_P and v_S and density σ as functions of depth. Figure 2.31a indicates that the asthenospheric layer (a layer of lower seismic wave velocities) in the PREM is not as well defined as in the Press model. This is apparently so because the Press model was not required to fit the global long-period data. The PREM distinctly reflects the 400-km discontinuity (also referred to as the 20°-discontinuity) with the density and seismic wave velocity jumps, as well as the discontinuity at a depth of 670 km. These discontinuities are caused by the phase transitions between the solid phases of the mantle material (Birch 1964, 1969). Ringwood (1969) interprets the 20°-discontinuity as being due to the pyroxene-garnate phase transitions and by the transition of spinel-structure olivine to β-phase olivine. The discontinuity at 670 km is caused by the transition of the β-phase. Anderson (1991) gives an overview of current opinions of the material composition of the boundary surfaces of the mantle. Recently, new models have been derived for seismic wave velocities within the Earth using tomographic methods (Woodhouse and Dziewonski 1984).

2.11 Lateral Density Variations

We shall now deal with the problem of determining the constraints imposed on the fine three-dimensional structure of the density distribution within the Earth. We shall assume that the external gravitational field, expressed in terms of

geopotential coefficients of a sufficiently high degree, is known in detail. We shall also assume that the shape of the Earth's surface is known in detail in terms of the coefficients of the spherical expansion of the radius-vector of the topographic surface. We shall draw on our knowledge of the mean radially symmetrical density model and impose the following conditions on the mean model:
1. Astro-geodetic: total Earth mass, mean moment of inertia, precession constant and zonal geopotential coefficient J_2.
2. Earth's natural oscillations and long-period surface waves.
3. Travel-time curves of seismic P and S waves.

2.11.1 Integral Density Equations

We shall start with Eq. (2.227) which expresses the complex geopotential coefficient A_{jm} in terms of density in integral form. The system of equations (2.227) for $j = 0, 1, 2, \ldots$; $m = -j, -j + 1, \ldots, j$, is equivalent to Eq. (2.225), which we shall consider to be the integral equation in density σ, and potential function $V(P)$ we shall assume to be known. In Eq. (2.227) we shall substitute for volume element $d\tau = \varrho'^2 \, d\varrho' \, d\Omega'$, and also put $\varrho' = x\varrho_s(\Omega'), 0 \leqslant x \leqslant 1$, where x is the geocentric radius-vector normalized to unity for points on the surface. We shall obtain $d\varrho' = \varrho_s(\Omega') \, dx$, $\varrho_s(\Omega')$ as the distance from the origin to the surface point along Ω'. After performing some algebra, Eq. (2.227) will read

$$A_{jm} = \frac{1}{Ma^j} \frac{4}{2j + 1} \int \varrho_s^{j+3}(\Omega') Y_{jm}(\Omega') \int\limits_0^1 \sigma(x, \varrho') x^{j+2} dx \, d\Omega',$$

$$j = 0, 1, 2, \ldots; m = -j, -j + 1, \ldots, j. \tag{2.260}$$

The system of equations (2.260) is equivalent to integral equation (2.225).

2.11.2 Analytical Density Model for a Spherically Asymmetrical Earth

We shall express density $\sigma(x, \Omega)$ as a function of normalized radius-vector x and angular coordinates (Pěč and Martinec 1984):

$$\sigma(x, \Omega) = \sigma(x) + \delta\sigma(x, \Omega). \tag{2.261}$$

Density model (2.261) emphasizes that the predominant part depends on density $\sigma(x)$, which is derived from the mean density model. The lateral variation of density, $\delta\sigma(x, \Omega)$ is small compared with $\sigma(x)$.

The lateral density variation is expressed as a series of orthogonal spherical harmonics, functions of x:

$$\delta\sigma(x, \Omega) = \sum_{j_2=1}^{\infty} \sum_{m_2=-j}^{j_2} \beta_{j_2 m_2}(x) Y_{j_2 m_2}(\Omega) \,. \tag{2.262}$$

The convergence of series (2.262) is guaranteed by condition

$$\int_{\Omega_0} |\delta\sigma(x, \Omega)|^2 \, d\Omega < \infty \,, \tag{2.263}$$

which is certainly satisfied for the Earth. Coefficients $\beta_{j_2 m_2}(x)$ can be developed into a series of shifted Legendre polynomials $\tilde{P}_n(x)$, orthogonal in interval $0 \leqslant x \leqslant 1$:

$$\beta_{j_2 m_2}(x) = \sum_{n=0}^{\infty} F_{j_2 m_2}^{(n)} \tilde{P}_n(x) \,, \quad 0 \leqslant x \leqslant 1 \,. \tag{2.264}$$

Series (2.264) converges if condition

$$\int_0^1 |\beta_{j_2 m_2}(x)|^2 \, dx < \infty \tag{2.265}$$

is satisfied. Finally, the analytical density model will read

$$\sigma(x, \Omega) = \sigma(x) + \sum_{j_2=1}^{\infty} \int_{m_2=-j_2}^{j_2} \sum_{n=0}^{\infty} F_{j_2 m_2}^{(n)} \tilde{P}_n(x) Y_{j_2 m_2}(\Omega) \,. \tag{2.266}$$

2.11.3 Powers x^n Developed into a Series of Shifted Legendre Polynomials

It is convenient to develop power x^{n+2}, occurring in integral equation (2.260), into a series of polynomials orthogonal in interval $\langle 0, 1 \rangle$. Shifted Legendre polynomials with the properties required will be adopted as the orthogonal functions. The shifted Legendre polynomials $\tilde{P}_n(x)$ can be expressed in terms of Legendre polynomials $P_n(\xi)$, $-1 \leqslant \xi \leqslant 1$:

$$\tilde{P}_n(x) = P_n(1 - 2x) = P_n(\xi); \quad \xi = 1 - 2x; \quad 0 \leqslant x \leqslant 1 \,;$$
$$-1 \leqslant \xi \leqslant 1 \,. \tag{2.267}$$

The following recurrent relations hold for the shifted Legendre polynomials:

$$(n + 1)\tilde{P}_{n+1}(x) = (2n + 1)(1 - 2x)\tilde{P}_n(x) - n\tilde{P}_{n-1}(x) \,. \tag{2.268}$$

In particular,

$$\tilde{P}_0(x) = 1 \,; \quad \tilde{P}_1(x) = 1 - 2x \,. \tag{2.269}$$

The shifted Legendre polynomials are orthogonal in interval $\langle 0, 1 \rangle$ and their norm follows from the expression

$$\int_0^1 \tilde{P}_n(x)\tilde{P}_m(x)\,dx = \delta_{nm}/(2n+1)\,. \tag{2.270}$$

Power x^n, $0 \leqslant x \leqslant 1$, can be developed into a series:

$$x^n = \sum_{k=0}^{n} c_k^{(n)} \tilde{P}_k(x)\,. \tag{2.271}$$

The coefficients of the series, $c_k^{(n)}$, satisfy these recurrent relations:

$$c_k^{(n+1)} = -c_{k-1}^{(n)} \frac{k}{2(2k-1)} + \frac{1}{2} c_k^{(n)} - \frac{1}{2} \frac{k+1}{2k+3} c_{k+1}^{(n)}\,. \tag{2.272}$$

The initial values in recurrent relations (2.272) are

$$c_0^{(0)} = 1\,;\quad c_0^{(1)} = 1/2\,;\quad c_1^{(1)} = -1/2 \tag{2.273}$$

with the constraints

$$c_{-k}^{(n)} = 0,\quad c_{n+k}^{(n)} = 0,\quad k = 1, 2, 3, \ldots\,. \tag{2.274}$$

2.11.4 Systems of Algebraic Equations for the Density Model Coefficients. Compatibility Conditions for the Mean Spherical Model

2.11.4.1 Algebraic Equations for Coefficients $F_{jm}^{(k)}$ of the Density Variations Model

Integral equations (2.260) express the relation between the geopotential coefficients of the external gravitational field, A_{jm}, and the density distribution. By substituting for $\varrho_s^{j+3}(\Omega')$ from (2.159) and for $\sigma(x, \Omega')$ from (2.266), Eq. (2.260) becomes

$$A_{jm} = \frac{4\pi}{2j+1} \frac{a_1^3}{M} \left(\frac{a_1}{a_0}\right)^j \sum_{j_1 m_1} E_{j_1 m_1}^{(j+3)} \int_{\Omega_0} Y_{j_1 m_1}(\Omega') Y_{jm}(\Omega')$$

$$\times \int_0^1 \left[\sigma(x) x^{j+2} + \sum_{\substack{j_2 m_2 \\ j_2 \geqslant 1}} \sum_{n=0}^{} F_{j_2 m_2}^{(n)} Y_{j_2 m_2} \sum_{k=0}^{j+2} c_k^{(j+2)} \tilde{P}_n(x)\tilde{P}_k(x) \right] dx\,d\Omega'\,,$$

$$j = 0, 1, 2, \ldots\,;\quad m = -j, -j+1, \ldots, j\,. \tag{2.275}$$

We shall introduce normalized spherical moments $S^{(j)}$:

$$S^{(j)} = \int_0^1 \sigma(x) x^j\,dx\,. \tag{2.276}$$

Using Eqs. (2.79) and (2.78) we now modify the integral of the triple product of spherical harmonics to read

$$\int_{\Omega_0} Y_{jm}^*(\Omega')Y_{j_1 m_1}(\Omega')Y_{j_2 m_2}(\Omega')\,d\Omega'$$

$$= \sqrt{\left[\frac{(2j_1+1)(2j_2+1)}{4\pi(2j+1)}\right]} C_{j_1 0 j_2 0}^{j0} C_{j_1 m_1 j_2 m_2}^{jm}; \tag{2.277}$$

$$(2j+1)^{3/2}\frac{M}{a_1^3}\left(\frac{a_0}{a_1}\right)^j \frac{A_{jm}}{2\sqrt{\pi}} - 2\sqrt{[(2j+1)]}\,E_{jm}^{(j+3)}S^{(j+2)}$$

$$= \sum_{\substack{j_2 m_2 \\ j_2 \geqslant 1}} \sum_{k=0}^{j+2} \frac{c_k^{(j+2)}}{2k+1}\,F_{j_2 m_2}^{(k)} \sum_{j_1 m_1} E_{j_1 m_1}^{(j+3)} \sqrt{[(2j_1+1)(2j_2+1)]}$$

$$\times C_{j_1 0 j_2 0}^{j0} C_{j_1 m_1 j_2 m_2}^{jm}, \quad j=0,1,2,\ldots; \quad m=-j,-j+1,\ldots,j. \tag{2.278}$$

Equations (2.278) may be expressed in a slightly different form:

$$\sqrt{(2j+1)}\frac{M}{a_1^3}\left(\frac{a_0}{a_1}\right)^j \frac{A_{jm}}{2\sqrt{\pi}} - \frac{2\sqrt{\pi}}{2j+1}\,E_{jm}^{(j+3)}S^{(j+2)}$$

$$= (-1)^m \sum_{\substack{j_2 m_2 \\ j_2 \geqslant 1}} \sum_{k=0}^{j+2} \frac{c_k^{(j+2)}}{2k+1}\,F_{j_2 m_2}^{(k)} \sum_{j_1 m_1} E_{j_1 m_1}^{(j+3)} \sqrt{\left(\frac{2j_2+1}{2j_1+1}\right)}$$

$$\times C_{j0 j_2 0}^{j_1 0} C_{j-m j_2 m_2}^{j_1 m_1}, \quad j=0,1,2,\ldots; \quad m=-j,-j+1,\ldots,j. \tag{2.279}$$

The identity of systems of equations (2.278) and (2.279) implies that

$$C_{j_1 m_1 j_2 m_2}^{jm} = (-1)^{j_2+m_2} \sqrt{\left(\frac{2j+1}{2j_1+1}\right)} C_{j-m j_2 m_2}^{j_1 -m_1}, \quad m=m_1+m_2, \tag{2.280}$$

$$E_{j_1 -m_1}^{(p)} = (-1)^{m_1} E_{j_1 m_1}^{(p)*}. \tag{2.281}$$

The reason for this modification of Eq. (2.279) will become clear in Section 2.11.5.

Equations (2.278) or (2.279) represent a system of linear algebraic equations for the coefficients of the density model, $F_{jm}^{(k)}$. The lhs of these equations contain the geopotential coefficients of the gravitational field and coefficients of the mean density model in the form of moments $S^{(j+2)}$ separately. The geometric parameters of the body, $E_{jm}^{(j+3)}$, occur in the matrix of the system and on the lhs of the equations. The coefficients of the density model, $F_{jm}^{(k)}$, interrelate the field and geometric parameters. The real density model of the Earth must satisfy other conditions besides the condition that its field effects agree with the external field observed, which is expressed by Eqs. (2.278) and (2.279). These other conditions may be formulated as follows:

1. The total mass of the Earth, M, must be invariant in any density model. For practical purposes, however, we must take into account that the Earth's mass is only known with an accuracy to four significant figures.

2. The computed, averaged, density model must be identical with the mean model for a spherically symmetrical Earth.

3. The principal moments of inertia, derived from the density model, must yield the observed value of precession constant H. This condition is identical with the condition of invariance of the trace of the tensor of inertia (see Sect. 2.11.5).

2.11.4.2 Total Mass of the Earth

The total mass of the Earth can be calculated from a known density distribution, $\sigma(P')$, as follows:

$$M = \int_\tau \sigma(P')\,d\tau' . \tag{2.282}$$

By substituting for volume element $d\tau' = \varrho'^2 d\varrho' d\Omega'$ and introducing $\varrho' = x\varrho_s(\Omega')$, Eq. (2.235) becomes

$$M = \int_0^1 x^2\,dx \int_{\Omega_0} \sigma(x,\Omega')\varrho_s^3(\Omega')\,d\Omega'$$

$$= \int_0^1 x^2\,dx \int_{\Omega_0} [\sigma(x) + \delta\sigma(x,\Omega')]\,\varrho_s^3(\Omega')\,d\Omega' . \tag{2.283}$$

By substituting into Eq. (2.283) for density variation $\delta\sigma(x,\Omega')$ from (2.262) and for $\varrho_s^3(\Omega')$ from (2.159) we arrive at

$$M = a_1^3 E_{00}^{(3)} S^{(2)} 2\sqrt{\pi} + a_1^3 \sum_{\substack{j_2 m_2 \\ j_2 \geqslant 1}} \sum_{n=0}^\infty F_{j_2 m_2}^{(n)} \sum_{j_1 m_1} E_{j_1 m_1}^{(3)*}$$

$$\times \int_0^1 x^2 \tilde{P}_n(x)\,dx \int_{\Omega_0} Y_{j_1 m_1}^*(\Omega') Y_{j_2 m_1}(\Omega')\,d\Omega' . \tag{2.284}$$

In view of (2.271) $x^2 = \sum_{k=0}^2 c_k^{(2)} \tilde{P}_k(x)$, and making use of the orthogonality of spherical harmonics $Y_{jm}(\Omega)$ and $\tilde{P}_n(x)$, the formula for the total Earth's mass can be modified to read

$$\frac{M}{a_1^3} = 2\sqrt{(\pi)} E_{00}^{(3)} S^{(2)} + W_2^{(1)} , \tag{2.285}$$

where

$$W_2^{(1)} = \sum_{\substack{j_2 m_2 \\ j_2 \geqslant 1}} E_{j_2 m_2}^{(3)*} \sum_{k=0}^2 \frac{c_k^{(2)}}{2k+1} F_{j_2 m_2}^{(k)} . \tag{2.286}$$

Equation (2.285) expresses the total mass of the Earth in terms of parameters $S^{(2)}$ of the mean spherical model and of coefficients $F_{jm}^{(k)}$ of the density model, and also of coefficients E_{jm} which reflect the Earth's topography.

We shall now prove that Eq. (2.285) is identical with the equation of system (2.278) for $j = m = 0$. We shall first express the latter equation explicitly:

$$\frac{M}{a_1^3}\frac{A_{00}}{2\sqrt{(\pi)}} - 2\sqrt{\pi}E_{00}^{(3)}S^{(2)} = \sum_{j_2 m_2}\sum_{k=0}^{2}\frac{c_k^{(2)}}{2k+1}F_{j_2 m_2}^{(k)}$$

$$\times \sum_{j_1 m_1} E_{j_1 m_1}^{(3)}\sqrt{[(2j_1+1)(2j_2+1)]}\,C_{j_1 0 j_2 0}^{00}\,C_{j_1 m_1 j_2 m_2}^{00}\,. \tag{2.287}$$

In view of (2.8) $A_{00} = 2\sqrt{\pi}$ and

$$C_{j_1 m_1 j_2 m_2}^{00} = (-1)^{j_1 - m_1}\delta_{j_1 j_2}\delta_{m_1 - m_2}/\sqrt{[2j_1+1]}\,, \tag{2.288}$$

$$E_{j_1 m_1}^{(3)*} = (-1)^{m_1}E_{j_1 - m_1}^{(3)}\,. \tag{2.289}$$

By substituting (2.288) and (2.289) into (2.287) it is easy to see that Eqs. (2.287) and (2.285) are identical.

The Mean Spherical Density Model

We shall briefly discuss the properties of the mean model $\sigma(x)$ for a radially symmetrical Earth.

The mean spherical density model is required to yield the total mass of the Earth, M. Equation (2.285) for the spherically symmetrical model of the Earth ($W_2^{(1)} = 0$, $E_{00} = 2\sqrt{\pi}R/a_1$, $E_{00}^{(3)} = E_{00}^3/4\pi$), where R is the Earth's mean radius, is very simple:

$$S^{(2)} = \frac{1}{4\pi}\frac{M}{R^3} = \frac{2\sqrt{\pi}}{E_{00}^3}\frac{M}{a_1^3}\,. \tag{2.290}$$

This equation expresses the second reduced moment $S^{(2)}$ of the mean spherical model in terms of quantity M/R^3. It represents a condition which must be satisfied by every mean density model, i.e. to be compatible with the theory of three-dimensional density distribution. We shall refer to Eq. (2.290) as the first equation of compatibility. By substituting into Eq. (2.285) for $S^{(2)}$ from Eq. (2.290) we get

$$\left(1 - 4\pi\frac{E_{00}^{(3)}}{E_{00}^3}\right)\frac{M}{a_1^3} = W_2^{(1)}\,. \tag{2.291}$$

2.11.5 Moments of Inertia

Symbols A, B, C will be used to denote the principal moments of inertia referred to axes x_1, x_2, x_3. According to (2.232) they are defined as follows:

$$C = \int(x_1^2 + x_2^2)\sigma(P)\,d\tau; \quad A + B = \int(x_1^2 + x_2^2 + 2x_3^2)\sigma(P)\,d\tau\,. \tag{2.292}$$

The square of distance expressed in spherical coordinates reads

$$x_1^2 + x_2^2 = \varrho^2 \sin^2 \vartheta = \frac{4\sqrt{\pi}}{3} \varrho^2 \left[Y_{00}(\Omega) - \frac{1}{\sqrt{5}}(\Omega) \right], \tag{2.293}$$

$$x_1^2 + x_2^2 + 2x_3^2 = \varrho^2(1 + \cos^2 \vartheta) = \frac{4\sqrt{\pi}}{3} \varrho^2 \left[2Y_{00}(\Omega) + \frac{1}{\sqrt{5}} Y_{20}(\Omega) \right]. \tag{2.294}$$

Substituting $\varrho = x\varrho_s(\Omega)$ into Eq. (2.292) for the moment of inertia yields

$$C = \frac{4\sqrt{\pi}}{3} \int_{\Omega_0} \left[Y_{00}(\Omega) - \frac{1}{\sqrt{5}} Y_{20}(\Omega) \right] \varrho_s^5(\Omega) \int_0^1 x^4 \sigma(x, \Omega) \, dx \, d\Omega. \tag{2.295}$$

By substituting for density $\sigma(x, \Omega)$ from (2.266) and developing $\varrho_s^5(\Omega)$ and x^4 into orthogonal series, after some minor algebra we obtain the final expression for the principal moment of inertia C (Pěč and Martinec 1984):

$$C = \tfrac{2}{3} a_1^5 \left\{ 2\sqrt{\pi} S^{(4)} \left[E_{00}^{(5)} - \frac{1}{\sqrt{5}} E_{20}^{(5)} \right] + W_4^{(1)} - W_4^{(2)} \right\}, \tag{2.296}$$

where

$$W_4^{(1)} = \sum_{\substack{j_2 m_2 \\ j_2 \geqslant 1}} E_{j_2 m_2}^{(5)} \sum_{k=0}^{4} \frac{c_k^{(4)}}{2k+1} F_{j_2 m_2}^{(k)}, \tag{2.297}$$

$$W_4^{(2)} = \sum_{\substack{j_2 m_2 \\ j_2 \geqslant 1}} \sum_{k=0}^{4} \frac{c_k^{(4)}}{2k+1} F_{j_2 m_2}^{(k)} \sum_{j_1 m_1} E_{j_1 m_1}^{(5)} \sqrt{\left[\frac{2j_2 + 1}{2j_1 + 1} \right]} C_{20 j_2 0}^{j_1 0} C_{20 j_2 m_2}^{j_1 m_1}. \tag{2.298}$$

In view of the properties of the Clebsch–Gordan coefficients, the summation over j_1, m_1 in Eq. (2.298) reduces to only three terms: $j_1 = j_2$, $j_1 = j_2 \pm 2$, $m_1 = m_2$. Equation (2.298) can be simplified to

$$W_4^{(2)} = \sum_{\substack{j_2 m_2 \\ j_2 \geqslant 1}} \sum_{k=0}^{4} \frac{c_k^{(4)}}{2k+1} F_{j_2 m_2}^{(k)} \left\{ E_{j_2 m_2}^{(5)} C_{j_2 0 20}^{j_2 0} C_{j_2 m_2 20}^{j_2 m_2} \right.$$

$$+ E_{j_2 - 2 m_2}^{(5)} \sqrt{\left(\frac{2j_2 + 1}{2j_2 - 3} \right)} C_{j_0 20}^{j_2 - 2 0} C_{j_2 m_2 20}^{j_2 - 2 m_2}$$

$$\left. + E_{j_2 + 2 m_2}^{(5)} \sqrt{\left(\frac{2j_2 + 1}{2j_2 + 5} \right)} C_{j_2 0 20}^{j_2 + 2 0} C_{j_2 m_2 20}^{j_2 + 2 m_2} \right\}. \tag{2.299}$$

The Clebsch–Gordan coefficients in Eq. (2.299) read

$$C_{j_2 m_2 20}^{j_2 m_2} = \frac{3m_2^2 - j_2(j_2 + 1)}{j_2(j_2 + 1)(2j_2 - 1)(2j_2 + 3)}, \quad j_2 \geqslant 2$$

$$C_{j_2 m_2 20}^{j_2 - 2 m_2} = \sqrt{\left[\frac{3(j_2 + m_2)(j_2 + m_2 - 1)(j_2 - m_2)(j_2 - m_2 - 1)}{2j_2(j_2 - 1)(2j_2 - 1)(2j_2 + 1)} \right]},$$

$$C_{j_2 m_2 20}^{j_2 + 2 m_2} = \sqrt{\left[\frac{3(j_2 + m_2 + 1)(j_2 + m_2 + 2)(j_2 - m_2 + 1)(j_2 - m_2 + 2)}{2(j_2 + 1)(j_2 + 2)(2j_2 + 1)(2j_2 + 3)} \right]}.$$

$$\tag{2.300}$$

By substituting (2.300) into (2.299) the expression for $W_4^{(2)}$ becomes

$$
\begin{aligned}
W_4^{(2)} = \sum_{\substack{j_2 m_2 \\ j_2 \geqslant 1}} \sum_{k=0}^{4} \frac{c_k^{(4)}}{2k+1} F_{j_2 m_2}^{(k)} \Bigg\{ & E_{j_2 m_2}^{(5)*} \frac{j_2(j_2+1) - 3m_2^2}{(2j_2-1)(2j_2+3)} \\
& + \frac{3}{2(2j_2-1)} E_{j_2-2\,m_2}^{(5)*} \\
& \times \sqrt{\left[\frac{(j_2+m_2)(j_2+m_2-1)(j_2-m_2)(j_2-m_2-1)}{(2j_2+1)(2j_2-3)} \right]} \\
& + \frac{3}{2(2j_2+3)} E_{j_2+2\,m_2}^{(5)*} \\
& \times \sqrt{\left| \frac{(j_2+m_2+1)(j_2+m_2+2)(j_2-m_2+1)(j_2-m_2+2)}{(2j_2+1)(2j_2+5)} \right|} \Bigg\} .
\end{aligned}
$$

$$(2.301)$$

In the way outlined above we get

$$
A + B = \tfrac{2}{3} a_1^5 \left\{ 2\sqrt{\pi} S^{(4)} \left[2E_{00}^{(5)} + \frac{1}{\sqrt{5}} E_{20}^{(5)} \right] + 2W_4^{(1)} + W_4^{(2)} \right\} . \tag{2.302}
$$

Equations (2.296) and (2.302) now yield the expression for the trace of moment of inertia I which is the basic invariant:

$$
Tr\,I = A + B + C = 2a_1^5 \left\{ 2\sqrt{\pi} S^{(4)} E_{00}^{(5)} + W_4^{(1)} \right\} . \tag{2.303}
$$

The mean density model must yield the same moment of inertia as that of the real non-spherical Earth. All that has been said about the total mass of the Earth also applies to the tensor of inertia. The trace of the tensor of inertia is invariant with respect to any density model.

Equation (2.303) reduced to the mean spherical model $[W_4^{(1)} = 0,\ E_{00} = 2\sqrt{\pi}\,R/a_1,\ E_{00}^{(5)} = E_{00}^5/16\pi^2]$ reads

$$
S^{(4)} = \frac{4\pi\sqrt{\pi}}{(a_1 E_{00})^5}\, Tr\,I . \tag{2.304}
$$

This equation represents the second condition which has to be satisfied for the mean spherically symmetrical model to be compatible with the general density model. We shall refer to it as the second equation of compatibility. It determines the relation between the trace of the tensor of inertia and the reduced moment of the mean spherical model, $S^{(4)}$.

Substituting Eq. (2.304) into Eq. (2.303) yields

$$
\left(1 - 16\pi^2 \frac{E_{00}^{(5)}}{E_{00}^5} \right) \frac{Tr\,I}{a_1^5} = 2W_4^{(1)} . \tag{2.305}
$$

Quantity $E_{00}^{(5)}$ is defined by Eq. (2.165). Constant H, (2.252), is defined as

$$H = 1 - \frac{1}{2}\frac{A+B}{C}, \quad C - \frac{1}{2}(A+B) = HC.$$
(2.306)

Constant H has been derived with satisfactory accuracy from astronomical observations (Kinoshita and Souchay 1990). Equations (2.296) and (2.302) imply that

$$C - \tfrac{1}{2}(A+B) = -a_1^5 \left\{ 2\sqrt{\pi}S^{(4)}\frac{E_{20}^{(5)}}{\sqrt{5}} + W_4^{(2)} \right\}.$$
(2.307)

Equation (2.307) enables (2.306) to be transformed to

$$\frac{H}{3}\frac{Tr\,I}{a_1^5} + (1 - \tfrac{2}{3}H)2\sqrt{\left(\frac{\pi}{5}\right)}S^{(4)}E_{20}^{(5)} = -(1 - \tfrac{2}{3}H)W_4^{(2)}.$$
(2.308)

Terms similar to those in Eq. (2.308) also appear in the equation of system (2.279) for $j = 2$, $m = 0$ which reads

$$\frac{1}{2}\sqrt{\left(\frac{5}{\pi}\right)}\frac{M}{a_1^3}\frac{a_0}{a_1}A_{20} - 2\sqrt{\left(\frac{\pi}{5}\right)}E_{20}^{(5)}S^{(4)} = W_4^{(2)}.$$
(2.309)

Eliminating $W_4^{(2)}$ from Eqs. (2.309) and (2.308) yields

$$\frac{1}{3}\frac{Tr\,I}{a_1^3} + \left(1 - \frac{2}{3}H\right)\frac{1}{2}\sqrt{\left(\frac{5}{\pi}\right)}\frac{M}{a_1^3}\left(\frac{a_0}{a_1}\right)^2\frac{A_{20}}{H} = 0.$$
(2.310)

On considering the relation between complex geopotential coefficient A_{20} and zonal geopotential coefficient J_2,

$$J_2 = \frac{1}{2}\sqrt{\left(\frac{5}{\pi}\right)}A_{20},$$
(2.311)

we immediately see that Eq. (2.311) can be expressed in an equivalent form:

$$\frac{1}{3}\frac{Tr\,I}{a_1^5} - \left(1 - \frac{2}{3}H\right)\frac{M}{a_1^3}\left(\frac{a_0}{a_1}\right)^2\frac{J_2}{H} = 0.$$
(2.312)

Equations (2.310) and (2.312) are familiar, e.g. Nakiboglu (1982). Equation (2.310) defines the trace of the tensor of inertia in terms of the basic observed Earth data (total Earth's mass M, second zonal geopotential coefficient J_2, constant H and semimajor axis a_1). Equation (2.312) is universal and holds regardless of the assumptions of hydrostatic equilibrium of the Earth. Moreover, it eliminates the conflict between Eq. (2.309), belonging to system of linear equations (2.279), and Eq. (2.310), which contains the dynamic flattening. Equation (2.312) must also hold for the density model of the mean spherically symmetrical Earth. At first glance it may seem a paradox that the spherically symmetrical model is being described using the dynamic flattening of the real Earth. However, considering that the general condition of the moment of inertia of the mean radially symmetrical Earth being the same as that of the real flattened Earth is being imposed on the mean density model, then including the dynamic flattening in the mean radial model is equally justified here as in the case of the moment of inertia.

For the mean spherical model ($C = \frac{1}{3} Tr\, I$) Eq. (2.312) is modified to read

$$\frac{C}{MR^2} = (1 - \tfrac{2}{3}H)\left(\frac{a_0}{R}\right)^2 \frac{J_2}{H}. \tag{2.313}$$

Substituting the values of parameters $H = 0.003\,273\,965$, $J_2 = 0.001\,082\,63$, $a_0 = 6\,378\,137$ m, $a_1 = 6\,378\,171$ m, $R = a_1 E_{00}/2\sqrt{\pi} = 0.998\,874\,441$, $a_1 = 6\,370\,994$ m yields

$$\frac{C}{MR^2} = 0.330\,69. \tag{2.314}$$

The PREM model (Dziewonski and Anderson 1981) considers $C/MR^2 = 0.3308$ which agrees well with (2.314).

Equation (2.279) for the coefficients of the 3-D density distribution together with the equation for the trace of the tensor of inertia (2.310) represent the conditions which the density model must necessarily satisfy if the external gravitational field it generates is to correspond with the field observed. The constraints related to the internal gravitational field have been discussed by Matyska (1989). Martinec and Pěč (1986) used Eqs. (2.279) and (2.310) to derive their density model. They required the 3-D density model to differ least from the mean spherically symmetrical model PREM in the least-squares sense.

The application of the equations derived does not eliminate the ambiguity of the solution of the inverse gravimetric problem, but only restricts it in accordance with the assumptions adopted. The constraints mentioned have been expanded should the non-negligible effect of the irregularities of the core-mantle boundary (Čadek and Matyska 1991) be considered, besides the effect of the topography of the Earth's surface, and should the compensation of masses on the undulating Mohorovičić discontinuity, which strongly correlates with the distribution of continents and oceans (Martinec 1991) be also considered.

3 Fundamentals of the Earth's Rotation Dynamics

3.1 Introduction

Space methods of determining the positions of points on the Earth's surface and the external gravitational field have already achieved an accuracy which enables us to determine quantitatively the time variations of the quantities which define position and the field. This is the reason why the work in geodynamics has been developing intensively in recent years. This scientific field covers the dynamics of the vector of the Earth's rotation, the dynamics of the ellipsoid of inertia of the Earth, tidal deformations of equipotential surfaces of the geopotential, atmosphere, oceans and the Earth's crust, other variations of the geopotential, deformations of the Earth of non-tidal origin and, finally, also relativistic cosmogonic phenomena.

We shall only discuss in more detail the geodynamic phenomena which can reliably be described theoretically, and which are already being observed as part of international cooperation. In this chapter we shall describe the fundamentals of the Earth's rotation dynamics, as well as the current results and problems of determining the variations of the modulus and direction of the vector of instantaneous Earth rotation within the Earth itself. Special attention will be devoted to deriving the Earth–Moon–Sun force function which is the basis for solving precession and nutation.

3.2 Fundamental Relations of the Earth's Rotation Dynamics, Euler's Dynamic and Kinematic Equations

First, let us consider the Earth to be perfectly rigid. We shall then be able to apply to it Euler's dynamic theory of the gyroscope with one fixed point, based on the second moment theorem:

$$\frac{d\mathbf{L}}{dt} = \mathbf{G} ; \tag{3.1}$$

$\mathbf{G}$ is the vector of the total moment of the external forces acting on the body and $\mathbf{L}$ is the vector of the total angular momentum,

$$\mathbf{L} = \int_M [\mathbf{r} \times \mathbf{v}] \, dm , \tag{3.2}$$

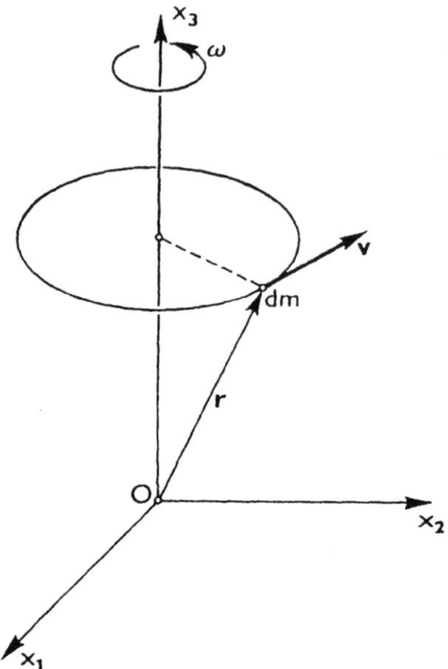

Fig. 3.1. Gyroscope with one fixed point

where **r** is (Fig. 3.1) the radius-vector of element dm of the body whose total mass is M, and **v** is its velocity.

We shall first operate in geocentric coordinate system x'_j whose origin is in mass centre O, but its axes are in general not identical with the axes of the ellipsoid of inertia.

We shall assume that axes x'_j are fixed to the body and that, consequently, their positions relative to the mass particles of the Earth do not change. This means that moments of inertia A', B', C' relative to axes x'_1, x'_2, x'_3, products of inertia D, E, F and, therefore, also tensor of inertia **I** [(2.231), (2.232)],

$$\mathbf{I} = \begin{pmatrix} A' & -F & -E \\ -F & B' & -D \\ -E & -D & C' \end{pmatrix} \tag{3.3}$$

are constant in time. Moments **L** and **G** will be reckoned relative to the Earth's centre of mass, O.

According to Resal's theorem (1884) the time derivative of vector (3.2) can be expressed as the sum of its relative time variation $(dL/dt)_r$ in system x'_j and rotation component $[\boldsymbol{\omega} \times \mathbf{L}]$, i.e.

$$\left(\frac{d\mathbf{L}}{dt}\right)_r + [\boldsymbol{\omega} \times \mathbf{L}] = \mathbf{G}, \tag{3.4}$$

$\boldsymbol{\omega}$ being the vector of instantaneous angular velocity of the Earth's rotation.

Since $\mathbf{v} = [\boldsymbol{\omega} \times \mathbf{r}]$, vector (3.2) now becomes

$$\mathbf{L} = \int\limits_M (\mathbf{r} \times [\boldsymbol{\omega} \times \mathbf{r}])\,dm = \int\limits_M [\boldsymbol{\omega}(\mathbf{r}\cdot\mathbf{r}) - \mathbf{r}(\mathbf{r}\cdot\boldsymbol{\omega})]\,dm$$

$$= \sum_{j=1}^{3} \omega_j \mathbf{e}_j \left(\sum_{j=1}^{3} \int\limits_M x_j'^2\,dm \right) - \int\limits_M \sum_{j=1}^{3} x_j' \mathbf{e}_j \left(\sum_{j=1}^{3} x_j'\omega_j \right) dm \,; \tag{3.5}$$

ω_j are the components of vector $\boldsymbol{\omega}$ along axes x_j' and $\mathbf{e}_j$ are the unit vectors along these axes. Components L_j then read

$$L_1 = \omega_1 A' - \omega_3 E - \omega_2 F \,,$$

$$L_2 = \omega_2 B' - \omega_1 F - \omega_3 D \,,$$

$$L_3 = \omega_3 C' - \omega_2 D - \omega_1 E \tag{3.6}$$

or, in other words,

$$\mathbf{L} = \mathbf{I}\cdot\boldsymbol{\omega}\,, \quad L_j = I_{ij}\omega_i \tag{3.7}$$

Equation (3.4) in terms of components in system x_j' now reads

$$\frac{dL_j}{dt} + \varepsilon_{jik}\omega_i L_k = G_j \tag{3.8}$$

or

$$A'\dot{\omega}_1 - E\dot{\omega}_3 - F\dot{\omega}_2 + (C' - B')\omega_2\omega_3 - E\omega_1\omega_2 + F\omega_1\omega_3$$
$$+ D(\omega_3^2 - \omega_2^2) = G_1 \,,$$
$$B'\dot{\omega}_2 - F\dot{\omega}_1 - D\dot{\omega}_3 + (A' - C')\omega_1\omega_3 - F\omega_2\omega_3$$
$$+ D\omega_2\omega_1 + E(\omega_1^2 - \omega_3^2) = G_2 \,,$$
$$C'\dot{\omega}_3 - D\dot{\omega}_2 - E\dot{\omega}_1 + (B' - A')\omega_1\omega_2 - D\omega_3\omega_1 + E\omega_3\omega_2$$
$$+ F(\omega_2^2 - \omega_1^2) = G_3 \,; \tag{3.9}$$

G_j are components of the moment of external forces, $\mathbf{G}$, also in system x_j'; $i, j,$ $k = 1, 2, 3$, ε_{ijk} is Levi-Civita's epsilon, a unit tensor of the third order defined as follows: $\varepsilon_{ijk} = 0$ if any two indices are equal, $\varepsilon_{ijk} = +1$ if the indices form an even permutation and $\varepsilon_{ijk} = -1$ if the indices form an odd permutation. The dotted quantities indicate their time derivative.

We shall now simplify the problem by putting $D = E = F = 0$, and axes x_j' will thus identify with the axes of the ellipsoid of inertia; for the practical realization of this ideal system, see Section 3.8. One should, however, bear in mind that axis x_1 of this geocentric system is deflected by an angle of about 14.9° to the west of axis x_1' of the commonly used Greenwich reference geocentric system x_j' (Fig. 3.2) and the results must be interpreted with this in view once the problem has been solved in ideal system x_j.

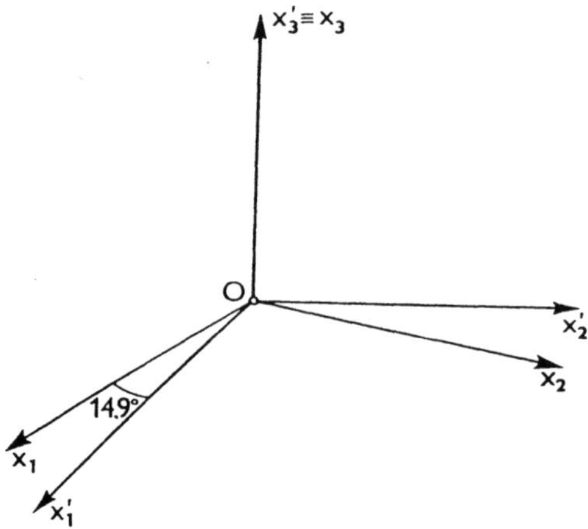

Fig. 3.2. System of axes of the Earth's mean ellipsoid of inertia, x_j, as referred to the Greenwich coordinate system, x_j'

This simplification yields Euler's classical dynamic equations of motion of the gyroscope (Euler 1765),

$$A\dot{\omega}_1 + (C - B)\omega_2\omega_3 = G_1 \,,$$

$$B\dot{\omega}_2 + (A - C)\omega_3\omega_1 = G_2 \,,$$

$$C\dot{\omega}_3 + (B - A)\omega_1\omega_2 = G_3 \,, \tag{3.10}$$

representing a non-linear system of three first-order differential equations; A, B, C are the principal moments of inertia relative to axes x_j of the ellipsoid of inertia.

They describe the motion of the vector of instantaneous angular velocity of rotation, $\boldsymbol{\omega}$, in the x_j-system of axes of the Earth's ellipsoid of inertia, fixed with the body. Therefore, they reflect its relative motion. To be able to describe it 'absolutely' (in Newton's sense), it is necessary to use a system which is inertial with respect to the rotation of the body involved, and to transform the components of the vector of instantaneous rotation velocity, ω_j, into this system. We shall denote this system $\bar{x}_j$, and describe the relation between x_j and $\bar{x}_j$ by orthogonal transformation

$$x_i = \sum_{j=1}^{3} l_{ji}\bar{x}_j \,, \quad i = 1, 2, 3 \,; \tag{3.11}$$

l_{ij} are the appropriate direction cosines. Systems x_j and $\bar{x}_j$ are orthogonal and, therefore,

$$\sum_{j=1}^{3} l_{ji}l_{jk} = \delta_{ik} \,, \tag{3.12}$$

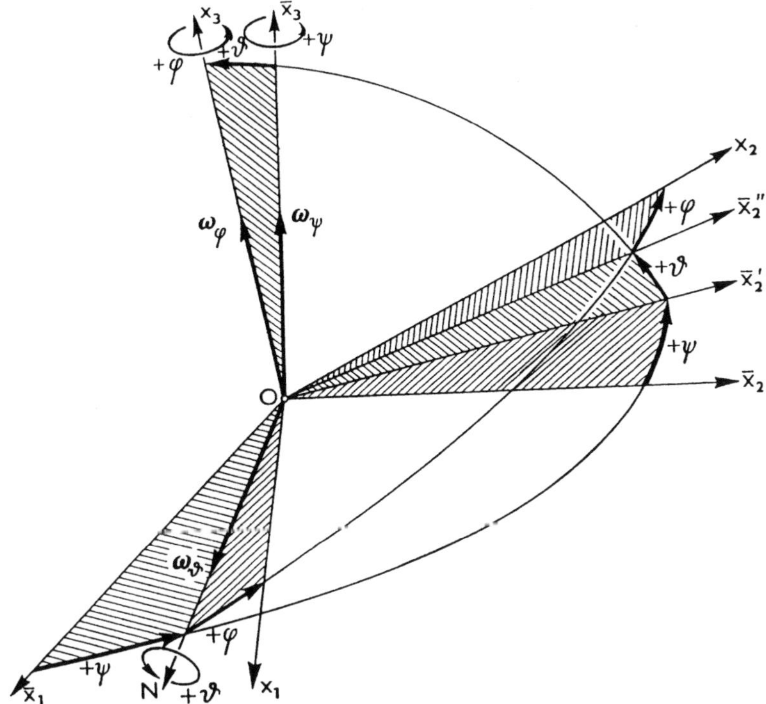

Fig. 3.3. Euler's angles: precession ψ, nutation ϑ and proper rotation φ (*shaded areas represent planes of rotation*)

δ_{ik} being Kronecker's delta. We shall denote Euler's angles (precession, nutation, and proper rotation) ψ, ϑ, φ (Fig. 3.3). The corresponding orthogonal rotation matrices are:

$$\boldsymbol{\psi} = \begin{pmatrix} \cos\psi & \sin\psi & 0 \\ -\sin\psi & \cos\psi & 0 \\ 0 & 0 & 1 \end{pmatrix},$$

$$\boldsymbol{\vartheta} = \begin{pmatrix} 1 & 0 & 0 \\ 0 & \cos\vartheta & \sin\vartheta \\ 0 & -\sin\vartheta & \cos\vartheta \end{pmatrix},$$

$$\boldsymbol{\varphi} = \begin{pmatrix} \cos\varphi & \sin\varphi & 0 \\ -\sin\varphi & \cos\varphi & 0 \\ 0 & 0 & 1 \end{pmatrix} \tag{3.13}$$

and

$$\mathbf{r} = \boldsymbol{\varphi}\boldsymbol{\vartheta}\boldsymbol{\psi}\bar{\mathbf{r}}, \tag{3.14}$$

where

$$
\boldsymbol{\varphi\vartheta\psi} =
\begin{pmatrix}
\cos\varphi\cos\psi - \sin\varphi\cos\vartheta\sin\psi & \cos\varphi\sin\psi + \sin\varphi\cos\vartheta\cos\psi & \sin\varphi\sin\vartheta \\
- \sin\varphi\cos\psi - \cos\varphi\cos\vartheta\sin\psi & - \sin\varphi\sin\psi + \cos\varphi\cos\vartheta\cos\psi & \cos\varphi\sin\vartheta \\
\sin\vartheta\sin\psi & - \sin\vartheta\cos\psi & \cos\vartheta
\end{pmatrix} .
$$

$$(3.15)$$

We shall now apply the principal of superposition of rotations:
1. If the body undergoes n instantaneous partial rotations ω_i, $i = 1, 2, \ldots, n$, about axes which intersect at a single point, its resultant rotational motion is defined by the vector of instantaneous rotation,

$$
\boldsymbol{\omega} = \sum_{i=1}^{n} \boldsymbol{\omega}_i .
\tag{3.16}
$$

2. The rotational motion can be resolved into a number of partial rotations about axes which intersect at a single point.

In view of theorem 2,

$$
\boldsymbol{\omega} = \boldsymbol{\omega}_\varphi + \boldsymbol{\omega}_\vartheta + \boldsymbol{\omega}_\psi ,
\tag{3.17}
$$

$$
\boldsymbol{\omega}_\varphi = \dot{\boldsymbol{\varphi}}, \quad \boldsymbol{\omega}_\vartheta = \dot{\boldsymbol{\vartheta}}, \quad \boldsymbol{\omega}_\psi = \dot{\boldsymbol{\psi}} .
\tag{3.18}
$$

The components of vectors $\boldsymbol{\omega}_\varphi$, $\boldsymbol{\omega}_\vartheta$, $\boldsymbol{\omega}_\psi$ along axes x_j (Fig. 3.3) are

$$
(\omega_\varphi)_1 = \dot{\varphi}\cos(x_3, x_1) = 0 ,
$$

$$
(\omega_\varphi)_2 = \dot{\varphi}\cos(x_3, x_2) = 0 ,
$$

$$
(\omega_\varphi)_3 = \dot{\varphi} ;
\tag{3.19}
$$

$$
(\omega_\vartheta)_1 = \dot{\vartheta}\cos(N, x_1) = \dot{\vartheta}\cos\varphi ,
$$

$$
(\omega_\vartheta)_2 = \dot{\vartheta}\cos(N, x_2) = - \dot{\vartheta}\sin\varphi ,
$$

$$
(\omega_\vartheta)_3 = \dot{\vartheta}\cos(N, x_3) = 0 ;
\tag{3.20}
$$

$$
(\omega_\psi)_1 = \dot{\psi}\cos(\bar{x}_3, x_1) = \dot{\psi}\sin\varphi\sin\vartheta ,
$$

$$
(\omega_\psi)_2 = \dot{\psi}\cos(\bar{x}_3, x_2) = \dot{\psi}\cos\varphi\sin\vartheta ,
$$

$$
(\omega_\psi)_3 = \dot{\psi}\cos(\bar{x}_3, x_3) = \dot{\psi}\cos\vartheta .
\tag{3.21}
$$

Equations (3.19) and (3.20) follow from the elementary matrix of rotation φ (3.13) and Eqs. (3.21) follow from the third column of matrix (3.15).

In view of (3.17) and (3.19)–(3.21) components ω_j of resultant vector $\boldsymbol{\omega}$ along the axes of the ellipsoid of inertia, x_j, can immediately be expressed as

$$
\omega_1 = (\omega_\varphi)_1 + (\omega_\vartheta)_1 + (\omega_\psi)_1 = \dot{\vartheta}\cos\varphi + \dot{\psi}\sin\varphi\sin\vartheta ,
$$

$$
\omega_2 = (\omega_\varphi)_2 + (\omega_\vartheta)_2 + (\omega_\psi)_2 = - \dot{\vartheta}\sin\varphi + \dot{\psi}\cos\varphi\sin\vartheta ,
$$

$$
\omega_3 = (\omega_\varphi)_3 + (\omega_\vartheta)_3 + (\omega_\psi)_3 = \dot{\varphi} + \dot{\psi}\cos\vartheta ,
\tag{3.22}
$$

which are Euler's kinematic equations (Euler 1765); these enable the transition from describing rotation in the system of the body's ellipsoid of inertia to describing it in the inertial system, and vice versa. They link the components of the vector of instantaneous rotation velocity of the body, occurring in Euler's dynamic equations (3.10), with angles φ, ϑ, ψ and with their time derivatives. They also link dynamic equations (3.10) in the x_j-system of the ellipsoid of inertia with system $\bar{x}_j$ which is inertial relative to the rotation being studied. As regards the Earth's rotation, it is convenient to take the ecliptical system, in which $\bar{x}_1$ points to the vernal equinox, Υ, of a particular epoch, as the inertial system. Components $\bar{\omega}_j$ of vector $\boldsymbol{\omega}$ in the inertial system, $\bar{x}_j$, can be expressed in very much the same way:

$$\bar{\omega}_1 = \dot{\varphi} \sin \vartheta \sin \psi + \dot{\vartheta} \cos \psi \,,$$

$$\bar{\omega}_2 = -\dot{\varphi} \sin \vartheta \cos \psi + \dot{\vartheta} \sin \psi \,,$$

$$\bar{\omega}_3 = \dot{\varphi} \cos \vartheta + \dot{\psi} \,. \tag{3.23}$$

The modulus

$$\omega = [\dot{\psi}^2 + \dot{\vartheta}^2 + \dot{\varphi}^2 + 2\dot{\psi}\dot{\varphi} \cos \vartheta]^{1/2} \tag{3.24}$$

and the direction cosines of the vector of instantaneous rotation, $\boldsymbol{\omega}$, relative to the axes x_j of the ellipsoid of inertia, are $a_\omega = \omega_1/\omega$, $b_\omega = \omega_2/\omega$, $c_\omega = \omega_3/\omega$. Euler's kinematic equations (3.22) can also be expressed as

$$\dot{\psi} \sin \vartheta = \omega_1 \sin \varphi + \omega_2 \cos \varphi \,,$$

$$\dot{\vartheta} = \omega_1 \cos \varphi - \omega_2 \sin \varphi \,,$$

$$\dot{\varphi} = \omega_3 - \dot{\psi} \cos \vartheta = \omega_3 - (\omega_1 \sin \varphi + \omega_2 \cos \varphi) \cot \vartheta \,. \tag{3.25}$$

They express the time variation of the direction to the vernal equinox, defined as the direction of the line of intersection of the plane of the instantaneous equator; the plane of the 'fixed' ecliptic $(\bar{x}_1 \bar{x}_2)$ of the epoch involved; and the time variation of the angle between the plane of the instantaneous equator and the $(\bar{x}_1 \bar{x}_2)$ plane. Quantity $\dot{\varphi}$ represents the instantaneous angular velocity of the Earth's proper rotation, the reference direction being that to Υ, i.e. the direction of nodal line N.

By substituting (3.22) into (3.6) we get (assuming that $D = E = F = 0$, $A' = A$, $B' = B$, $C' = C$) the components of the resultant vector of the angular momentum along the axes of the ellipsoid of inertia, x_j,

$$L_1 = A\omega_1 = A(\dot{\vartheta} \cos \varphi + \dot{\psi} \sin \varphi \sin \vartheta) \,,$$

$$L_2 = B\omega_2 = B(-\dot{\vartheta} \sin \varphi + \dot{\psi} \cos \varphi \sin \vartheta) \,,$$

$$L_3 = C\omega_3 = C(\dot{\varphi} + \dot{\psi} \cos \vartheta) \tag{3.26}$$

and along the axes of the inertial system, $\bar{x}_j$,

$$\bar{L}_1 = L_1(\cos\varphi\cos\psi - \sin\psi\cos\vartheta\sin\varphi)$$
$$- L_2(\sin\varphi\cos\psi + \sin\psi\cos\vartheta\cos\varphi) + L_3\sin\psi\sin\vartheta,$$
$$\bar{L}_2 = L_1(\sin\psi\cos\varphi + \cos\psi\cos\vartheta\sin\varphi)$$
$$- L_2(\sin\varphi\sin\psi - \cos\psi\cos\vartheta\cos\varphi) - L_3\cos\psi\sin\vartheta,$$
$$\bar{L}_3 = L_1\sin\vartheta\sin\varphi + L_2\sin\vartheta\cos\varphi + L_3\cos\vartheta \tag{3.27}$$

whose time variations, in the absence of external forces, are zero, i.e.

$$\dot{\bar{L}}_j = 0, \quad j = 1, 2, 3, \tag{3.28}$$

which is easy to prove by substituting (3.26) into (3.27) and differentiating the resultant relation with respect to time.

3.3 The Earth's Rotation Dynamics in the Absence of External Moments; Euler's Free Nutation

We have still not defined the resultant vector of external forces, $\mathbf{G}$, on the rhs of Eq. (3.1). Before doing so we shall consider the case in which $\mathbf{G} = 0$. Then

$$\frac{d\mathbf{L}}{dt} = 0, \tag{3.29}$$

i.e. the vector of the angular momentum remains unchanged (in the inertial system), and we obtain a system of homogeneous equations with no rhs:

$$\dot{\omega}_1 + \alpha\omega_2\omega_3 = 0,$$
$$\dot{\omega}_2 + \beta\omega_3\omega_1 = 0,$$
$$\dot{\omega}_3 + \gamma\omega_1\omega_2 = 0; \tag{3.30}$$

$$\alpha = \frac{C-B}{A}, \quad \beta = \frac{A-C}{B}, \quad \gamma = \frac{B-A}{C},$$

$$\alpha = 1/305.461, \quad \beta = 1/303.426, \quad \gamma = 1/45\,540. \tag{3.31}$$

The solution of system (3.30) is known in theoretical mechanics (see, e.g., Landau and Lifshits 1965) and leads to Legendre's (Jacobi's) elliptical functions. The two first integrals of system (3.30) read

$$A\omega_1^2 + B\omega_2^2 + C\omega_3^2 = T = \int_M \sum_{j=1}^{3} [\boldsymbol{\omega}\times\boldsymbol{\varrho}]_j^2 \, dm = C\omega^2 + (A-C)\omega_1^2$$
$$+ (B-C)\omega_2^2 \tag{3.32}$$

and

$$A^2\omega_1^2 + B^2\omega_2^2 + C^2\omega_3^2 = L^2 = C^2\omega^2 + (A^2 - C^2)\omega_1^2 + (B^2 - C^2)\omega_2^2 \; ; \quad (3.33)$$

they reflect the law of kinetic energy conservation, $T/2$, and of the modulus of the vector of angular momentum L; $CT > L^2$ and $L^2 > AT$ always, if $C > B > A$. If we put $A\omega_1 = \bar{x}_1$, $B\omega_2 = \bar{x}_2$, $C\omega_3 = \bar{x}_3$, (3.32) and (3.33) yield

$$\frac{\bar{x}_1^2}{AT} + \frac{\bar{x}_2^2}{BT} + \frac{\bar{x}_3^2}{CT} = 1 , \tag{3.34}$$

$$\sum_{j=1}^{3} \bar{x}_j^2 = L^2 . \tag{3.35}$$

Systems (3.34) and (3.35) can be interpreted geometrically as defining the curve created by a sphere of radius L, intersecting the surface of a triaxial ellipsoid, semi-axes $(AT)^{1/2}, (BT)^{1/2}, (CT)^{1/2}$. The first integrals of (3.32) and (3.33) yield, for example,

$$\omega_1^2 = \frac{\omega^2 BC + L^2 - T(B + C)}{(B - A)(C - A)} ,$$

$$\omega_2^2 = -\frac{\omega^2 AC + L^2 - T(C + A)}{(B - A)(C - B)} ,$$

$$\omega_3^2 = -\frac{\omega^2 AB + L^2 - T(A + B)}{(C - B)(C - A)} . \tag{3.36}$$

If the body is perfectly rigid ($A = $ const, $B = $ const, $C = $ const) the dynamics of vector $\boldsymbol{\omega}$ are described by the first integrals T, L and by its modulus ω; the latter can also be expressed as

$$\omega^2 = \frac{1}{BC} [T(B + C) - L^2] - \beta\gamma\omega_1^2$$

$$= \frac{1}{CA} [T(C + A) - L^2] - \gamma\alpha\omega_2^2$$

$$= \frac{1}{AB} [T(A + B) - L^2] - \alpha\beta\omega_3^2 . \tag{3.37}$$

System (3.30) can be solved, for example, by altering (3.32) and (3.33) to yield

$$\omega_1^2 = \frac{TC - L^2}{A(C - A)} - \omega_2^2 \frac{B(C - B)}{A(C - A)} ,$$

$$\omega_3^2 = \frac{L^2 - TA}{C(C - A)} - \omega_2^2 \frac{B(B - A)}{C(C - A)} \tag{3.38}$$

and substituting into the second equation of (3.30):

$$\dot{\omega}_2 = B^{-1}(AC)^{-1/2}[TC - L^2 - \omega_2^2 B(C - B)]^{1/2}$$
$$\times [L^2 - TA - \omega_2^2 B(B - A)]^{1/2} . \tag{3.39}$$

After performing some algebra we arrive at

$$(c\dot{\omega}_2)^2 = k_0^2 [1 - (c\omega_2)^2][1 - k^2(c\omega_2)^2] ;$$

$$c^2 = \frac{B(C - B)}{TC - L^2}, \quad k_0^2 = (L^2 - AT)(C - B)(ABC)^{-1} ,$$

$$k^2 = \frac{(B - A)(TC - L^2)}{(C - B)(L^2 - TA)} . \tag{3.40}$$

Solution (3.40) leads to an elliptical integral of the first kind, which also applies to the solutions for ω_1 and ω_3. The motion of vector ω is periodic with period

$$T_0 = 4k_0 \int_0^1 [(1 - c^2\omega_2^2)(1 - k^2 c^2\omega_2^2)]^{-1/2} \, \mathrm{d}(c\omega_2) \tag{3.41}$$

and its components are

$$\omega_1 = k_1 \, \mathrm{cn}(k_0 t - \beta_0) ,$$
$$\omega_2 = k_2 \, \mathrm{sn}(k_0 t - \beta_0) ,$$
$$\omega_3 = k_3 \, \mathrm{dn}(k_0 t - \beta_0) ; \tag{3.42}$$

$$k_1^2 = \frac{TC - L^2}{A(C - A)}, \quad k_2^2 = \frac{TC - L^2}{B(C - B)}, \quad k_3^2 = \frac{L^2 - TA}{C(C - A)} ; \tag{3.43}$$

β_0 is the third integration constant, and

$$\mathrm{cn}(k_0 t - \beta_0) = \cos(k_0 t - \beta_0) + 4k^2[(k_0 t - \beta_0)^4/4!$$
$$- (11 + 4k^2)(k_0 t - \beta_0)^6/6!$$
$$+ (102 + 228k^2 + 16k^4)(k_0 t - \beta_0)^8/8! + \cdots] ,$$
$$\mathrm{sn}(k_0 t - \beta_0) = \sin(k_0 t - \beta_0)$$
$$- k^2\{(k_0 t - \beta_0)^3/3! + (14 + k^2)(k_0 t - \beta_0)^5/5!$$
$$- [135(1 + k^2) + k^4](k_0 t - \beta_0)^7/7! + \cdots \} ,$$
$$\mathrm{dn}(k_0 t - \beta_0) = 1 - k^2[(k_0 t - \beta_0)^2/2!$$
$$+ (4 + k^2)(k_0 t - \beta_0)^4/4!$$
$$- (16 + 44k^2 + k^4)(k_0 t - \beta_0)^6/6!$$
$$+ (64 + 912k^2 + 408k^4 + k^6)(k_0 t - \beta_0)^8/8! + \cdots] . \tag{3.44}$$

Solution (3.42) defines the position of the vector of instantaneous rotation of the Earth in the system of axes x_j of the central ellipsoid of inertia as a function of time and of integration constants L, T, β_0, which must be determined by observation [from the initial conditions, e.g., from $(\omega_1)_0, (\omega_2)_0, (\omega_3)_0$ for $t = t_0$]. We can now express them as

$$\omega_1 = \left[\frac{TC - L^2}{A(C - A)}\right]^{1/2} \cos(k_0 t - \beta_0) + f_1(k^2),$$

$$\omega_2 = \left[\frac{TC - L^2}{B(C - B)}\right]^{1/2} \sin(k_0 t - \beta_0) + f_2(k^2),$$

$$\omega_3 = \left[\frac{L^2 - TA}{C(C - A)}\right]^{1/2} + f_3(k^2); \tag{3.45}$$

terms $f_1(k^2), f_2(k^2), f_3(k^2)$, obvious from comparing (3.42), (3.45) and (3.44), are functions of the Earth's equatorial flattening ($\alpha_1 = 1/92\,000$) due to the difference in the equatorial moments of inertia (2.250): $B - A = 4Ma_0^2 J_2^{(2)} = 1.764 \times 10^{33}\,\mathrm{kg\,m^2}$. Since Eqs. (3.32) and (3.33) imply

$$TC - L^2 = A(C - A)\omega_1^2 + B(C - B)\omega_2^2, \tag{3.46}$$

$$L^2 - TA = B(B - A)\omega_2^2 + C(C - A)\omega_3^2, \tag{3.47}$$

the order-of-magnitude estimate of $(TC - L^2)/(L^2 - TA)$ is 10^{-11}. Also, $(B - A)/(C - B) = 6.730 \times 10^{-3}$, and the order of magnitude of parameter k^2 is thus less than 10^{-13}. This indicates that the effect of terms $f_1(k^2), f_2(k^2), f_3(k^2)$ on the Earth is small compared with the effect of the main terms $\cos(k_0 t - \beta_0)$ and $\sin(k_0 t - \beta_0)$.

If the Earth were rotationally symmetrical, i.e. if $B = A$, these terms would be exactly equal to zero, and

$$\omega^2 = [T(A + C) - L^2]/(AC) = \mathrm{const}, \quad \dot{\omega} = 0,$$

$$\omega_3^2 = \frac{L^2 - TA}{C(C - A)} = \mathrm{const}, \quad \dot{\omega}_3 = 0, \tag{3.48}$$

$$\omega_1 = \left[\frac{TC - L^2}{A(C - A)}\right]^{1/2} \cos(k_0 t - \beta_0),$$

$$\omega_2 = \left[\frac{TC - L^2}{A(C - A)}\right]^{1/2} \sin(k_0 t - \beta_0). \tag{3.49}$$

Equations (3.49) represent the parametric equation of a circle of radius $(TC - L^2)^{1/2}[A(C - A)]^{-1/2}$. The motion of the end of vector ω is thus circular with period $2\pi/k_0$ being the familiar Euler period $T_E = 304.44$ days $\sim 1.18°/\mathrm{day}$;

$$k_0 = \left[\frac{(L^2 - TA)(C - A)}{A^2 C}\right]^{1/2} = \frac{C - A}{A}\,\omega_3 \tag{3.50}$$

is the circular frequency. The direction of the free nutation of the vector of rotation in the system of axes of the ellipsoid of inertia is identical with the direction of rotation of the body.

Moreover, vector $\boldsymbol{\omega}$ is located on the side of $\mathbf{L}$ opposite to that of axis x_3 of the ellipsoid of inertia (Fig. 3.4) because

$$\frac{\omega_1^2 + \omega_2^2}{\omega^2} > \frac{A^2\omega_1^2 + B^2\omega_2^2}{L^2},$$

which is easy to prove. If the preceding relation holds, the following must also be true:

$$\frac{\omega^2 - \omega_3^2}{\omega^2} > \frac{L^2 - C^2\omega_3^2}{L^2},$$

$$\frac{\omega_3^2}{\omega^2} < \frac{C^2\omega_3^2}{L^2},$$

$$\frac{\omega_3^2}{\omega^2} < \frac{C^2\omega_3^2}{A^2\omega_1^2 + B^2\omega_2^2 + C^2(\omega_3^2 - \omega^2) + C^2\omega^2},$$

$$\frac{\omega_3^2}{\omega^2} < \frac{C^2\omega_3^2}{\omega_1^2(A^2 - C^2) + \omega_2^2(B^2 - C^2) + C^2\omega^2},$$

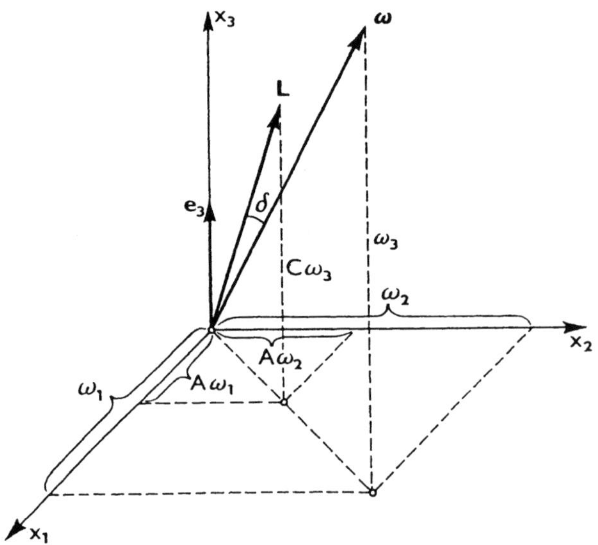

Fig. 3.4. Vector of instantaneous rotation $\boldsymbol{\omega}$ and angular momentum $\mathbf{L}$

$$\frac{\omega_3^2}{\omega^2} < \frac{\omega_3^2}{\omega^2}\left[1 + \frac{\omega_1^2}{\omega^2}\frac{A^2 - C^2}{C^2} + \frac{\omega_2^2}{\omega^2}\frac{B^2 - C^2}{C^2}\right]^{-1},$$

$$1 - \frac{\omega_1^2}{\omega^2}\frac{C^2 - A^2}{C^2} - \frac{\omega_2^2}{\omega^2}\frac{C^2 - B^2}{C^2} < 1.$$

The proof is obvious because $C > B > A$.

If $B = A$, vectors $\boldsymbol{\omega}$, $\mathbf{L}$ and axis x_3 of the ellipsoid of inertia are coplanar (Fig. 3.5) because

$$[\boldsymbol{\omega} \times \mathbf{L}] \cdot \mathbf{e}_3 = \begin{vmatrix} 0 & 0 & 1 \\ \omega_1 & \omega_2 & \omega_3 \\ A\omega_1 & A\omega_2 & C\omega_3 \end{vmatrix} = 0.$$

The plane defined by this relation rotates about vector $\mathbf{L}$, because $\mathbf{L} = \text{const}$ under free nutation. Vector $\boldsymbol{\omega}$ thus moves along a circle about $\mathbf{L}$ [herpolhode, small cone (1) in Fig. 3.6] and about the smallest axis of the ellipsoid of inertia, i.e. vector $\mathbf{e}_3$ [polhode, large cone in Fig. 3.5 and (2) in Fig. 3.6]. The large cone rotates (rolls) about the generating lines (vector of instantaneous rotation $\boldsymbol{\omega}$) of the small cone (Poinsot 1851), which has a constant position in the inertial (ecliptical) system. The centre of the large cone (the pole of the ellipsoid of inertia) moves along a circle about vector $\boldsymbol{\omega}$ (dashed line in Fig. 3.6).

If $B \neq A$, which is the case for the real Earth, Eq. (3.45) holds true, of course, and the motion of the end-point of vector $\boldsymbol{\omega}$ is more complicated, although its projection into plane (x_1, x_2) does circumscribe an ideal ellipse, semi-axes k_1, k_2,

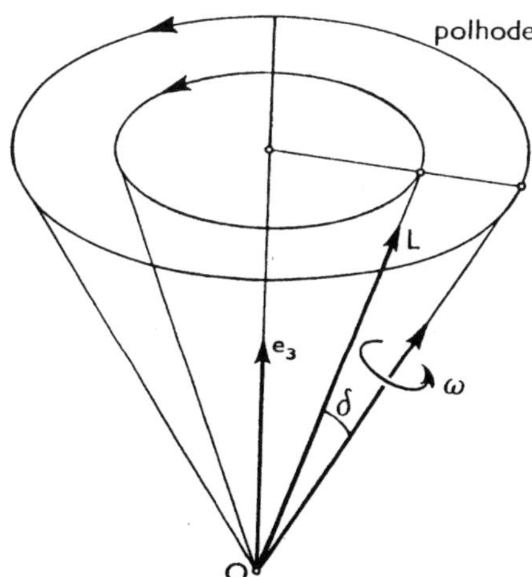

Fig. 3.5. Coplanarity of the vector of instantaneous rotation, of the angular momentum vector, and of the smallest axis of the Earth's ellipsoid of inertia (given $B = A$). *Polhode* is the trace of the pole on the ellipsoid of inertia

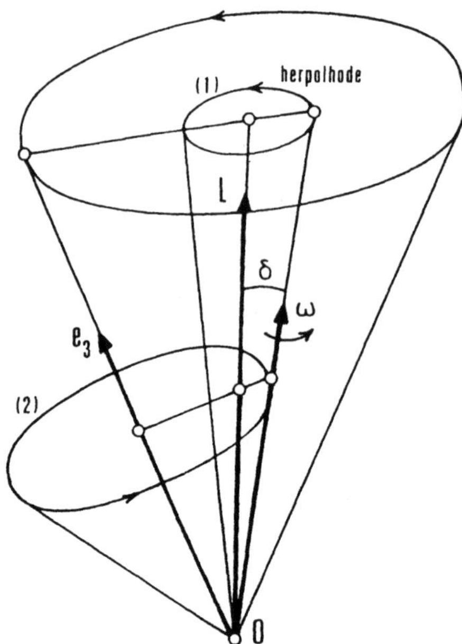

Fig. 3.6. *Herpolhode*, trace of the pole on an (inertial) plane fixed in space

because, in view of (3.42),

$$\left(\frac{\omega_1}{k_1}\right)^2 + \left(\frac{\omega_2}{k_2}\right)^2 = 1 \,,$$

as $\mathrm{sn}^2 x + \mathrm{cn}^2 x = 1$. In general, free nutation differs from ideal circular motion less the larger the difference $(B - A)$. However, the end-point of vector $\boldsymbol{\omega}$ can never move along an ideal ellipse. If $B = A$ ($k = 0$) its motion is circular; if $B \neq A$, $k \neq 0$ and it is described by Eqs. (3.45). Only the projection into plane (x_1, x_2) of the trajectory which the end-point of vector $\boldsymbol{\omega}$ circumscribes is elliptical. However, it is frequently considered to be elliptical; this occurs if terms $f_1(k^2)$, $f_2(k^2)$ and $f_3(k^2)$ in Eqs. (3.45) are omitted, but $B \neq A$ retained in the amplitudes:

$$\omega_1 = \left[\frac{TC - L^2}{A(C - A)}\right]^{1/2} \cos(k_0 t - \beta_0) \,,$$

$$\omega_2 = \left[\frac{TC - L^2}{B(C - B)}\right]^{1/2} \sin(k_0 t - \beta_0) \,,$$

$$\omega_3 = \left[\frac{L^2 - TA}{C(C - A)}\right]^{1/2} = \text{const.} \tag{3.51}$$

The first two equations in (3.51) indeed define an ellipse, semi-axes

$$a_0 = \left[\frac{TC - L^2}{B(C - B)}\right]^{1/2}, \quad b_0 = \left[\frac{TC - L^2}{A(C - A)}\right]^{1/2}. \tag{3.52}$$

However, if $(B - A)$ is small, and terms $f_1(k^2), f_2(k^2), f_3(k^2)$ can be neglected, difference $a_0 - b_0$ is also small and, therefore, the flattening of this ellipse, $\varepsilon = (a_0 - b_0)/a_0$, is also small:

$$\varepsilon = \frac{(B - A)(B + A - C)}{A(C - A) + [AB(C - A)(C - B)]^{1/2}}$$

$$= \frac{1}{2}\frac{B - A}{C - A} = -2\frac{J_{2,2}}{J_2^{(0)}} = 3 \times 10^{-3}, \tag{3.53}$$

where

$$J_{2,2} = [(J_2^{(2)})^2 + (S_?^{(2)})^2]^{1/2}.$$

Given an amplitude of about 10 m, the difference between the semi-axes of the ellipse being considered is then about 0.03 m, which can be proved by observation only with difficulty.

According to Guinot (1982), the whole ensemble of observations from 1900–1980 yields an ellipse flattening of 0.035 ± 0.015 and the west longitude of its axis $135° \pm 15°$. Note that this is not the ellipse of forced motion due to the dynamics of the atmosphere (Sect. 3.5). Angle δ (Figs. 3.4 and 3.5) between the vector of instantaneous rotation and the vector of the resultant angular momentum is small; it obviously follows from

$$\cos \delta = \cos(\boldsymbol{\omega}, \mathbf{L}) = \frac{A\omega_1^2 + B\omega_2^2 + C\omega_3^2}{\omega L} = \frac{T}{\omega L}. \tag{3.54}$$

If $B = A$,

$$\cos \delta = \frac{A(\omega_1^2 + \omega_2^2) + C\omega_3^2}{\omega[A^2(\omega_1^2 + \omega_2^2) + C^2\omega_3^2]^{1/2}}$$

$$= \frac{C\omega^2 - (\omega_1^2 + \omega_2^2)(C - A)}{\omega[C^2\omega^2 - (\omega_1^2 + \omega_2^2)(C^2 - A^2)]^{1/2}}$$

$$= \frac{1 - \dfrac{\omega_1^2 + \omega_2^2}{\omega^2}\dfrac{C - A}{C}}{\left[1 - \dfrac{\omega_1^2 + \omega_2^2}{\omega^2}\dfrac{C^2 - A^2}{C^2}\right]^{1/2}}, \tag{3.55}$$

i.e.

$$\delta = \frac{1}{\omega}(\omega_1^2 + \omega_2^2)^{1/2}\frac{C - A}{C} = \frac{1}{\omega}\left[\frac{TC - L^2}{A(C - A)}\right]^{1/2}\frac{C - A}{C}; \tag{3.56}$$

given a free nutation amplitude of about $0.3''$, $\dfrac{C-A}{C} = 0.00327$, $\delta = 0.001''$, i.e.

~ 3 cm at the Earth's surface.

The kinetic energy, $T/2$, of the actual rotation (3.32) is larger than the kinetic energy of the ideal rotation

$$\tfrac{1}{2} T_0 = \tfrac{1}{2} C \omega_0^2 \,,$$

when the axis of rotation is identical with the smallest axis of the ellipsoid of inertia. Since the angular momentum must be the same ($L = L_0$) in both cases, i.e. necessarily

$$L_0 = C\omega_0 = [\omega_1^2 A^2 + \omega_2^2 B^2 + \omega_3^2 C^2]^{1/2}$$
$$= C\omega \left[1 - \frac{C^2 - A^2}{C^2} \frac{\omega_1^2 + \omega_2^2}{\omega^2} \right]^{1/2} ,$$

the angular velocity of the ideal (i.e. with zero free nutation) rotation (simplified for $B = A$) comes out as

$$\omega_0 = \omega \left[1 - \frac{C^2 - A^2}{C^2} \frac{\omega_1^2 + \omega_2^2}{\omega^2} \right]^{1/2} .$$

If $\omega_1 \neq 0$, $\omega_2 \neq 0$, $\omega > \omega_0$, i.e. under free nutation the Earth rotates faster than in the ideal case, in the absence of free nutation. The difference

$$C(\omega - \omega_0) = C\omega - L_0 = \frac{1}{2} \frac{C^2 - A^2}{C} \frac{\omega_1^2 + \omega_2^2}{\omega} = \frac{1}{2} \frac{C^2 - A^2}{C} \omega a_0^2$$

is always positive, which also applies to the difference

$$\tfrac{1}{2}(T - T_0) = \tfrac{1}{2} [A(\omega_1^2 + \omega_2^2) + C(\omega_3^2 - \omega_0^2)]$$
$$= \tfrac{1}{2} A \frac{C - A}{C} (\omega_1^2 + \omega_2^2) = \tfrac{1}{2} A \frac{C - A}{C} \omega^2 a_0^2 \,,$$

representing the kinetic energy of the Earth's free nutation. Given $a_0 = (\omega_1^2 + \omega_2^2)^{1/2} \omega^{-1} = 0.3''$ and $\tfrac{1}{2}(T - T_0) = 1.5 \times 10^{15}$ kg m^2 s^{-2}, $C(\omega - \omega_0) = 4.1 \times 10^{19}$ kg m^2 s^{-1}.

3.4 Liouville's Equations

If transfer or deformation of terrestrial masses occurs, Euler's dynamic equations (3.10), or in the more general case (3.9), no longer hold exactly because all the Earth's moments of inertia, A', B', C', D, E, F, and, consequently, also the tensor of inertia (3.3) become functions of time.

Angular momentum **L** will be changed not only because its components (3.6) will now contain the effect of the changes in the configuration (geometry) of the positions of mass elements, which will be reflected in the instantaneous values (for a particular instant) of moments $A', \ldots, F$, but also because of the additional part

$$\delta\mathbf{L} = \int\limits_M [\mathbf{r} \times \mathbf{u}]\, dm, \quad \mathbf{u} = \frac{d\mathbf{r}}{dt}, \tag{3.57}$$

which is generated by the velocity field $\mathbf{u}$ of the motion of mass particles, e.g. atmospheric particles, and which may formally be expressed in terms of its components as

$$\delta L_j = \int\limits_M \varepsilon_{jik} x_i u_k\, dm,$$

$$\delta L_1 = \int\limits_M (x_2 u_3 - x_3 u_2)\, dm = \int\limits_M \varrho^2 (\dot\phi \sin \Lambda - \dot\Lambda \sin\phi \cos\phi \cos\Lambda)\, dm,$$

$$\delta L_2 = \int\limits_M (x_3 u_1 - x_1 u_3)\, dm = - \int\limits_M \varrho^2 (\dot\phi \cos \Lambda + \dot\Lambda \sin\phi \cos\phi \sin\Lambda)\, dm,$$

$$\delta L_3 = \int\limits_M (x_1 u_2 - x_2 u_1)\, dm = \int\limits_M \varrho^2 \dot\Lambda \cos^2\phi\, dm; \tag{3.58}$$

$x_1 = \varrho \cos\phi \cos \Lambda$, $x_2 = \varrho \cos\phi \sin \Lambda$, $x_3 = \varrho \sin\phi$ are the geocentric coordinates of mass element dm, $\varrho = |\mathbf{r}|$ is its radius-vector and M is the total mass.

Let us again consider the case when $\mathbf{G} = 0$ and, for the sake of simplicity, let us assume that in the initial epoch $t = t_0$, $\delta L = 0$, $D_0 = E_0 = F_0 = 0$, i.e. that axes $x_j' = x_j$ are the axes of the ellipsoid of inertia, and that $A_0' = A_0$, $B_0' = B_0$, $C_0' = C_0$ are the principal moments of inertia. Equations (3.42) would then represent the solution at time $t = t_0$.

If the Earth is deformed, the dynamic equations determining components ω_j will read

$$\frac{d}{dt}(\mathbf{I} \cdot \boldsymbol{\omega} + \delta\mathbf{L}) + [\boldsymbol{\omega} \times (\mathbf{I} \cdot \boldsymbol{\omega} + \delta\mathbf{L})] = \mathbf{G}$$

or

$$\frac{d}{dt}(I_{ji}\omega_i + \delta L_j) + \varepsilon_{jik}\omega_i(I_{kl}\omega_l + \delta L_k) = G_j. \tag{3.59}$$

These are Liouville's equations (Liouville 1856) which differ from (3.8) in vector $\delta\mathbf{L}$ and in the time variable tensor of inertia $\mathbf{I}$. They can be expressed as

$$A_0 \dot\omega_1 + (C_0 - B_0)\omega_2\omega_3 = f_1(t),$$

$$B_0 \dot\omega_2 + (A_0 - C_0)\omega_3\omega_1 = f_2(t),$$

$$C_0 \dot\omega_3 + (B_0 - A_0)\omega_1\omega_2 = f_3(t). \tag{3.60}$$

The rhs of Liouville's equations (3.60) are called excitation functions, and read

$$f_1(t) = -\omega_1 \frac{dA'}{dt} + \frac{d(F\omega_2)}{dt} + \frac{d(E\omega_3)}{dt} + E\omega_1\omega_2 + D\omega_2^2$$

$$- F\omega_1\omega_3 - D\omega_3^2 - \omega_2\delta L_3 + \omega_3\delta L_2 - \frac{d\delta L_1}{dt},$$

$$f_2(t) = -\omega_2 \frac{dB'}{dt} + \frac{d(F\omega_1)}{dt} + \frac{d(D\omega_3)}{dt} - E\omega_1^2 - D\omega_1\omega_2$$

$$+ F\omega_2\omega_3 + E\omega_3^2 + \omega_1\delta L_3 - \omega_3\delta L_1 - \frac{d\delta L_2}{dt},$$

$$f_3(t) = -\omega_3 \frac{dC'}{dt} + \frac{d(E\omega_1)}{dt} + \frac{d(D\omega_2)}{dt} + F\omega_1^2 + D\omega_1\omega_3$$

$$- F\omega_2^2 - E\omega_2\omega_3 - \omega_1\delta L_2 + \omega_2\delta L_1 - \frac{d\delta L_3}{dt}. \tag{3.61}$$

Compared with the principal terms of the homogeneous equations, i.e. with (3.30) without the rhs, they have the character of perturbations. Lagrange's method of varying constants can be used to advantage in solving (3.60).

We are thus seeking a solution formally identical with (3.42), but with integration constants L, T and β_0 functions of time,

$$L = L(t), \quad T = T(t), \quad \beta_0 = \beta_0(t) \tag{3.62}$$

and general solutions

$$\omega_j = \omega_j[t; L(t), T(t), \beta_0(t)], \quad j = 1, 2, 3. \tag{3.63}$$

Since

$$\dot{\omega}_j = \frac{\partial \omega_j}{\partial t} + \frac{\partial \omega_j}{\partial L}\dot{L} + \frac{\partial \omega_j}{\partial T}\dot{T} + \frac{\partial \omega_j}{\partial \beta_0}\dot{\beta}_0, \tag{3.64}$$

and in view of the initial equations, (3.60), it must hold that

$$\frac{\partial \omega_1}{\partial t} + \frac{\partial \omega_1}{\partial L}\dot{L} + \frac{\partial \omega_1}{\partial T}\dot{T} + \frac{\partial \omega_1}{\partial \beta_0}\dot{\beta}_0 + \alpha\omega_2\omega_3 = \frac{1}{A_0}f_1(t),$$

$$\frac{\partial \omega_2}{\partial t} + \frac{\partial \omega_2}{\partial L}\dot{L} + \frac{\partial \omega_2}{\partial T}\dot{T} + \frac{\partial \omega_2}{\partial \beta_0}\dot{\beta}_0 + \beta\omega_1\omega_3 = \frac{1}{B_0}f_2(t),$$

$$\frac{\partial \omega_3}{\partial t} + \frac{\partial \omega_3}{\partial L}\dot{L} + \frac{\partial \omega_3}{\partial T}\dot{T} + \frac{\partial \omega_3}{\partial \beta_0}\dot{\beta}_0 + \gamma\omega_1\omega_2 = \frac{1}{C_0}f_3(t). \tag{3.65}$$

This yields three natural conditions, viz.

$$\frac{\partial \omega_1}{\partial L}\dot{L} + \frac{\partial \omega_1}{\partial T}\dot{T} + \frac{\partial \omega_1}{\partial \beta_0}\dot{\beta}_0 = \frac{1}{A_0}f_1(t),$$

$$\frac{\partial \omega_2}{\partial L}\dot{L} + \frac{\partial \omega_2}{\partial T}\dot{T} + \frac{\partial \omega_2}{\partial \beta_0}\dot{\beta}_0 = \frac{1}{B_0}f_2(t),$$

$$\frac{\partial \omega_3}{\partial L}\dot{L} + \frac{\partial \omega_3}{\partial T}\dot{T} + \frac{\partial \omega_3}{\partial \beta_0}\dot{\beta}_0 = \frac{1}{C_0}f_3(t). \tag{3.66}$$

By solving system (3.66) we get $\dot{L}$, $\dot{T}$, $\dot{\beta}_0$ as functions of $f_1(t)$, $f_2(t)$ and $f_3(t)$.

In this general case, however, the solution is quite complicated because Jacobi's elliptical functions (3.44), apart from being functions of argument $(\mu t - \beta_0)$, are also functions of modulus k and, consequently, derivatives with respect to k have to be considered in the variations. Therefore, to simplify matters, we shall put $B = A$ and thus $k^2 = 0$. Then

$$\partial \omega_1/\partial L = -L[A(C - A)(CT - L^2)]^{-1/2}\cos(\mu t - \beta_0)$$
$$\qquad - tLA^{-3/2}C^{-1/2}(CT - L^2)^{1/2}(L^2 - AT)^{-1/2}\sin(\mu t - \beta_0),$$

$$\partial \omega_1/\partial T = \tfrac{1}{2}C[A(C - A)(CT - L^2)]^{-1/2}\cos(\mu t - \beta_0)$$
$$\qquad + \tfrac{1}{2}t(CT - L^2)^{1/2}[AC(L^2 - AT)]^{-1/2}\sin(\mu t - \beta_0),$$

$$\partial \omega_1/\partial \beta_0 = [A(C - A)]^{-1/2}(CT - L^2)^{1/2}\sin(\mu t - \beta_0),$$

$$\partial \omega_2/\partial L = -L[A(C - A)(CT - L^2)]^{-1/2}\sin(\mu t - \beta_0)$$
$$\qquad + tLA^{-3/2}C^{-1/2}(CT - L^2)^{1/2}(L^2 - AT)^{-1/2}\cos(\mu t - \beta_0),$$

$$\partial \omega_2/\partial T = \tfrac{1}{2}C[A(C - A)(CT - L^2)]^{-1/2}\sin(\mu t - \beta_0)$$
$$\qquad - \tfrac{1}{2}t(CT - L^2)^{1/2}[(AC)(L^2 - AT)]^{-1/2}\cos(\mu t - \beta_0),$$

$$\partial \omega_2/\partial \beta_0 = -[A(C - A)]^{-1/2}(CT - L^2)^{1/2}\cos(\mu t - \beta_0),$$

$$\partial \omega_3/\partial L = L[C(C - A)(L^2 - AT)]^{-1/2},$$

$$\partial \omega_3/\partial T = -\tfrac{1}{2}A[C(C - A)(L^2 - AT)]^{-1/2},$$

$$\partial \omega_3/\partial \beta_0 = 0; \tag{3.67}$$

$\mu = A^{-1}C^{-1/2}[(C - A)(L^2 - AT)]^{1/2} = k_0$, provided that $B = A$ (3.50). Using this simplification we may calculate determinant D of system (3.66),

$$D = -\frac{\partial \omega_3}{\partial L}\left(\frac{\partial \omega_1}{\partial \beta_0}\frac{\partial \omega_2}{\partial T} - \frac{\partial \omega_1}{\partial T}\frac{\partial \omega_2}{\partial \beta_0}\right) - \frac{\partial \omega_3}{\partial T}\left(\frac{\partial \omega_1}{\partial L}\frac{\partial \omega_2}{\partial \beta_0} - \frac{\partial \omega_1}{\partial \beta_0}\frac{\partial \omega_2}{\partial L}\right), \tag{3.68}$$

for $t = 0$:

$$D = -\tfrac{1}{2}LA^{-1}[C(C - A)(L^2 - AT)]^{-1/2} < 0. \tag{3.69}$$

If $D \neq 0$ also in the general case, $\dot{L}$, $\dot{T}$, $\dot{\beta}_0$ can be determined as functions of the rhs, $f_1(t)$, $f_2(t)$ and $f_3(t)$, i.e. of parameters generated by the transfer of terrestrial masses, by solving (3.66). However, since time derivatives $\dot{L}$, $\dot{T}$, $\dot{\beta}_0$ and, consequently, also their time variations, can be determined by observation, i.e. from the time variation in components ω_1, ω_2, ω_3 of the vector of instantaneous rotation, there is a way of practically determining at least some of the parameters in $f_1(t)$, $f_2(t)$ and $f_3(t)$ from experimental data.

In a first approximation one could assume that $\omega_3 \gg \omega_1$, $\omega_3 \gg \omega_2$, put $\omega_3 = \omega_0$, and only retain the principal terms,

$$f_1(t) = \omega_0 \dot{E} - D\omega_0^2 + \omega_0 \delta L_2 - \frac{\mathrm{d}\delta L_1}{\mathrm{d}t},$$

$$f_2(t) = \omega_0 \dot{D} + E\omega_0^2 - \omega_0 \delta L_1 - \frac{\mathrm{d}\delta L_2}{\mathrm{d}t},$$

$$f_3(t) = -\omega_0 \dot{C} - \frac{\mathrm{d}\delta L_3}{\mathrm{d}t}; \tag{3.70}$$

then resolve functions $f_j(t)$ into a Fourier series,

$$f_j(t) = \sum_{m=1}^{\infty} (a_j^{(m)} \sin \alpha_j^{(m)} t + b_j^{(m)} \cos \beta_j^{(m)} t) \tag{3.71}$$

and seek amplitudes a_j, b_j and frequencies α_j, β_j of the resultant perturbations.

Liouville's equations (3.60) will simplify if they are expressed in the system of Tisserand's mean coordinate axes (Tisserand 1891), defined so that

$$\int_{M_\oplus} u^2 \, \mathrm{d}m = \min. \tag{3.72}$$

and

$$\delta \mathbf{L} = 0.$$

Then ($\omega_0 = \mathrm{const}$, $G_j = 0$)

$$A\dot{\omega}_1 + (C - B)\omega_0\omega_2 - \dot{E}\omega_0 + D\omega_0^2 = 0,$$

$$B\dot{\omega}_2 + (A - C)\omega_0\omega_1 - \dot{D}\omega_0 - E\omega_0^2 = 0,$$

$$C\dot{\omega}_3 + \dot{C}\omega_0 = 0. \tag{3.73}$$

Under this simplification and given $B = A$, the solution can be expressed in the following form (Jeffreys 1959):

$$\omega_1 = c \cos \omega \frac{t - t_0}{T_{CH}},$$

$$\omega_2 = c \sin \omega \frac{t - t_0}{T_{CH}}, \tag{3.74}$$

where c and t_0 are integration constants and T_{CH} is Chandler's period (Chandler 1892), which theoretically comes out as (Larmor 1909; Jeffreys 1959)

$$T_{CH} = \frac{A + \frac{1}{3}k_2\omega^2\bar{a}^5/G}{C - A - \frac{1}{3}k_2\omega^2\bar{a}^5/G},\tag{3.75}$$

where k_2 is the Love number (see Chap. 4) and $\bar{a}$ the mean equatorial radius-vector of the geoid.

We shall modify this expression to contain constants which have been determined reliably, namely GM instead of constant G which is known but only with relatively low accuracy. Now [A is replaced by $\frac{1}{2}(A + B)$],

$$\begin{aligned}
T_{CH} &= \frac{1}{2}\frac{A + B}{C - (A + B)/2}\frac{1 + \frac{1}{3}k_2\dfrac{\omega^2\bar{a}^3}{GM}\dfrac{2M\bar{a}^2}{A + B}}{1 - \frac{1}{3}k_2\dfrac{\omega^2\bar{a}^3}{GM}\dfrac{M\bar{a}^2}{C - (A + B)/2}} \\
&= \frac{1 - H}{H}\frac{1 + \frac{1}{3}k_2qH[J_2^{(0)}(H - 1)]^{-1}}{1 + \frac{1}{3}k_2q/J_2^{(0)}},
\end{aligned}\tag{3.76}$$

$$T_{CH} = T_E\frac{1 - k_2H[k_s(H - 1)]^{-1}}{1 - k_2/k_s} = T_E\frac{k_s + k_2/T_E}{k_s - k_2};$$

$$\omega^2\bar{a}^3/(GM) = q,\tag{3.77}$$

the latter being a dimensionless (Helmert's) parameter in the potential of centrifugal forces which is known with relatively high accuracy:

$$q = (3461.390 \pm 0.002)\times 10^{-6},$$

$$2M\bar{a}^2/(A + B) = H[J_2^{(0)}(H - 1)]^{-1} = (303399 \pm 1)\times 10^{-5},\tag{3.78}$$

$$M\bar{a}^2[C - \tfrac{1}{2}(A + B)]^{-1} = 1/J_2^{(0)} = -(923.6715 \pm 0.0006),\tag{3.79}$$

$$\tfrac{1}{2}(A + B)[C - \tfrac{1}{2}(A + B)]^{-1} = (1 - H)/H = 304.4408 \pm 0.0005 = T_E;\tag{3.80}$$

$$H = [C - \tfrac{1}{2}(A + B)]/C = 0.003\,273\,9567 \pm 2\times 10^{-9}\tag{3.81}$$

is the precession constant (Kinoshita and Souchay 1990) and $T_E = 2\pi/k_0$ is Euler's period [see (3.50)]. Parameter

$$k_s = -3\frac{J_2^{(0)}}{q}\tag{3.82}$$

characterizes the state of the body relative to the ideal state of hydrostatic equilibrium and is referred to as the secular Love number (Sect. 4.9). Numerically

$$k_s = 0.938\,32 \pm 0.000\,01.\tag{3.83}$$

Table 3.1. Mean values of parameters of Chandler's motion

Reference	Period	T_{CH} [day]	Q
Jeffreys (1972)	1899–1967	433.15	60
Yatskiv (1974)	1846–1971	433.54	40–60
Currie (1974)	1900–1973	432.95	36

Using numerical values (3.78)–(3.81), if $k_2 = 0.29$, we get

$$T_{CH} = 441.1 \, , \tag{3.84}$$

and if $k_2 = 0.30$,

$$T_{CH} = 448.0 \, .$$

However, these values differ quite considerably from that observed which, based on the observations of the past period as a whole (roughly from 1900), comes out at

$$T_{CH} = 433.2 \pm 0.2 \, . \tag{3.85}$$

Some of the pertinent values are given in Table 3.1. Chandler's period, determined by observation, fluctuates in the interval of 425–440 days, and its amplitude also varies quite considerably. Its mean value is $\sim 0.15''$, and variable damping is 10–70 years. One may assume, of course, that Chandler's period itself is constant, and that the changes observed are due to correlation with the atmosphere.

The actual period thus differs quite considerably from Euler's ideal period, being about 4 months longer (~ 14 months); Chandler (1892) had already derived this from observations in 1891. He used observations of geographical latitude at 45 stations throughout the world. In the same year Newcomb explained why it was 4 months longer: three-quarters of the time was due to the elastic properties of the Earth and one-quarter was due to the motion of the waters of oceans and seas (Newcomb 1891). At the same time Newcomb proved that this phenomenon could be used in deriving the physical properties of the Earth.

3.5 Polar Motion: Variations in the Angular Velocity of the Earth's Rotation. Numerical Results

The theory presented in the preceding sections by no means reflects the actual motion of the Earth's vector of instantaneous rotation, ω; this motion is considerably more complicated. This is so because the models being considered, for which this theory holds exactly, differ substantially from the actual Earth. In addition, it is necessary to consider the effect of transfer of atmospheric masses which have not been included even in Chandler's model.

We have not yet been able to derive the excitation function comprehensively to reflect sufficiently the whole spectrum of observed variations of vector ω, called Earth polar motion. The nature of this complicated motion is illustrated, for example, in Fig. 3.7a, which schematically depicts the relative motion of the end-point of vector ω relative to the conventional origin in the time interval 1962–1968; this figure has been adopted from the annual report of the International Polar Motion Service (IPMS) (see below) (Yumi 1969). Figure 3.7b illustrates this motion in the interval 1985.0–1991.0. The trajectory of polar motion is thus, generally speaking, an open curve, the angle between vector ω and axis x_3 always being $< 0.3''$. The motion's period fluctuates between 425 and 440 days and its mean value is approximately 433 days. Figure 3.8, adopted from Dickey (1984), shows the observed polar motion (curve a) and the calculated motion generated by atmospheric effects (curve b). The comparison of the two curves proves how important atmospheric effects are for explaining polar motion. The polar motion phenomenon thus contains two periodic quasi-circular components: (a) ~ 14 month (Chandler's) and (b) annual (see below).

Coordinates x, y of the instantaneous rotation pole, P, relative to a particular 'mean' position, P_0, are defined as shown in Fig. 3.9 (the y-axis is positive to the west). If x_j ($j = 1, 2, 3$) are the axes of the Earth's ellipsoid of inertia, P_0 should be a pole of the ellipsoid of inertia whose position, however, has still not been determined accurately. In current practice P_0 is a conventional origin (Conventional International Origin; CIO) essentially defined by a set of conventional latitudes of five stations of the International Latitude Service (ILS) system for epoch 1900.0 (Table 3.2); however, this is only a provisional solution.

The transformation from components ω_j of rotation vector to pole coordinates x, y, adopted internationally, can evidently be achieved by using relation (R is the mean radius of the Earth)

$$x = \overline{P_0 P} \cos \lambda_{\mathrm{w}} = R(\varphi - \varphi_0) \cos \lambda_{\mathrm{w}} ,$$

$$y = \overline{P_0 P} \sin \lambda_{\mathrm{w}} = R(\varphi - \varphi_0) \sin \lambda_{\mathrm{w}} . \tag{3.86}$$

The difference $\varphi - \varphi_0 = \delta\varphi$ in the geographical latitudes of poles P and P_0 is then

$$\delta\varphi = x \cos \lambda_{\mathrm{w}} + y \sin \lambda_{\mathrm{w}} , \tag{3.87}$$

where x and y are angular quantities.

The ILS was founded in 1895 and regular observations with visual zenith telescopes at selected stations (Table 3.2) of approximately the same geographical latitude, $39°08'\,$N, based on an identical programme and the same set of stars, began in 1899. Using the same stars enables one to eliminate errors due to their proper motions, still not known accurately enough. This is the advantage

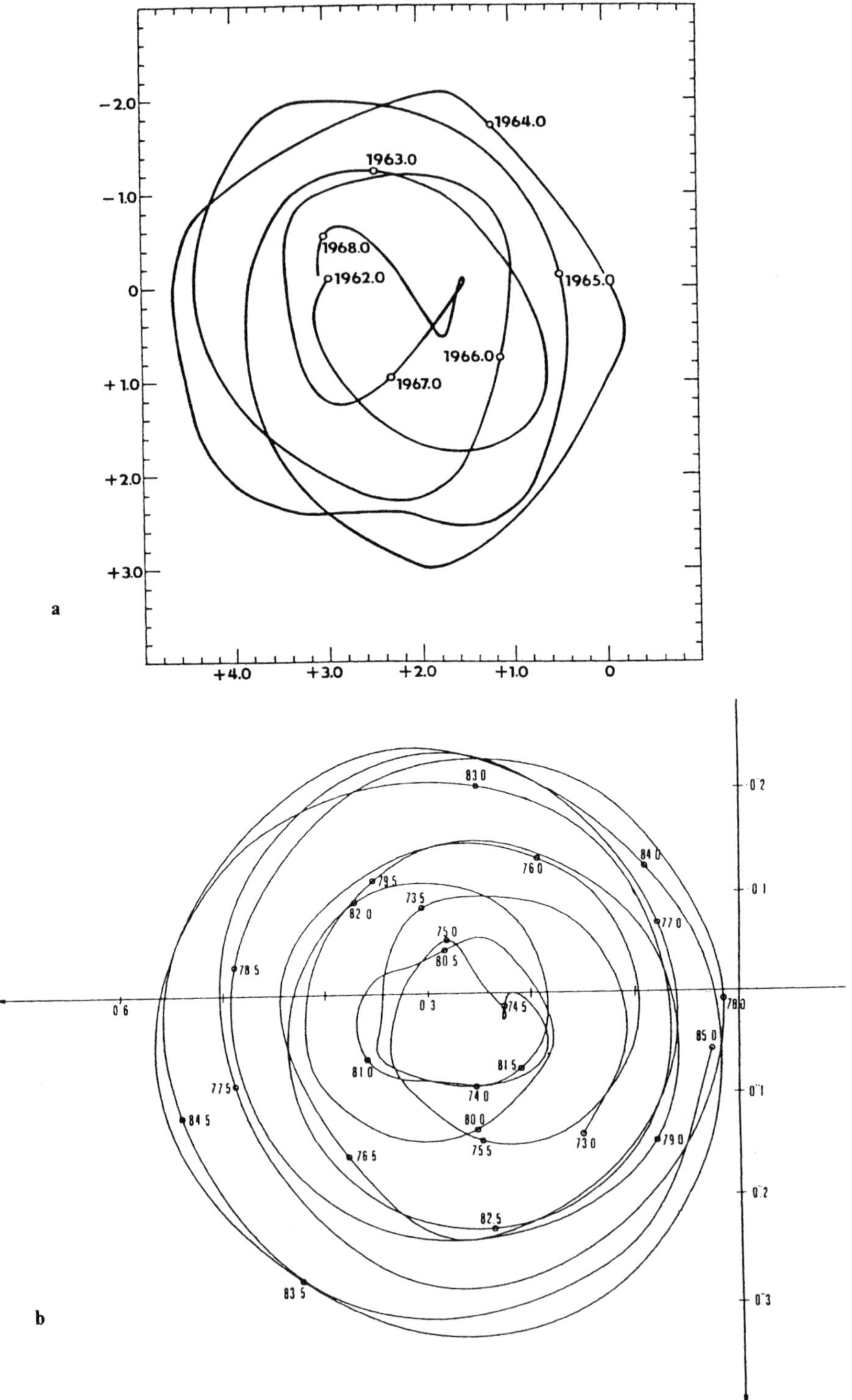
-2.0
-1.0
0
+1.0
+2.0
+3.0
1964.0
1963.0
1968.0
1962.0
1965.0
1966.0
1967.0
+4.0   +3.0   +2.0   +1.0   0
a
83 0
84 0
76 0
79 5
73 5
82 0
77 0
75 0
80 5
78 5
74 5
85 0
78 0
81 5
0 6
0 3
81 0
77 5
74 0
80 0
84 5
73 0
79 0
76 5
75 5
82 5
83 5
0 2
0 1
0"1
0"2
0"3
b

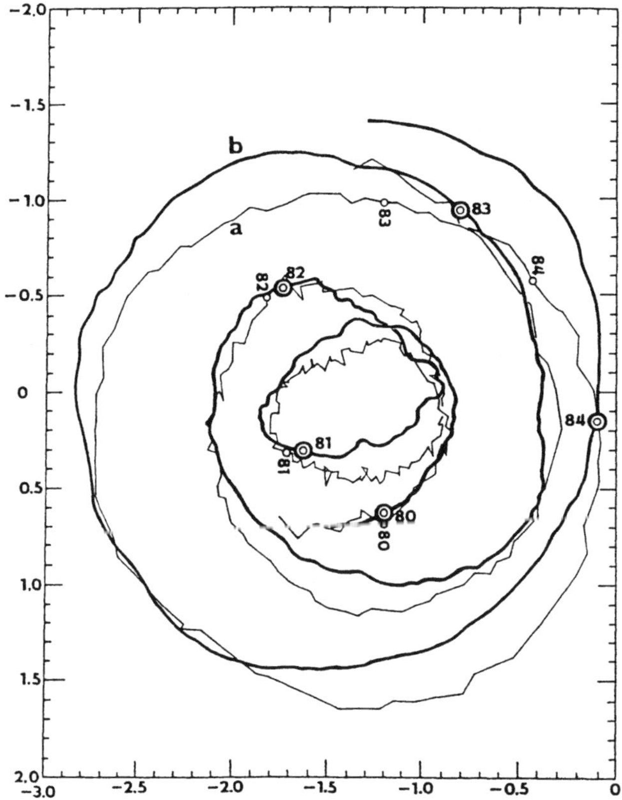

Fig. 3.8. Comparison of observed pole wandering (*curve a*) and of pole wandering calculated from meteorological data (*curve b*). Scales in tenths of a second of arc; the points on the curves are marked with the two last digits of the year. (After Dickey 1984)

and motivation behind the founding of this worldwide system. Nevertheless, its disadvantages are considerable and they were the reason why this system was abandoned as of 1 January 1988. The problem lies in the dynamics of the Earth's crust, especially in the motion of lithospheric plates. This motion, observed in the vicinity of station Ukiah, affects the results of the observations to such an extent that, considering the small set of stations and the current high accurary of astronomical observations, the resultant mean values of the ILS system suffer.

In 1962 the IPMS was founded, which then consisted of 56 observatories in 26 different countries with a total of 89 astronomical instruments of the highest precision, the most important of which are photographic zenith telescopes. This represents a substantially larger set, so that the mean values of the pole

Fig. 3.7. a Earth's pole wandering curve 1962.0–1968.0. Scale in tenths of a second of arc. (After Yumi 1969). **b** Pole wandering curve 1985.0–1991.0. Compiled from IPMS and IERS data

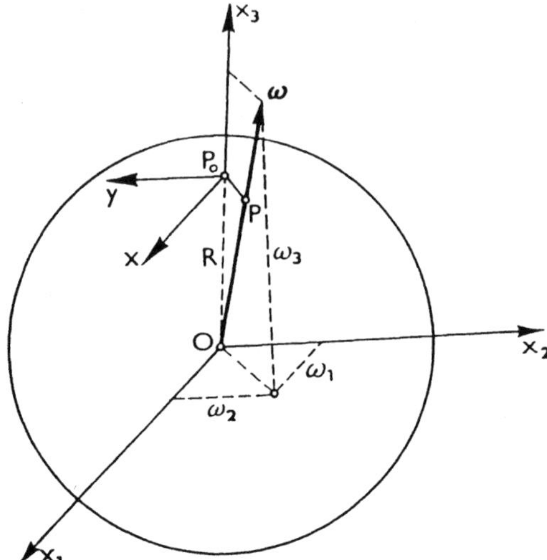

Fig. 3.9. Pole coordinates

Table 3.2. International longitude stations of the ILS astrometric system

i	Station	φ_i (1900.0)
1	Mizusawa	39°08′03.602″
2	Kitab	39°08′01.850″
3	Carloforte	39°08′08.941″
4	Gaithersburg	39°08′13.202″
5	Ukiah	39°08′12.096″

coordinates derived in this system were much less affected by motions of the Earth's crust than were the values of the ILS system. However, the IPMS values were, in general, relatively more affected by the proper motions of the observed stars, which were not known with sufficient accuracy. Nevertheless, the situation should improve substantially in this respect once the accuracy of stars' positions and that of their proper motions are improved by a whole order of magnitude ($\pm 0.002''$ and $0.002''$ per year, respectively) by using space methods, i.e. space telescopes orbiting outside the dense layers of the atmosphere (the HIPPARCOS project: High Precision Parallax Collecting Satellite) (Kovalevsky 1982). However, the IPMS was also abandoned on 1 January 1988 because optical astrometry cannot compete in accuracy with modern techniques, such as satellite laser ranging (SLR), lunar laser ranging (LLR) and very long baseline interferometry (VLBI).

Concurrently with IPMS, pole coordinates were also determined in the international Bureau International de l'Heure (BIH) centre, founded in 1911, originally exclusively for time standardizing on the basis of universal time (UT). After the introduction of atomic time (1955), whose uniform scale is guaranteed physically, electronically, i.e. independently of astronomical phenomena, the BIH's main programme was to determine time variations in UT, i.e. variations in the modulus of vector ω; however, BIH determined the variations of vector ω as a whole, inclusive of its direction. Stations which operated in the IPMS also took part in this programme. The BIH system was also abandoned as of 1 January 1988.

The astronomical observations of the highest accuracy in the systems mentioned, namely with the 16 photographic zenith telescopes and other astronomical instruments at a total of about 80 stations at various locations on the Earth's surface, were used to determine the variations in direction of the vector of instantaneous rotations, ω, with an accuracy of ~ 0.4 m, and variations in modulus ω, expressed in terms of the length of day with an accuracy of $\pm 2 \times 10^{-4}$ s; these were mean values over intervals of 5 days. All this, however, is now history.

Concurrently with classical methods of optical astrometry used to determine the variation in the vector of the Earth's instantaneous rotation, a very promising development of dynamic space methods has arisen in recent years. In the period 1982–1988 these made dramatic progress. The accuracy which can be achieved with them was studied as part of the international MERIT project (Monitor Earth Rotation and Intercompare the Techniques of observation and analysis). The main worldwide campaign took place from 1 September 1983 to 31 October 1984, covering one Chandler period. Three methods are involved:

1. Radio interferometry from VLBI founded on interference radio-telescopic ranging of distant space radio sources (quasars).
2. Laser ranging of geodynamic artificial Earth satellites (SLR) from a network of satellite stations distributed over the whole Earth. The LAGEOS (Laser Geodynamics) satellite of relatively large mass (411 kg; a sphere 60 cm in diameter with 426 corner reflectors on its surface) launched for this purpose in 1976 to orbit outside the dense layers of the atmosphere (perigee height 5800 km, $i = 109.9°$, $e = 0.004$) is particularly important in this respect. Its orbit is being determined with an accuracy of a few centimeters. Apart from the LAGEOS-type satellites (LAGEOS-1, LAGEOS-2, LAGEOS-3) there are a number of other satellites with laser reflectors (as of 1990): STARLETTE, AJISAI, STELLA, ETALON-1, ETALON-2, SALT, ERS-1.
3. Laser ranging of corner laser reflectors on the Moon's surface (LLR), as yet from a relatively small number of observatories. Five laser reflectors have been placed on the Moon's surface for this purpose, and two are being used (at the Apollo 15 landing site and on the Lunochod 2).

These three modern methods constitute the basis of the new international system for determining the variations in the Earth's rotation vector, International Earth Rotation Service (IERS) which began operating on 1 January 1988.

At the beginning of the MERIT project compaign, Doppler ranging of artificial Earth satellites was also being contemplated; however, in the course of the campaign it was found that the accuracy of this method was so low that this satellite technique was abandoned in the IERS. The real accuracy of determining the direction of the rotation vector (polar motion) using the methods mentioned is $\pm 0.001''$. This is an improvement by a whole order of magnitude over the accuracy provided by the IPMS and the BIH system. The new international service IERS has its centre in Paris.

IERS has completely replaced the earlier IPMS and the Earth Rotation Section of the BIH. The BIH Time Section now operates within the International Bureau of Measures and Weights, BIPM (Bureau International des Poids et Measures). The IERS cooperates with BIPM in producing the Coordinate Universal Time (UTC) time scale, and with the International Association of Geodesy in monitoring polar motion due to the transfer of atmospheric masses, which forms its substantial component.

IERS has the following basic centres: Central Bureau – Observatoire de Paris, France; Rapid Service Bureau – US Naval Observatory, Washington, United States; Atmospheric Data Bureau – National Meteorological Center, Washington, United States; VLBI Coordinating Centre – National Geodetic Survey, Rockville, United States; Coordinating Centre for Satellite Laser Ranging (SLR) – Center for Space Research, University of Texas at Austin, United States; Coordinating Centre for Lunar Laser Ranging (LLR) – Centre d'Etudes et de Recherches Géodynamiques et Astronomiques (CERGA), Grasse, France; and Coordinating Center for Global Positioning System (GPS) – Jet Propulsion Laboratory, Austin, Texas, United States. Approximately 50 observatories, forming five specialized worldwide networks (as of 1988), participate in the IERS: three for VLBI and one each for SLR and LLR.

It is expected that within the next few years the variations in the vector of the Earth's instantaneous rotation will be determined more accurately and in time intervals equal to fractions of a day. In fact, by the year 2000 we should be able to determine the position of the Earth's poles with an accuracy of a few centimetres at 6-h intervals. This will enable us to interpret this phenomenon in terms of geophysics, i.e. with regard to the dynamics of the Earth's internal structure.

The disagreement between quantities (3.84) and (3.85) alone is evidence that the Earth is not perfectly elastic. If the Love number k_2 is expressed using (3.77), i.e.

$$k_2 = \frac{T_{CH} - (1 - H)/H}{T_{CH} + 1} k_s = 0.278 \pm 0.002 , \qquad (3.88)$$

the value we get differs considerably from the value derived from tide and tilt measurements (~ 0.30). The disagreement between (3.84) and (3.85) will diminish considerably if Chandler's motion is solved for a model of the Earth consisting of a liquid core and elastic mantle.

Assuming that core J and mantle P (together with the Earth's crust) rotate separately, and that the interaction of both components, reflected by friction, is proportional to the difference between the angular velocities of rotation of the core and mantle, the equations of motion can be simplified to read

$$\dot{\mathbf{L}}_P + [\boldsymbol{\omega} \times \mathbf{L}_P] = \varkappa(\boldsymbol{\omega}_J - \boldsymbol{\omega}_P)\,,$$

$$\dot{\mathbf{L}}_J + [\boldsymbol{\omega} \times \mathbf{L}_J] = \varkappa(\boldsymbol{\omega}_J - \boldsymbol{\omega}_P)\,; \tag{3.89}$$

$\mathbf{L}_P$ and $\boldsymbol{\omega}_P$ stand for the vectors of the angular momentum and of the rotation of the mantle, and $\mathbf{L}_J$ and $\boldsymbol{\omega}_J$ for the vectors of the angular momentum and rotation of the core:

$$\mathbf{L}_P = \mathbf{I}_P \cdot \boldsymbol{\omega}_P\,,$$

$$\mathbf{L}_J = \mathbf{I}_J \cdot \boldsymbol{\omega}_J\,, \tag{3.90}$$

where $\mathbf{I}_P$ and $\mathbf{I}_J$ are tensors of inertia of the mantle and core, respectively, and $\varkappa$ is a constant of proportionality.

The solution is a periodic damped motion described by equations of the type (Jeffreys 1949; Bondi and Gold 1956)

$$\omega_1 = k_\beta e^{-\lambda t} \cos(\beta t + \gamma_0)\,,$$

$$\omega_2 = k_\beta e^{-\lambda t} \sin(\beta t + \gamma_0)\,; \tag{3.91}$$

parameters β and λ can be determined from the spectrum of the observed motion of vector $\boldsymbol{\omega}$.

The relation (3.77) between Chandler's (3.85) and Euler's (3.80) period can also be expressed as

$$T_{CH} = T_E(1 + \bar{\mu})/\bar{\mu}\,, \tag{3.92}$$

where $\bar{\mu}$ is a dimensionless coefficient characterizing the rigidity of the Earth and proportional to the shear modulus μ (Table 2.13). For a perfectly rigid body $\mu \to \infty$. Since, numerically,

$$T_{CH} = (1.448 \pm 0.001)T_E\,, \tag{3.93}$$

theoretically, coefficient $\bar{\mu}$ comes out as

$$\bar{\mu} = 2.228 \pm 0.015\,; \tag{3.94}$$

its accuracy is constrained by the error in k_2. This value yields the following shear modulus:

$$\mu = (2/19)\,W_0 \bar{\sigma} \bar{\mu} = 8.10 \times 10^{10}\,\mathrm{N\,m^{-2}}\,,$$

$(W_0 = 62\,636\,857\,\mathrm{m^2\,s^{-2}}, \bar{\sigma} = 5515\,\mathrm{kg\,m^{-3}})$.

By comparing (3.92) and (3.77) we get

$$\bar{\mu} = \frac{k_s - k_2}{k_2}\left(1 + \frac{1}{T_E}\right)^{-1} \tag{3.95}$$

or

$$\frac{k_2}{k_s} = \frac{1}{1 + \bar{\mu} + \bar{\mu} T_E^{-1}} . \tag{3.96}$$

Chandler's period, correlated with the effect of the atmosphere, however, is not constant but varies within the interval of 1.10–1.25 years (Table 3.1). Its amplitude also varies considerably. This involves a complicated phenomenon of energy transfers between excitation and dissipation processes.

It may be assumed that Chandler's polar motion is generated by random and irregular phenomena, the relevant dissipation period being between 15 and 25 years (maximum 70 years). The dissipation can be characterized by the dimensionless factor (Lambeck 1980)

$$Q = \pi \frac{f_0}{\tilde{\alpha}} , \tag{3.97}$$

where f_0 is the frequency of Chandler's motion and $\tilde{\alpha}$ the dissipation factor ($1/\tilde{\alpha}$ is the dissipation period). The value of factor Q comes out between 40 and 60 (Table 3.1). This value is apparently too low to be able to explain the dissipation as due to the inelasticity of the Earth's mantle. The short-period free oscillations of the Earth indicate that $Q > 100$. The difference could perhaps be explained by the dissipation of the energy of Chandler's motion in oceans and seas.

The reciprocal value of Chandler's frequency comes out as (Yatskiv 1974)

$$f_0^{-1} = 1.187 \pm 0.005 . \tag{3.98}$$

The question as to which geophysical phenomena, within and at the surface of the Earth or in outer space, could excite the observed Chandler's motion remains unanswered.

Apart from Chandler's polar motion (period ~ 433 days) there also exists a forced motion of shorter periods due to the transfer of atmospheric masses. This motion is quite considerable. Dynamic atmospheric phenomena account for about 90% of the annual polar motion (Fig. 3.8). The study of the forced motion of vector ω, due to seasonal transfers of atmospheric masses, indicates that this motion is elliptical and has an annual and semi-annual period. The amplitude of the annual forced component is $\sim 0.09''$; the ellipse along which the forced motion of vector ω takes place relative to the axis of inertia, x_3, is quite flat with semi-axes of $\sim 0.103''$ (3.1 m) and $0.084''$ (2.5 m), the flattening being ~ 0.2. The size of the semi-axes fluctuates over a period of 90 years as follows: 3.4–2.7 m; 2.5–1.8 m. The direction of the semimajor axis varies from 205°E to 145°E.

In the course of the year particles of air, water, snow and ice move between oceans and continents, on the one hand, and between the Northern and Southern hemispheres, on the other hand. For example, in January the weight of the air above the Eurasian continent is about 6×10^{15} kg greater than in July. About 4×10^{15} kg of atmospheric masses move from the Northern to the Southern Hemisphere between January and July. During the winter snow accumulates at the north of the Eurasian continent and North America. In the spring the snow melts and the water runs into the oceans and seas. All these transfers of masses change the Earth's tensor of inertia and, consequently, the vector of the Earth's rotation varies too.

The component with the semi-annual period displays a relatively small amplitude, $\sim 0.01''$, and, strictly speaking, its origin still lacks a precise explanation. Zonal tidal deformations of the Earth's crust due to the Sun, and which also display a semi-annual period (Chap. 4), are apparently also involved.

Monthly and fortnightly variations with theoretical amplitudes of $\sim 0.001''$ and approximately diurnal variations with amplitudes $\leq 0.02''$ are assumed.

Polar motion has also been observed to display irregular 'jumps' in its amplitude variations; examples occurred in 1907, 1942, 1948, 1960, 1967 and 1980 (Guinot 1982). Their origin remains unexplained. According to Rochester and Smylie (1965) the electromagnetic forces at the core-mantle boundary are at least several orders of magnitude smaller than the forces required to explain it.

The effects of earthquakes, the solar wind and volcanic activity on polar motion remain an open question. Individual earthquakes can cause polar motion of considerable amplitude, but no proof is available that the ensemble of all earthquake phenomena could explain the observed polar motion (Rochester 1973).

The secular term, whose existence has not yet been proved beyond doubt but has been assumed since 1891, is problematic. According to Mikhailov (1971), for example, it amounts to 0.102 m/year along meridian 76°W; according to Markowitz (1960) it is 0.10 m/year in the direction towards Greenland (60°W) and the western shores of Australia; and Vondrák (1985) claims 0.099 m/year along meridian 78.2°W (Table 3.3). However, there are studies, e.g. by the Kiev astrometric school led by Fedorov and Yatskiv, indicating that no conclusion may as yet be drawn on this subject (Fedorov and Rasulov 1981). If the problem is being solved using observations from only five ILS stations observing the same stars, the results have insufficient statistical weight and, besides, they are apparently affected by variations of the verticals and deformations of the Earth's crust in general at the locations of these five ILS stations. If the observations of all IPMS and BIH stations are employed, the results are affected by errors in the proper motions of the stars and by the systematic errors in stellar catalogues, which may be of the same magnitude as the phenomenon in question. Even a catalogue as accurate as the FK4, with more than 1500 stars of welldefined positions, has systematic errors of the order of 0.1''. Not until the accuracy of the positions is improved by means of the space methods mentioned, combined with new satellite lunar and radio astronomic methods, will it be possible to solve reliably the problem of secular polar motion. If proved, it should be interpreted as the consequence of lithospheric motion rather than the motion of the axis of rotation, because it has no part in the theory of rotational motion. This motion cannot be explained by external perturbing forces. The moment of inertia of the Earth's crust relative to axis x_3 is two orders of magnitude smaller than the principal moment C of the Earth as a whole, and it is more probable that it is

Table 3.3. Rate and direction of secular polar motion

Author(s)	Rate of secular motion ($''\text{year}^{-1}$)	Direction of motion ($°W$)	Observation period
Lambert	0.0066	83	1900–1918
Kimura	0.0058	67	1900–1924
Wanach	0.0047	42	1900–1925
Wanach and Mahnkopf	0.0063	84	1900–1911
Wanach and Mahnkopf	0.0051	62	1900–1923
Hattori	0.0045	73	1900–1939
Hattori	0.0036	65.2	1900–1947
Orlov	0.0042	69	1900–1950
Sekiguchi	0.0050	60	1900–1950
Markowitz	0.0032	60	1900–1959
Proverbio et al.	0.0029	65.6	1900–1962
Proverbio et al.	0.0028	60.7	1900–1962
Yumi and Wako	0.0023	77	1900–1965
Mikhailov	0.0043	88	1900–1965
Yumi and Wako	0.0027	72.9	1933–1966
Stoyko	0.0032	69.8	1890–1966
Proverbio and Quesada	0.0031	69.6	1900–1969
Mikhailov	0.0033	76	1900–1969
Dickman	0.0035	80.1	1900–1978
Pejović	0.0036	80.5	1900–1978
Vondrák	0.0033	78.2	1900–1984

only the lithosphere 'sliding' over the asthenosphere in the opposite direction. The principal moment of inertia of the Earth as a whole relative to axis x_3 is $\sim 8.04 \times 10^{37}\ \text{kg m}^2$; the moment of inertia of the Earth's crust is only $\sim 8 \times 10^{35}\ \text{kg m}^2$. According to Mikhailov (1970) this type of motion can be explained as due to the systematic effect of the horizontal components of the 'anomalous centrifugal forces' acting on parts of continents with large sea-level heights, i.e. the mountain range of Tibet whose effect is quite dominant. According to Migal' (Migal' and Markovich 1977) vertical motions of the Earth's crust also come to bear.

As regards the time variations of the modulus of vector $\boldsymbol{\omega}$, i.e. the changes in the Earth's angular velocity of rotation, $\delta\omega$, the existence of the secular term has been proved (Lambeck 1977; Pariiskii 1978) and amounts to (cy, hereinafter denotes century)

$$\frac{\dot{\omega}}{\omega} = -(2.4 \pm 0.2) \times 10^{-8}\ \text{cy}^{-1}, \quad \dot{\omega} = -(5.4 \pm 0.5) \times 10^{-22}\ \text{rad s}^{-2}, \quad (3.99)$$

or in the length of the day (rotation period)

$$\dot{T} = -\frac{\dot{\omega}}{\omega} T = (2.1 \pm 0.2)10^{-3}\,\text{s} \qquad (3.100)$$

per century, i.e. about 1 s per 50 000 years.

Figure 3.10 shows the long-term variations in the Earth's angular rotation in the interval 1775–1975, essentially adopted from Stoyko (1970), (see also Sect. 3.6). Between the beginning of the eighteenth century and the middle of the nineteenth century, the Earth's rotation velocity changed relatively little, but observations in this interval were, of course, sporadic. Quite irregular changes have been observed over an interval of 60 to 80 years since the middle of the nineteenth century. In this interval the Earth rotated fastest in around 1870 when the length of the day was roughly 3 ms less than the nominal length (86 400 s). In contrast, the Earth rotated at its lowest rate in around 1910 when the length of the day was 4 ms more. Roughly from 1910 to 1934 the Earth's rotation accelerated, but from then to 1972 its rate of rotation decreased again. Three well-defined minima (1840, 1910, 1970) correlate with geomagnetic activity (Le Mouël et al. 1979).

However, the periodic variations with the annual and semi-annual period (Sidorenkov 1975) (Fig. 3.11) and the short-period variations (Poma and Proverbio 1981) (Fig. 3.12) are much larger (by an order of magnitude). Seasonal variations were discovered in 1936 by Scheibe and Adelsberger, using a quartz-crystal clock, and 1 year later by Stoyko, using a pendulum clock. Of the annual periodic variations with an amplitude of approx. 0.5 ms in the length of the day, 90% can be attributed to seasonal variations of zonal atmospheric circulation.

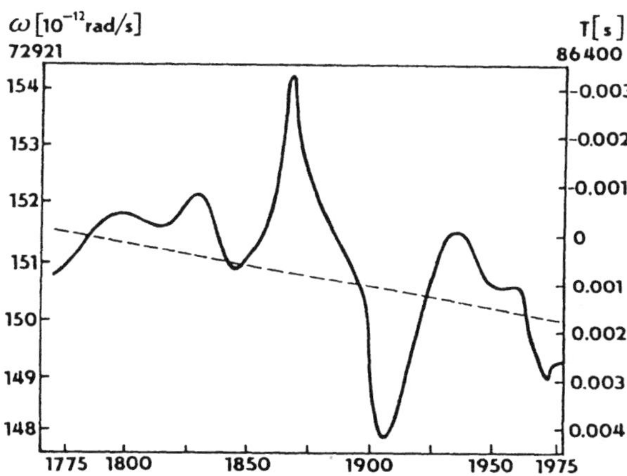

Fig. 3.10. Secular and long-term variations in the Earth's angular velocity of rotation ω (*left-hand scale*) and in the length of day T (*right-hand scale*). *Dashed line* represents secular drift

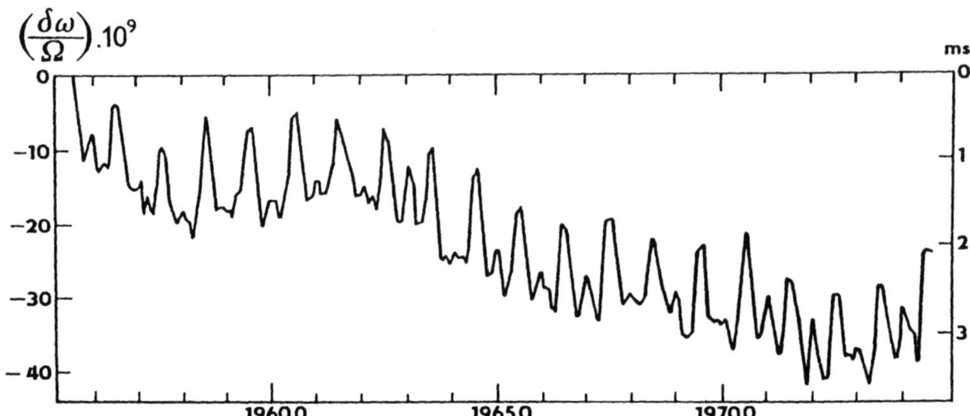

Fig. 3.11. Variations in the Earth's angular velocity of rotation ω (*left-hand scale*) and in the length of day T (*right-hand scale*) with annual and semi-annual periods; $\Omega = 2\pi/86\,400$ s. (After Sidorenkov 1975)

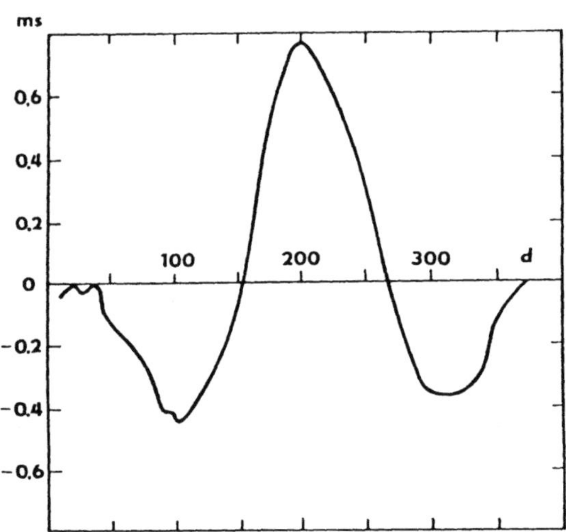

Fig. 3.12. Mean annual curve of seasonal diurnal variations in the length of day calculated for the period 1962–1978. (After Poma and Proverbio 1981)

The origin of the periodic semi-annual variations with an amplitude of approx. 0.3 ms is partly due to atmospheric variations and partly to the tidal effects of the Sun (tidal variations in the principal moment of inertia, C). The Earth always rotates more slowly in the first half of the year (the length of the day in April is as much as 0.5 ms longer) and faster in the second half (in July the length of the day is as much as 0.8 ms shorter). The secondary rotation velocity minimum occurs in November and the secondary maximum at the end of January. In other words, the angular velocity of the Earth's rotation is maximal in July and August and minimal in March. The length of the day, T, varies within an interval of about 1 ms in the course of the year:

$$T = 86\,400^{s} + 0^{s}.0004 \cos(nt + \varphi_1) + 0^{s}.0003 \cos(2nt + \varphi_2)\,, \tag{3.101}$$

where n is the mean value of the angular orbital velocity (around the Sun) and φ_1 and φ_2 are the phases of the annual and semi-annual harmonic terms; the amplitudes and phases vary with time. The variations are caused by exchange of angular momentum between the atmosphere and the solid Earth, in which the sum of the moments of the Earth plus atmosphere system must remain unchanged.

One has to bear in mind that the atmosphere does not rotate with the same angular velocity as the Earth itself. Satellite observations indicate that the atmosphere rotates faster, the ratio being roughly 71 : 70 (the atmosphere rotates 71 times over the same interval that the Earth rotates 70 times).

Other periodic variations, e.g. the fortnightly (period 13.66 days, amplitude 0.36 ms) and monthly (27.55 days, amplitude 0.19 ms) variations apparently have their origin in the tidal deformations of the Earth. Variations with periods of 1–4 years, however, are mainly caused by atmospheric circulation, while those with a period of roughly 10 years are only partly due to this circulation (10 to 20%).

Much larger are irregular variations (Table 3.4) whose origin has not been explained yet; it should be sought in the phenomena occurring within the Earth. Their existence was first proved by de Sitter (1927a, b). Sudden jumps over intervals of several weeks may be larger by as much as two orders of magnitude. Jumps in the angular velocity of the Earth's rotation of relative magnitude

$$\frac{\delta\omega}{\omega} = 4 \times 10^{-8} \tag{3.102}$$

over a short interval of 3 years, i.e. about 3.4 ms in the length of the day (first line in Table 3.4), are not rare. Such large irregular changes in the Earth's rotation

Table 3.4. Examples of irregular rapid changes in the angular velocity of rotation of the Earth

Period	$\mathrm{d}\omega/\mathrm{d}t$ [rad s^{-2}]
1897–1900	-3×10^{-20}
1958–1961	2×10^{-21}
1963–1966	-8×10^{-21}
1967–1970	-2×10^{-21}

velocity cannot be explained by phenomena occurring on and close to the Earth's surface, i.e. by atmospheric phenomena, precipitation, melting of icebergs, meteorite falls, etc. The origin of these irregular changes is apparently in the phenomena taking place within the Earth, namely in the changes of the electromagnetic interactions at the core and lower mantle boundary. They are certainly not due to tidal effects.

The modulus of the rotation vector is thus constant only to about 10^{-8}, i.e. in terms of numbers

$$\omega = 7.292\,115 \times 10^{-5}\ \mathrm{rad\,s^{-1}} = 15.041\,081''\mathrm{s}^{-1} = 15.\dot{0}41\,081°\mathrm{h}^{-1}\,; \quad (3.103)$$

no further digits can be given without giving the time to which they would apply. The corresponding value of the sidereal rotation period $T = 2\pi/\omega$ is $T = 86\,164.10\ \mathrm{s}$.

In interpreting variations $\mathrm{d}\omega/\mathrm{d}t$, one has to take into account the theorem of preserving the total angular momentum of the Earth–Moon system, provided, of course, that short-term variations, due to changes in the Earth's moment of inertia caused by the transfer of terrestrial and atmospheric masses, are involved. If the partial angular momentum, defined by Eq. (3.2), can be considered constant over a sufficiently short interval of time, i.e.

$$\mathbf{L} = \mathbf{I} \cdot \boldsymbol{\omega} = \begin{pmatrix} A' & -F & -E \\ -F & B' & -D \\ -E & -D & C' \end{pmatrix} \cdot \boldsymbol{\omega} = \mathrm{const}\,, \qquad (3.104)$$

then

$$\mathbf{I} \cdot \frac{\mathrm{d}\boldsymbol{\omega}}{\mathrm{d}t} = -\boldsymbol{\omega} \cdot \frac{\mathrm{d}\mathbf{I}}{\mathrm{d}t}\,. \qquad (3.105)$$

This relation can be used to calculate the partial variations in $\dot{\omega}$ due to the variations in the Earth's moments of inertia (Sect. 4.4). However, if a more precise solution is required, it is necessary to consider also the effects of the Moon and Sun on $\mathbf{L}$, i.e. to take into account the total moment of the Earth–Moon–Sun system. Assuming that the Earth is a perfectly elastic body, the tidal variations of the Earth's rotation velocity are (Sect. 4.3)

$$\frac{1}{\omega}\frac{\mathrm{d}\omega}{\mathrm{d}t} = 2\frac{\mathrm{d}J_2^{(0)}}{\mathrm{d}t} = 2\frac{M_{\mathbb{D}}}{M_{\oplus}} k_2 \left(\frac{R}{a_0}\right)^2 \frac{\mathrm{d}}{\mathrm{d}t}\left[\left(\frac{R}{\varDelta_{\oplus\mathbb{D}}}\right)^3 \mathrm{P}_2^{(0)}(\sin\delta_{O'})\right]$$
$$+ 2\frac{M_{\odot}}{M_{\oplus}} k_2 \left(\frac{R}{a_0}\right)^2 \frac{\mathrm{d}}{\mathrm{d}t}\left[\left(\frac{R}{\varDelta_{\oplus\odot}}\right)^3 \mathrm{P}_2^{(0)}(\sin\delta_{O''})\right]\,; \qquad (3.106)$$

$M_{\oplus}, M_{\mathbb{D}}, M_{\odot}$ are the total masses, $\varDelta_{\oplus\mathbb{D}} = \overline{OO'}$, $\varDelta_{\oplus\odot} = \overline{OO''}$, O, O', O'' are the centres of mass of the Earth, Moon and Sun respectively, $R = 6\,370\,997$ m is the Earth's mean radius, $a_0 = 6\,378\,140$ m is the mean equatorial radius-vector, $\delta_{O'}$ and $\delta_{O''}$ are the geocentric declination of the Moon's and Sun's centres of mass, and $k_2 = 0.3$ is the Love number (Sect. 4.7). The periodicity of the phenomenon is defined by the periodicity of functions $P_2^{(0)}(\sin \delta_{o'})$, $P_2^{(0)}(\sin \delta_{o''})$, $\Delta_{\oplus \mathrm{D}}$, $\Delta_{\oplus \odot}$. The maximum amplitude of these variations is about 10^{-8}. Formula (3.106) is approximate and holds under the assumption that $\omega C = \mathrm{const}$, which in fact is not satisfied exactly, and that condition (4.69) is satisfied. Jeffreys (1928) was the first to draw attention to the variations of the Earth's angular rotation velocity due to tides.

Changes of atmospheric origin are important. Fig. 3.13 shows variations $d\omega/dt$ expressed in terms of variations ΔT of the length of the day $(86\,400\,\mathrm{s} + \Delta T)$ and of variations of the angular momentum L (in $10^{26}\,\mathrm{kg\,m^2\,s^{-1}}$) due to the transfer of atmospheric masses; the correlation is quite evident.

The origin of the whole spectrum of variations of the Earth's rotation vector has not been established, and the mechanisms responsible for it have not been described to date. Atmospheric effects (transfer of atmospheric masses, wind pressure, atmospheric tides), oceanic effects (water friction against ocean and sea bottoms), interactions at the Earth's core-mantle boundary, the elasticity and viscosity of the Earth's mantle and apparently also the action of solar wind and earthquakes are all substantial. The frequency of earthquakes usually increases with the rotation velocity. This is a geodynamic phenomenon which provides

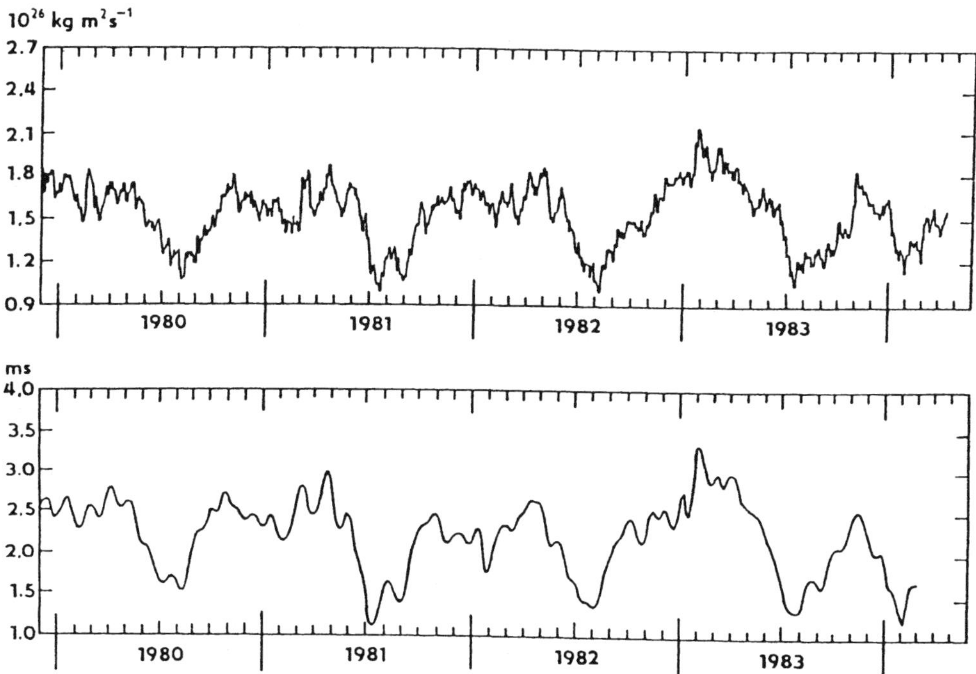

Fig. 3.13. Variations in the length of day T (*bottom*) and in the Earth's angular momentum L (*top*) due to movement of atmospheric masses. (After Dickey 1984)

unique information about the basic physical properties of the Earth, whose geophysical interpretation, however, is still in its infancy due to the complexity of the problem. Substantial progress may be expected from space geodynamic methods which are now being developed on a worldwide scale.

3.6 Dynamics of the Earth's Rotation and the Problem of Defining Time

While but little was known of the variations of the Earth's rotation, the Earth's rotation period was considered to be a constant unit of time. The time variations of the modulus and direction of the Earth's rotation vector, ω, required distinction to be made between the following:

UT0 – 'rotation' time derived directly from astronomical observations of star passages. It is determined from observations made at just one station, assuming that its astronomical longitude is constant, i.e. neglecting the effect of polar motion on longitude. Consequently, at every i-th station time UT0(i) is in general different at the same instant. This is the mean solar time reckoned from midnight on the prime meridian, derived from direct astronomical observation at the i-th station.

UT1 – time UT0 corrected for the changes in the direction of ω (for the fluctuation or position of the instantaneous Earth poles) at the i-th station.

UT2 – time UT1 corrected for the annual and semi-annual, periodic, seasonal variations in modulus ω, i.e. for the seasonal variations in the Earth's rotation.

The different UT1 $-$ UT0 is given by (3.146) and the difference UT2 $-$ UT1 is given by a periodic function of the type

$$\text{UT2} - \text{UT1} = a_1 \sin 2\pi t + b_1 \cos 2\pi t + a_2 \sin 4\pi t + b_2 \cos 4\pi t , \qquad (3.107)$$

where t is expressed in fractions of the year from its beginning, and a_1, b_1, a_2, b_2 are constants determined from observations [e.g. in the interval from 1967 to 1974 their mean values were (BIH) $a_1 = 0.022\,\text{s}, b_1 = -0.012\,\text{s}, a_2 = -0.006\,\text{s}, b_2 = 0.007\,\text{s}$].

However, not even time UT2 passes uniformly, because the corrections introduced in UT1 in no way reflect all the variations of the modulus of vector ω and it does not seem at all probable that they will do so in the near future. The variations of an irregular nature ('jumps' in the angular velocity of the Earth's rotation), which may be many times larger than the periodic variations, and the secular term have made the UT time scale a standard which is no longer satisfactory.

Since 1955 there has existed a physical time scale, reproducible with a high relative accuracy of 10^{-13}, referred to as Atomic Time (AT) and consisting of a set of 236 (in 1990) atomic frequency (caesium) standards (atomic clocks). These are synchronized by means of artificial satellites, the television method or by radio interferometry, which is the most effective method (VLBI; Sect. 3.5) for

establishing the mean atomic time, linked with UT2. Due to the high degree of
stability and high accuracy of the AT scale differences, UT1 — AT or UT2 — AT
can be considered practically absolute, i.e. the variations in the modulus of the
Earth's rotation vector, namely, the variations in the length of the day δT
(Fig. 3.14), can be monitored directly.

The third time scale, quite different in essence from UT and AT, is the
Ephemeris Time (ET) introduced in 1956. It passes uniformly as an independent
variable in Newton's dynamics, and is theoretically defined on the basis of the
annual motion of the Sun (Earth–Moon system). The Earth's orbiting about the
fictitious mean Sun displays a relatively constant period, the Earth's orbit
a constant semi-axis and, therefore, the ET scale is relatively uniform. However,
ET in general varies for different bodies and different theories of their motion,

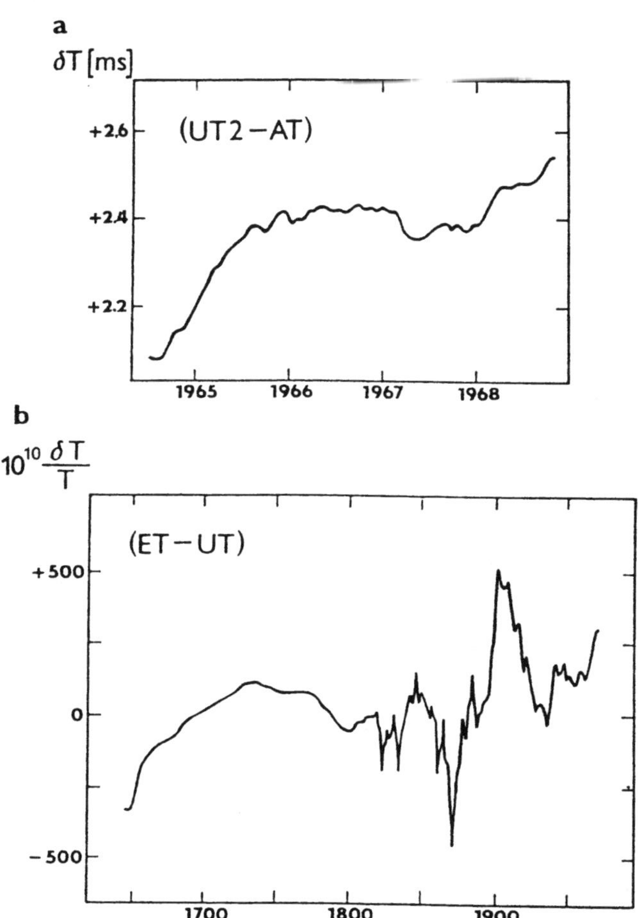

Fig. 3.14. Difference in times **a** UT_2–AT (after Guinot 1970) and **b** ET–UT (after Stoyko 1970)

and this is its principal disadvantage. In 1984 it was replaced by Dynamic Time (DT) which is linked to atomic time by definition: DT = AT + 32.184 s.

The fundamental parameter defining the ET scale is the mean tropical longitude of the Sun measured in the plane of the ecliptic from the mean position of the vernal equinox of the epoch in question, $\bar{T}_0$,

$$L = 279°41'48.04'' + 129\,602\,768.13'' \, \bar{T} + 1.089'' \, \bar{T}^2 \, ; \tag{3.108}$$

$\bar{T}$ is expressed in Julian centuries of 36 525 ephemeris days beginning with epoch $\bar{T}_0 \equiv 1900$, January 0, 12 h ET; at that instant the mean solar longitude $L_0 = 279°41'48.04''$. The instantaneous rate of change of L per ephemeris second is

$$\dot{L} = 0.041\,068\,638\,9744'' + 6.9017'' \times 10^{-10} \, \bar{T}, \tag{3.109}$$

i.e. the mean tropical longitude of the Sun increases at a rate of about $1''$ per 24.349 s.

In ET the mean solar longitude is

$$L(t_{\text{ephem}}) = L_0 + n_\odot t_{\text{ephem}} \, , \tag{3.110}$$

where $n_\odot = 0.985\,647° \, \text{d}^{-1}$ is the mean motion (d hereinafter stands for day). In UT

$$L(t_{\text{UT}}) = L_0 + n_\odot t_{\text{UT}} - \tfrac{1}{2} n_\odot \frac{\dot{\omega}}{\omega} t_{\text{UT}}^2 \, . \tag{3.111}$$

The (positive) coefficient with t_{UT}^2 is usually denoted as $v_\odot$; it is proportional to the mean motion of the Sun (T stands for the Earth's rotation period and T_0 for its value at epoch $\bar{T}_0$)

$$v_\odot = -\tfrac{1}{2} n_\odot \frac{\dot{\omega}}{\omega_0} = \tfrac{1}{2} n_\odot \frac{\dot{T}}{T_0} \tag{3.112}$$

or, with $n_\odot = 1.296\,0276'' \times 10^8 \, \text{cy}^{-1}$,

$$v_\odot = -6.480\,138'' \times 10^7 \frac{\dot{\omega}}{\omega_0} \, . \tag{3.113}$$

However, L is reckoned from the actual ♈ which moves as a result of precession (Sect. 3.9). Consequently,

$$\dot{L}_\Upsilon = -5026.6'' - 2.2'' t_{\text{ephem}} \, \text{cy}^{-1} \, . \tag{3.114}$$

This means that the longitude of the mean Sun is dynamically accelerated by $+1.1'' \, \text{cy}^{-2}$, and

$$L(t_{\text{ephem}}) = L_0 + n_\odot t_{\text{ephem}} + 1.1'' t_{\text{ephem}}^2 \, , \tag{3.115}$$

which corresponds to definition (3.108).

The length of the tropical year is defined as:

$$1 \text{ tropical year} = Y = t_{\text{ephem}}(L + 2\pi) - t_{\text{ephem}}(L) \, ,$$

where L is reckoned from the moving Υ;

$$Y = 360°/n_\odot = 365.25\,\mathrm{d} - \Delta\,,$$

$$n_\odot = [0.985\,626\,283\,367\,556\,468/(1 - \Delta/365.25)]°\,\mathrm{d}^{-1}\,,$$

$$\Delta = 0.008\,\mathrm{d}\ \text{(increases with time)}\,.$$

Due to precession the tropical year is about $50.3 \times 24.349\,\mathrm{s} = 1225\,\mathrm{s} = 20.4\,\mathrm{min}$ shorter than the sidereal year.

The acceleration of motion of Υ is responsible for the diminution of the length of the tropical year by $\dot{Y} = -0.54\,\mathrm{s\,cy}^{-1}$ [or $\dot{\Delta} = +6.3 \times 10^{-6}\,\mathrm{d\,cy}^{-1}$ (Wittmann 1982)].

The ephemeris second is defined as $1/31\,556\,925.9747$ ($86\,400 \times 365.242\,1988 = 31\,556\,925.9747$) of the tropical year at epoch 1900, January 0.5 UT (12 h ET), or as $1/86\,400$ of the mean day for the epoch mentioned. The complete independence of ET on the Earth's rotation has made it possible to derive the variations in the length of the day over a time interval longer than three centuries (Fig. 3.14).

As regards the relation between ET and AT, no sign of these scales diverging has been observed to date. In fact, observations of the Moon with special cameras, designed for the specific purpose of determining ET from lunar motion, were abandoned. Should differences exist at all, they would be caused by deficiencies in the theory of lunar motion or by non-Newtonian effects. Moreover, they would be so small that at least a 20-year interval would be required for them to reach measurable magnitude.

The ephemeris time (ET) is a time scale for Julian ephemeris days (JED), reckoned from the instant of mean noon at the Greenwich meridian for 1 January 4713 B.C.; the Julian century has $36\,525$ ephemeris days. The last so-called standard epoch J1950 is defined in the Julian system by the position of the mean equator and mean vernal equinox (differing from the actual positions only by a small correction for nutation) at instant January 1.0 ET = JED $2\,433\,282.5$. The new fundamental stellar catalogue FK5 has been compiled for standard epoch J2000, at which the position of the mean equator and mean vernal equinox is referred to instant January 1.5 ET = JED $2\,451\,545.0$.

Difference ET–UT contains long-term variations apparently generated by interactions at the core-mantle boundary. In general

$$\text{ET} = \text{UT2} + \Delta T;\tag{3.116}$$

correction ΔT is determined from the orbit dynamics of bodies of the Solar System, practically from those of the Sun and Moon. The diurnal variation $\delta\Delta T$ in seconds is approximately

$$\delta\Delta T = 0.001\,980 + 0.001\,640\,T,\tag{3.117}$$

where T is reckoned in Julian centuries (of $36\,525$ mean days) from epoch 1900.0. Figure 3.15 shows a diagram of ΔT in the interval from 1620 to 1978 relative to epoch 1900.0. Between 1903 and 1980 the difference reached $\sim 50\,\mathrm{s}$.

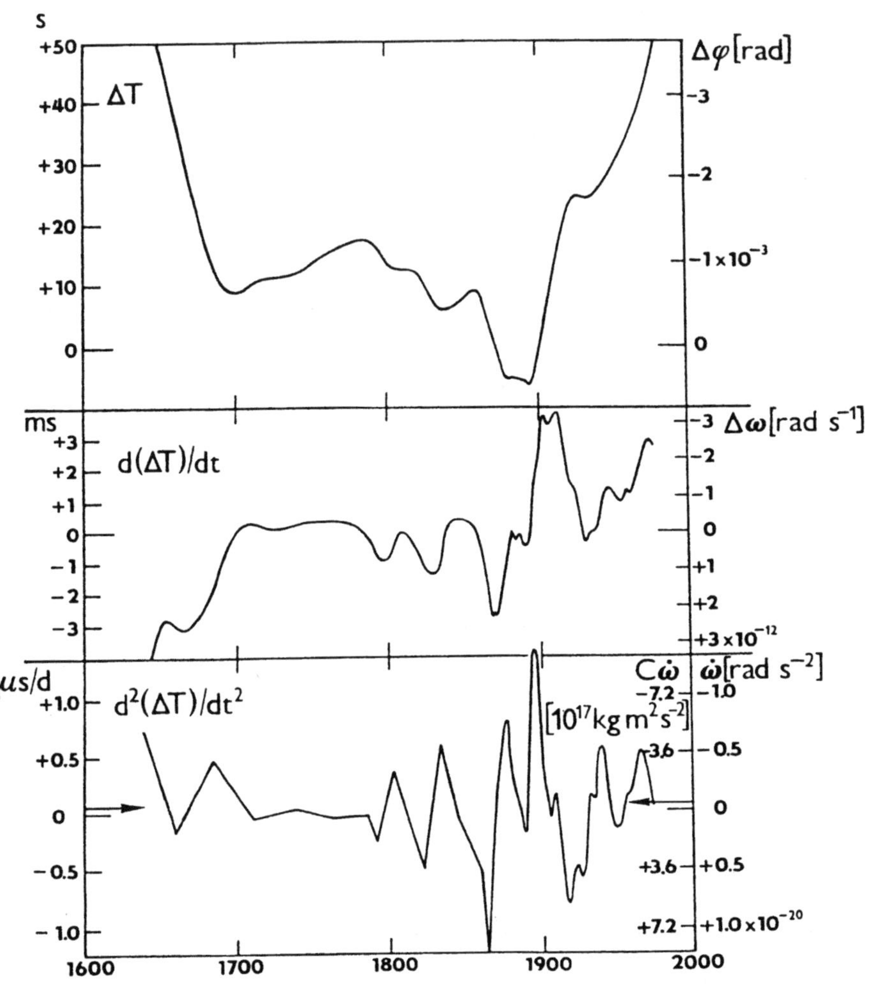

Fig. 3.15. Difference ΔT and its first and second derivatives since the middle of the seventeenth century (after Morrison and Stephenson 1981). *Top* ΔT (s), $\Delta\phi$ (rad); *middle* d(ΔT)/dt (ms), $\Delta\omega$ (rad/s); *bottom* d^2(ΔT)/dt^2 (μs/day), $C\dot{\omega}$(10^{17} kg m^2/s^2), $\dot{\omega}$ (rad/s^2). The horizontal time scale is in years. The *arrows* in the bottom diagram indicate the level of the tidal moment

In Morrison and Stephenson (1981) the main source of information was the catalogues of observed stellar eclipses by the Moon and lunar and solar eclipses. The ΔT-scale has been chosen so that difference ΔT would be equal to the reference value, $\Delta T = +47.52$ s, at epoch 1977, January 1, 0 h AT. The same information was used to plot the first derivative of difference ΔT in Fig. 3.15, i.e. dΔT/dt, which expresses the variations in the length of day and in angular rotation velocity ω (as in Figs. 3.10 and 3.14). The bottom part of Fig. 3.15 depicts the second derivative, d$^2\Delta T$/dt^2, i.e. the first derivative of the angular rotation velocity ω.

Quantity ΔT is caused by the variations of $d\omega/dt$, mostly due to the secular term (3.99). We shall denote the secular term in ΔT by $\overline{\Delta T}$; it is caused by the linear change of ω with time, i.e.

$$\omega(t) = \omega(t_0) + \frac{d\omega}{dt}(t - t_0).$$
(3.118)

Let us calculate difference $\overline{\Delta T}$ from epoch $t = t_0 = 0$:

$$\overline{\Delta T} = -\int_0^t \frac{d\omega}{dt} t \, dt = -\frac{1}{2}\frac{d\omega}{dt}t^2 \quad \text{(radians)}$$
(3.119)

or

$$\overline{\Delta T} = -\frac{1}{2}\frac{1}{\omega}\frac{d\omega}{dt}t^2 \quad \text{(units of time)}.$$
(3.120)

The observed value of the secular term is (3.99)

$$\frac{d\omega}{dt} = -(5.4 \pm 0.5) \times 10^{-22} \, \text{rad s}^{-2} = -(1.11 \pm 0.1)'' \times 10^{-16} \text{s}^{-2}$$

$$= -(5.4 \pm 0.5) \times 10^{-3} \, \text{rad cy}^{-2} = -(1.11 \pm 0.1)'' \times 10^3 \text{cy}^{-2}, \quad (3.121)$$

i.e. the length of day will increase by about 2 s every 100 000 years.

Using this value and expressing $t = T$ in Julian centuries of 36 525 days,

$$\overline{\Delta T} = 2.7 \times 10^{-3} \, T^2 \, \text{rad} = 37 \, T^2 \text{s}.$$
(3.122)

The deceleration of the Earth's rotation during one century thus causes a difference of roughly $\Delta T = 0.5$ min; in 1000 years the difference will already be about 1 h, in 2000 years it will be about 4 h, and in 3400 years about 12 h.

Since ET is independent of the Earth's rotation it cannot be used directly to calculate hour angles which are defined by the Earth's rotation. The part of the mediator is, in this case, played by the artificially introduced 'prime ephemeris meridian'. Its position corresponds to the position which the Greenwich meridian would assume if the Earth rotated uniformly with respect to ET. This auxiliary meridian is located $1.002\,738\,\Delta T_0$ [s] east of the Greenwich meridian, with $\Delta T_0 = \text{ET} - \text{UT0}$.

The deceleration of the Earth's rotation affects the determination of positions of celestial bodies (see Sect. 3.7) if the observations are being carried out in UT, which was the case before ET was introduced.

The mean longitude L of an observed object must be correct by the value corresponding to (3.120) and proportional to the mean motion of the body n:

$$\delta L = n\overline{\Delta L} = -\frac{1}{2}n\frac{1}{\omega}\frac{d\omega}{dt}t^2 = \frac{1}{2}\frac{dn}{dt}t^2,$$
(3.123)

$$\frac{dn}{dt} = -n\frac{1}{\omega}\frac{d\omega}{dt}, \quad \frac{1}{n}\frac{dn}{dt} = -\frac{1}{\omega}\frac{d\omega}{dt}.$$
(3.124)

If, for example, the observed object is the Sun,

$$n_\odot = 0.985\,647^\circ\,\mathrm{d}^{-1} = 1.296\,0276'' \times 10^8\,\mathrm{cy}^{-1} \tag{3.125}$$

and

$$\frac{\mathrm{d}n_\odot}{\mathrm{d}t} = 3.0''\,\mathrm{cy}^{-2}\,. \tag{3.126}$$

If the observed object is the Moon,

$$n_\leftmoon = 47\,435''\,\mathrm{d}^{-1} \tag{3.127}$$

and

$$\frac{\mathrm{d}n_\leftmoon}{\mathrm{d}t} = 40.5''\,\mathrm{cy}^{-2}\,. \tag{3.128}$$

The reader is reminded that value (3.128) represents the apparent acceleration of the mean motion of the Moon and has nothing to do with its tidal deceleration (5.18).

Since the observations made on the Earth's surface are linked to its rotation, the 'working time' commonly used should not differ much from the 'rotational terrestrial time' UT1. For this reason UTC, whose scale is defined by the AT system, was introduced on 1 January 1972. The unit is the 'atomic second' which is made to agree with UT1 by repeated translations so that the deviations do not exceed ± 0.7 s. The translation is effected by inserting either a positive or negative second on 31 December or 30 June in the UTC system as required, as soon as the value of difference UTC $-$ UT1 increases.

3.7 Effect of the Deceleration of the Earth's Rotation on the Observed Ephemerides of Orbiting Bodies

Due to tidal friction the angular velocity of the Earth's rotation is not constant. Denoting it ω_0 at time (epoch) $t = 0$,

$$\omega = \omega_0 + \frac{\mathrm{d}\omega}{\mathrm{d}t}\,t\,.$$

The change $\Delta\varphi$ in the angle of proper rotation $\varphi = \omega t$, due to angular acceleration $\dfrac{\mathrm{d}\omega}{\mathrm{d}t}$, is (in radians) (3.119)

$$\Delta\varphi = \frac{1}{2}\frac{\mathrm{d}\omega}{\mathrm{d}t}\,t^2 \tag{3.129}$$

or (in seconds if ω is in rad s^{-1})

$$\frac{\Delta\varphi}{\omega} = \frac{1}{2}\frac{1}{\omega}\frac{d\omega}{dt}t^2 .$$

If $\Delta T = \mathrm{ET} - \mathrm{UT}$ is the difference between the ephemeris and rotation time, difference $\Delta\varphi$ is negative because $d\omega/dt = -(5.4 \pm 0.5)\times 10^{-22}$ rad s$^{-2} < 0$, and [in seconds, (3.122)]

$$\Delta T = -\frac{\Delta\varphi}{\omega} = -\frac{1}{2}\frac{1}{\omega}\frac{d\omega}{dt}t^2 = 37.0\ T^2 , \tag{3.130}$$

if T is time t expressed in Julian centuries;

$$\Delta\varphi = -2.70\times 10^{-3}\ \mathrm{rad}\ T^2 = -557''\ T^2 . \tag{3.131}$$

The apparent difference ΔL in ecliptical longitudes $L = nt$ of observed bodies between their ephemeris position, calculated in rotation time, and actual position (or calculated in ephemeris time) is proportional to their mean motions n,

$$\Delta L = n\Delta T = -\frac{1}{2}\frac{n}{\omega}\frac{d\omega}{dt}t^2 , \tag{3.132}$$

if n is the mean motion of the observed orbiting body, e.g.

$$n_{\mathrm{D}} = 2\pi/(\text{sider. month}) = (2\pi/27.32)\ \mathrm{rad}\ \mathrm{d}^{-1}$$

$$= 2.6617\times 10^{-6}\ \mathrm{rad}\ \mathrm{s}^{-1} ,$$

$$n_{\odot} = 1.9910\times 10^{-7}\ \mathrm{rad}\ \mathrm{s}^{-1} ,$$

$$n_{\varphi} = 3.2366\times 10^{-7}\ \mathrm{rad}\ \mathrm{s}^{-1} ,$$

$$n_{\yen} = 8.2665\times 10^{-7}\ \mathrm{rad}\ \mathrm{s}^{-1} . \tag{3.133}$$

Consequently,

$$\Delta L_{\mathrm{D}} = 20.3''\ T^2 ,$$

$$\Delta L_{\odot} = 1.5''\ T^2 ,$$

$$\Delta L_{\varphi} = 2.5''\ T^2 ,$$

$$\Delta L_{\yen} = 6.3''\ T^2 . \tag{3.134}$$

Since $L = nt$,

$$n = n_0 + \frac{dn}{dt}t ,$$

and also

$$\Delta L = \frac{1}{2}\frac{dn}{dt}t^2 ; \tag{3.135}$$

by comparing (3.135) and (3.132) we arrive at (3.124)

$$-\frac{1}{\omega}\frac{d\omega}{dt} = \frac{1}{n}\frac{dn}{dt} \tag{3.136}$$

or

$$\frac{dn}{dt} = -\frac{n}{\omega}\frac{d\omega}{dt}. \tag{3.137}$$

Equation (3.137) implies that the apparent acceleration of the mean motion of orbiting bodies is proportional to their mean motion. In view of (3.132) the same applies to their apparent displacements in longitude. The apparent accelerations for the same bodies as in (3.134) are

$$\frac{dn_{\leftmoon}}{dt} = 40.5'' \, \mathrm{cy}^{-2}, \tag{3.138}$$

$$\frac{dn_{\odot}}{dt} = 3.0'' \, \mathrm{cy}^{-2}, \tag{3.139}$$

$$\frac{dn_{\female}}{dt} = 4.9'' \, \mathrm{cy}^{-2}, \tag{3.140}$$

$$\frac{dn_{\female}}{dt} = 12.6'' \, \mathrm{cy}^{-2}. \tag{3.141}$$

3.8 Problem of Realization of the Reference Coordinate System in the Earth's Rotation Dynamics

To be able to monitor the variations of the vector of the Earth's rotation requires the practical realization of a coordinate system which is fixed to the Earth but independent of the variations in question. The geocentric system, whose axes x_j are the axes of the central ellipsoid of inertia, is used in Euler's theory. However, the directions of these axes vary with time as a result of the changes of the tensor of inertia.

Transfers of terrestrial and atmospheric masses in general, including deformations of the Earth's crust, preclude an ideal geocentric system from being realized with sufficient accuracy. In these circumstances, in which we do not know whether there are 'fixed points' on the Earth's crust at all, we have to seek an auxiliary solution. First, it is necessary to assume that the position of the Earth's centre of mass O is invariable and that vector $\boldsymbol{\omega}$ passes through it at any time. It is also necessary to define suitably the direction of axis x_3, i.e. direction $\overline{OP_0}$ (Fig. 3.9), and the position of the prime geocentric meridian $(x_1 x_3)$.

Components ω_j of the vector of instantaneous rotation velocity relative to axes x_j of the Earth's ellipsoid of inertia occur in Euler's equations. As already mentioned, their positions have not been determined reliably yet, although satellite methods have, in principle, made this possible. Moreover, the initial conditions are not known, i.e. the pole coordinates (or parameters equivalent thereto) for the given epoch $t = t_0$.

The practical solution to this problem is the introduction of the so-called mean pole, a mean position calculated from the positions of the instantaneous poles over a 'long period'. It was assumed that this empirically determined mean pole would be identical with the pole of the smallest axis of the Earth's ellipsoid of inertia.

However, the analysis of an observation series covering more than 70 years showed that this 'mean' pole position was not constant. This phenomenon has not as yet been explained in terms of dynamics; however, it should be accepted that deformations of the Earth's crust, errors in the catalogue coordinates of stars and atmospheric effects are partly responsible.

The 13th General Assembly of the IAU (International Astronomical Union) and the 14th General Assembly of the IUGG (International Union of Geodesy and Geophysics) dealt with the problem in 1967. The practical solution was to establish conventional geographical latitudes of five international astrometric stations of the ILS system (Table 3.2). The latitudes of all other participating stations of the IPMS system were referred to this conventional system. The direction through the Earth's centre of mass, with adjusted angular distance $90° - \varphi_i$ from the tangents to the verticals running through five defined points of the ILS latitude stations mentioned, defines the position of the 'mean pole' P_0 (3.86); the reference direction, $\overline{OP_0}$, thus defined is referred to as the CIO (Sect. 3.5). In Figs. 3.9 and 3.16 the projection of this position on to an auxiliary sphere is marked P_0; P is the projection of the position of the instantaneous rotation pole. The position of instantaneous pole P relative to the CIO is defined by arc $P_0P = \gamma$ and by angle Γ, reckoned from the prime 'mean' astronomical Greenwich meridian. The position of vertical $\overline{OM}$ (astronomical zenith) of any point (point of observation) on the Earth's surface in the CIO system is defined by coordinates (φ_0, λ_0), and in the system of instantaneous pole P by coordinates (φ, λ). Longitudes λ, λ_0 are positive to the west, as generally accepted in BIH and IPMS. However, since 1984 longitude has been reckoned positive to the east in these centres, as is obligatory in geodesy and astrodynamics.

The difference in times UT1 $-$ UT0 depends on difference $\lambda - \lambda_0 = \delta\lambda$ which, like

$$\delta\varphi = \varphi - \varphi_0 = \gamma \cos(\lambda_0 - \Gamma), \tag{3.142}$$

is small and of the order of amplitude $P_0P = \gamma$, not exceeding 0.3″. The squares of quantities $\delta\varphi$, $\delta\lambda$, may practically be neglected completely. In this approximation

$$\sin \lambda = \sin \lambda_0 + \delta\lambda \cos \lambda_0,$$

$$\cos \lambda = \cos \lambda_0 - \delta\lambda \sin \lambda_0, \tag{3.143}$$

$$\delta\lambda = \lambda - \lambda_0 = \gamma \sin(\lambda_0 - \Gamma) \tan \varphi_0 + \gamma \sin \Gamma \tan \bar{\varphi}_0 \tag{3.144}$$

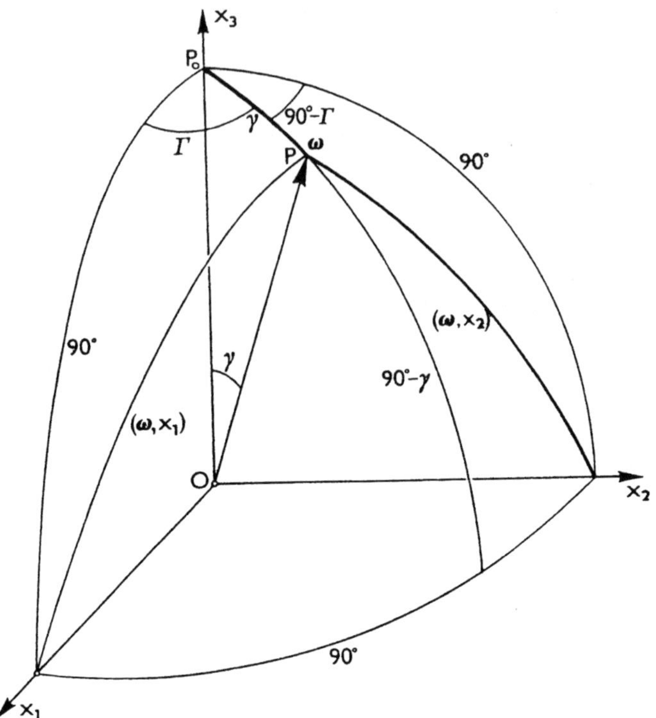

Fig. 3.16. Polar coordinates Γ, γ of the instantaneous pole

and, since

$$\gamma \cos \Gamma = x, \quad \gamma \sin \Gamma = y, \tag{3.145}$$

finally,

$$\lambda - \lambda_0 = (x \sin \lambda_0 - y \cos \lambda) \tan \varphi_0 + y \tan \bar{\varphi}_0 = \text{UT1} - \text{UT0}. \tag{3.146}$$

A more precise derivation, emphasizing terms of the order of γ^2, can be found in Zhongolovich (1971a).

In the international BIH centre, term $y \tan \bar{\varphi}_0$ is practically omitted in deriving the mean value of difference UT1 $-$ UT0 from the observations of all participating stations. The rotation time, therefore, is determined not for the plane of the instantaneous prime astronomical Greenwich meridian but for the plane of the prime geocentric meridian (Zhongolovich 1971b), the two planes being at angle $y \tan \bar{\varphi}_0 = y \tan 51°28.6'$, and the difference in rotation time for the two meridians amounting to $\frac{1}{15} y'' \tan \bar{\varphi}_0 = 0.0838$ s. This problem is analysed in detail in Zhongolovich (1971a, b).

In the near future the accuracy of the reference coordinate system for studying the dynamics of the Earth's rotation vector is sure to be improved, thanks to promising SLR, VLBI and LLR, and to the possibility of determining the positions of axes x_j of the Earth's ellipsoid of inertia and their time variations. The definition of the CIO is based on the system of astronomical zeniths of the five ILS stations (Table 3.2). The definition assumes that there are no relative changes of these selected zeniths (verticals) at the points of the Earth's surface in question. However, this does not agree with reality and only the results of the methods of space geodynamics will apparently be able to improve the accuracy of the solution of this fundamental geodynamic problem, viz. the definition of the reference system. Anyway, effort is being made to refer the observed component of rotation vector $\boldsymbol{\omega}$ to the axes of the ellipsoid of inertia, whose position is not constant (Sect. 5.4) but can be determined by satellite methods.

3.9 Fundamentals of the Dynamics of the Earth's Precession and Nutation

3.9.1 Force Function of the Earth–Moon–Sun System

In Sections 3.3 and 3.4 we dealt with solving Euler's and Liouville's dynamic homogeneous equations, describing the dynamics of rotational motion in which there are no external perturbations, i.e. the resultant moment of external forces is equal to zero ($\mathbf{G} = 0$).

The Earth's rotation dynamics is, however, strongly perturbed by the Moon and Sun; it is therefore necessary to derive the components G_j of the resultant moments of the forces acting due to these perturbing bodies. This could be done directly, but it is more convenient to do so via the force function of the Earth–Moon–Sun system,

$$V_{\oplus \mathbb{)}\odot} = V_{\oplus \mathbb{)}} + V_{\oplus \odot} + V_{\mathbb{)}\odot}, \qquad (3.147)$$

which, of course, has to be expressed in a particular form using the geopotential coefficients of the bodies mentioned. The expressions for the partial force functions of the separate pairs of bodies, $V_{\oplus\mathbb{)}}$, $V_{\oplus\odot}$, $V_{\mathbb{)}\odot}$, are given by (1.3). Component $V_{\mathbb{)}\odot}$ does not come to bear in the Earth's dynamics as its derivatives with respect to Euler's angles ψ, ϑ, φ are equal to zero; consequently, it will be omitted.

The relation between G_j and force function (3.147) in system x_j of the axes of the ellipsoid of inertia can best be obtained from Lagrange's equations of motion of the second kind (1.18), putting $q_1 = \psi$ (precession angle), $q_2 = \vartheta$ (nutation angle), and $q_3 = \varphi$ (angle of proper rotation). Angles ψ, ϑ, φ will be defined in terms of matrices of rotations ψ, ϑ, φ, as expressed generally by Eqs. (3.13) and used in transformation equation (3.14). System $\bar{x}_j$ is defined by the position of the ecliptic at a particular epoch (Figs. 3.17 and 3.3). The Lagrangian

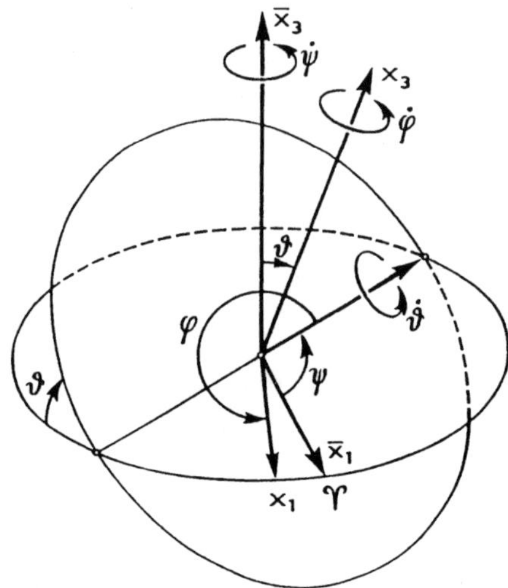

Fig. 3.17. System of the fixed ecliptic $\bar{x}_j$ as referred to the system of axis of the ellipsoid of inertia, x_j

of the problem reads

$$L = \frac{1}{2}T - V_{\oplus \mathbb{D} \odot} ; \tag{3.148}$$

T is double the kinetic energy, $T = I_{ij}\omega_i\omega_j$ (3.33). The equations of motion are now

$$\frac{1}{2}\left[\frac{d}{dt}\frac{\partial T}{\partial \dot{\psi}} - \frac{\partial T}{\partial \psi}\right] = \frac{\partial V_{\oplus \mathbb{D} \odot}}{\partial \psi},$$

$$\frac{1}{2}\left[\frac{d}{dt}\frac{\partial T}{\partial \dot{\vartheta}} - \frac{\partial T}{\partial \vartheta}\right] = \frac{\partial V_{\oplus \mathbb{D} \odot}}{\partial \vartheta},$$

$$\frac{1}{2}\left[\frac{d}{dt}\frac{\partial T}{\partial \dot{\varphi}} - \frac{\partial T}{\partial \varphi}\right] = \frac{\partial V_{\oplus \mathbb{D} \odot}}{\partial \varphi}. \tag{3.149}$$

Euler's kinematic equations (3.22) enable the kinetic energy to be expressed directly in terms of angles ψ, ϑ, φ and of their time derivatives (generalized coordinates and generalized impulses), since

$$\frac{\partial T}{\partial \dot{q}_r} = \sum_{j=1}^{3} \frac{\partial T}{\partial \omega_j}\frac{\partial \omega_j}{\partial \dot{q}_r}, \tag{3.150}$$

$$\frac{\partial T}{\partial q_r} = \sum_{j=1}^{3} \frac{\partial T}{\partial \omega_j}\frac{\partial \omega_j}{\partial q_r}. \tag{3.151}$$

In particular,

$$\frac{1}{2}\frac{\partial T}{\partial \dot\psi} = A\omega_1 \sin\varphi \sin\vartheta + B\omega_2 \cos\varphi \sin\vartheta + C\omega_3 \cos\vartheta,$$

$$\frac{1}{2}\frac{\partial T}{\partial \dot\vartheta} = A\omega_1 \cos\varphi - B\omega_2 \sin\varphi,$$

$$\frac{1}{2}\frac{\partial T}{\partial \dot\varphi} = C\omega_3,$$

$$\frac{1}{2}\frac{\partial T}{\partial \psi} = 0, \tag{3.152}$$

$$\frac{1}{2}\frac{\partial T}{\partial \vartheta} = A\omega_1\dot\psi \sin\varphi \cos\vartheta + B\omega_2\dot\psi \cos\varphi \cos\vartheta - C\omega_3\dot\psi \sin\vartheta,$$

$$\frac{1}{2}\frac{\partial T}{\partial \varphi} = A\omega_1(-\dot\vartheta \sin\varphi + \dot\psi \cos\varphi \sin\vartheta) - B\omega_2(\dot\vartheta \cos\varphi + \dot\psi \sin\varphi \sin\vartheta).$$

$$\tag{3.153}$$

We shall modify the last two equations by substituting the relevant expressions from Euler's kinematic equations (3.25) for $\dot\psi$ and $\dot\vartheta$:

$$\frac{1}{2}\frac{\partial T}{\partial \vartheta} = A\omega_1 \sin\varphi \cot\vartheta(\omega_1 \sin\varphi + \omega_2 \cos\varphi)$$
$$+ B\omega_2 \cos\varphi \cot\vartheta(\omega_1 \sin\varphi + \omega_2 \cos\varphi)$$
$$- C\omega_3(\omega_1 \sin\varphi + \omega_2 \cos\varphi),$$

$$\frac{1}{2}\frac{\partial T}{\partial \varphi} = (A - B)\omega_1\omega_2. \tag{3.154}$$

If we now substitute (3.152) and (3.154) into (3.149) and assume that the Earth is a perfectly rigid body ($A = \text{const}$, $B = \text{const}$, $C = \text{const}$), we get the equations of motion in the following form:

$$A\dot\omega_1 \sin\varphi \sin\vartheta + B\dot\omega_2 \cos\varphi \sin\vartheta + C\dot\omega_3 \cos\vartheta$$
$$+ \dot\vartheta(A\omega_1 \sin\varphi \cos\vartheta + B\omega_2 \cos\varphi \cos\vartheta - C\omega_3 \sin\vartheta)$$
$$+ \dot\varphi(A\omega_1 \cos\varphi \sin\vartheta - B\omega_2 \sin\varphi \sin\vartheta) = \frac{\partial V_{\oplus \mathrm{)}\odot}}{\partial \psi},$$

$$A\dot\omega_1 \cos\varphi - B\dot\omega_2 \sin\varphi - \dot\varphi(A\omega_1 \sin\varphi + B\omega_2 \cos\varphi)$$
$$- A\omega_1^2 \sin^2\varphi \cot\vartheta - B\omega_2^2 \cos^2\varphi \cot\vartheta - (A + B)$$
$$\times \omega_1\omega_2 \sin\varphi \cos\varphi - C\omega_3(\omega_1 \sin\varphi + \omega_2 \cos\varphi) = \frac{\partial V_{\oplus \mathrm{)}\odot}}{\partial \vartheta},$$

$$C\dot\omega_3 + (B - A)\omega_1\omega_2 = \frac{\partial V_{\oplus \mathrm{)}\odot}}{\partial \varphi}. \tag{3.155}$$

The next step is to make the lhs of the first two equations identical with the lhs of Euler's dynamic equations (3.10):

$$A\dot{\omega}_1 + (C - B)\omega_2\omega_3 = \frac{\sin\varphi}{\sin\vartheta}\left(\frac{\partial V_{\oplus\leftmoon\odot}}{\partial\psi} - \cos\vartheta\,\frac{\partial V_{\oplus\leftmoon\odot}}{\partial\varphi}\right) + \cos\varphi\,\frac{\partial V_{\oplus\leftmoon\odot}}{\partial\vartheta},$$

$$B\dot{\omega}_2 + (A - C)\omega_3\omega_1 = \frac{\cos\varphi}{\sin\vartheta}\left(\frac{\partial V_{\oplus\leftmoon\odot}}{\partial\psi} - \cos\vartheta\,\frac{\partial V_{\oplus\leftmoon\odot}}{\partial\varphi}\right) - \sin\varphi\,\frac{\partial V_{\oplus\leftmoon\odot}}{\partial\vartheta},$$

$$C\dot{\omega}_3 + (B - A)\omega_1\omega_2 = \frac{\partial V_{\oplus\leftmoon\odot}}{\partial\varphi}. \tag{3.156}$$

Therefore,

$$G_1 = \frac{\sin\varphi}{\sin\vartheta}\left(\frac{\partial V_{\oplus\leftmoon\odot}}{\partial\psi} - \cos\vartheta\,\frac{\partial V_{\oplus\leftmoon\odot}}{\partial\varphi}\right) + \cos\varphi\,\frac{\partial V_{\oplus\leftmoon\odot}}{\partial\vartheta},$$

$$G_2 = \frac{\cos\varphi}{\sin\vartheta}\left(\frac{\partial V_{\oplus\leftmoon\odot}}{\partial\psi} - \cos\vartheta\,\frac{\partial V_{\oplus\leftmoon\odot}}{\partial\varphi}\right) - \sin\varphi\,\frac{\partial V_{\oplus\leftmoon\odot}}{\partial\vartheta},$$

$$G_3 = \frac{\partial V_{\oplus\leftmoon\odot}}{\partial\varphi}. \tag{3.157}$$

It is now easy to express explicitly the derivatives of the force function with respect to Euler's angles:

$$\frac{\partial V_{\oplus\leftmoon\odot}}{\partial\psi} = G_1\sin\vartheta\sin\varphi + G_2\sin\vartheta\cos\varphi + G_3\cos\vartheta,$$

$$\frac{\partial V_{\oplus\leftmoon\odot}}{\partial\vartheta} = G_1\sin\vartheta\cos\varphi - G_2\sin\vartheta\sin\varphi,$$

$$\frac{\partial V_{\oplus\leftmoon\odot}}{\partial\varphi} = G_3. \tag{3.158}$$

We have simultaneously derived the components of the vector of the resultant moment of acting external forces as functions of force function $V_{\oplus\leftmoon\odot}$ (3.147) and of Euler's angles. These components depend on the density and shape anomalies of the body being studied and the perturbing body.

Components G_j of the moment of external forces can also be expressed in terms of the direction cosines of the direction to the perturbing body and of the derivatives of the force function with respect to them. If $\lambda_\leftmoon, \mu_\leftmoon, \nu_\leftmoon$ and $\lambda_\odot, \mu_\odot, \nu_\odot$ are the appropriate direction cosines in system x_j ($\lambda_\leftmoon = \cos\delta_{O'}, \cos T_{O'}$, $\mu_\leftmoon = -\cos\delta_{O'}, \sin T_{O'}, \nu_\leftmoon = \sin\delta_{O'}$, etc.), then

$$G_1 = \nu_\leftmoon\frac{\partial V_{\oplus\leftmoon\odot}}{\partial\mu_\leftmoon} - \mu_\leftmoon\frac{\partial V_{\oplus\leftmoon\odot}}{\partial\nu_\leftmoon} + \nu_\odot\frac{\partial V_{\oplus\leftmoon\odot}}{\partial\mu_\odot} - \mu_\odot\frac{\partial V_{\oplus\leftmoon\odot}}{\partial\nu_\odot},$$

$$G_2 = \lambda_\leftmoon\frac{\partial V_{\oplus\leftmoon\odot}}{\partial\nu_\leftmoon} - \nu_\leftmoon\frac{\partial V_{\oplus\leftmoon\odot}}{\partial\lambda_\leftmoon} + \lambda_\odot\frac{\partial V_{\oplus\leftmoon\odot}}{\partial\nu_\odot} - \nu_\odot\frac{\partial V_{\oplus\leftmoon\odot}}{\partial\lambda_\odot},$$

$$G_3 = \mu_\leftmoon\frac{\partial V_{\oplus\leftmoon\odot}}{\partial\lambda_\leftmoon} - \lambda_\leftmoon\frac{\partial V_{\oplus\leftmoon\odot}}{\partial\mu_\leftmoon} + \mu_\odot\frac{\partial V_{\oplus\leftmoon\odot}}{\partial\lambda_\odot} - \lambda_\odot\frac{\partial V_{\oplus\leftmoon\odot}}{\partial\mu_\odot}. \tag{3.159}$$

The system (3.156) of three Euler's differential dynamic equations together with the system of Euler's kinematic equations (3.22) describe completely the motion of a rigid body about its fixed centre of mass.

Deriving both partial force functions $V_{\oplus\mathrm{)}}$ and $V_{\oplus\odot}$, which add up to $V_{\oplus\mathrm{)}\odot}$, is formally identical. We shall now consider just the force function $V_{\oplus\mathrm{)}}$ of the Earth–Moon system and only after obtaining its final form will we write out the expression for $V_{\oplus\odot}$.

It is not necessary to know the density distributions of the bodies to determine $V_{\oplus\mathrm{)}}$ in concrete form. In this more general case it is also possible to use geopotential coefficients (1.72), which can be determined from quantities observed in outer space, to express the gravitational potential of a general body no matter how complicated its internal structure and shape. Of course, besides the geopotential coefficients of the Earth $J_n^{(k)}$, $S_n^{(k)}$ and of the Moon $(J_n^{(k)})'$, $(S_n^{(k)})'$, it is also necessary to know the mutual positions of the centres of mass of both bodies and the mutual orientation of the axes of their ellipsoids of inertia.

We shall first express function $\dfrac{1}{r}$ in integral

$$V_{\oplus\mathrm{)}} = G \int_{M_\oplus} \int_{M_\mathrm{)}} \frac{\mathrm{d}m_\oplus\,\mathrm{d}m_\mathrm{)}}{r}, \tag{3.160}$$

similarly as in (1.68),

$$\frac{1}{r} = \frac{1}{\varrho'_\oplus} \sum_{n=0}^{\infty} \left(\frac{\varrho'_\mathrm{)}}{\varrho'_\oplus}\right)^n \mathrm{P}_n^{(0)}(\cos\psi'), \tag{3.161}$$

where (the meaning of all quantities can be seen from Fig. 3.18)

$$\frac{1}{\varrho'_\oplus} = \frac{1}{\varDelta_{\oplus\mathrm{)}}} \sum_{m=0}^{\infty} \left(\frac{\varrho_\oplus}{\varDelta_{\oplus\mathrm{)}}}\right)^m \mathrm{P}_m^{(0)}(\cos\psi) \tag{3.162}$$

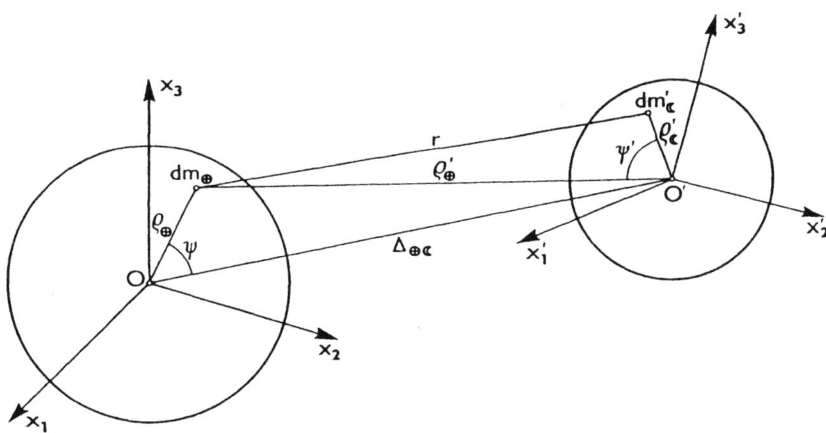

Fig. 3.18. Force function of the Earth–Moon system

or, after substituting (3.162) into (3.161),

$$\frac{1}{r} = \frac{1}{\Delta_{\oplus\mathbb{D}}} \sum_{n=0}^{\infty} \left(\frac{\varrho'_{\mathbb{D}}}{\Delta_{\oplus\mathbb{D}}}\right)^n P_n^{(0)}(\cos\psi') \left[\sum_{m=0}^{\infty} \left(\frac{\varrho_\oplus}{\Delta_{\oplus\mathbb{D}}}\right)^m P_m^{(0)}(\cos\psi)\right]^{n+1}, \qquad (3.163)$$

$$P_n^{(0)}(\cos\psi') = P_n^{(0)}(\sin\delta'_\oplus)\, P_n^{(0)}(\sin\phi'_{\mathbb{D}})$$

$$+\, 2 \sum_{i=1}^{n} \frac{(n-i)!}{(n+i)!} P_n^{(i)}(\sin\delta'_\oplus)\, P_n^{(i)}(\sin\phi'_{\mathbb{D}}) \cos i\,(T'_\oplus + \Lambda'_{\mathbb{D}}), \qquad (3.164)$$

$$P_m^{(0)}(\cos\psi) = P_m^{(0)}(\sin\phi_\oplus)\, P_m^{(0)}(\sin\delta_{O'}) +$$

$$+\, 2 \sum_{k=1}^{n} \frac{(m-k)!}{(m+k)!} P_m^{(k)}(\sin\phi_\oplus)\, P_m^{(k)}(\sin\delta_{O'}) \sin k\,(\Lambda_\oplus + T_{O'}); \qquad (3.165)$$

$\varrho_\oplus$, $\phi_\oplus$, $\Lambda_\oplus$ are the geocentric coordinates of mass element $dm_\oplus$; $\varrho'_{\mathbb{D}}$, $\phi'_{\mathbb{D}}$, $\Lambda'_{\mathbb{D}}$ are the selenocentric coordinates of $dm_{\mathbb{D}}$; $\Delta_{\oplus\mathbb{D}} = \overline{OO'}$ where O and O' are the centres of mass of the Earth and Moon, respectively; $\varrho'_\oplus$, $\delta'_\oplus$, $T'_\oplus$ are the selenocentric radius-vector, selenocentric declination and selenocentric hour angle, respectively, [relative to the prime meridian $(x'_1 x'_3)$] of mass element $dm_\oplus$; $\varrho_{O'}$, $\delta_{O'}$, $T_{O'}$ are the geocentric radius-vector, geocentric declination and hour angle, respectively, [relative to the prime meridian $(x_1 x_3)$] of the Moon's centre of mass; x_j and x'_j are respectively the geocentric and selenocentric coordinate systems whose axes are identical with the axes of the Earth's and Moon's ellipsoids of inertia.

Instead of raising the power of the series in (3.163), we can use to advantage ultraspheric (Gegenbauer's) polynomials $C_m^{(\lambda)}(\cos\psi)$ for which the generating function reads

$$\left[1 - 2\frac{\varrho_\oplus}{\Delta_{\oplus\mathbb{D}}}\cos\psi + \left(\frac{\varrho_\oplus}{\Delta_{\oplus\mathbb{D}}}\right)^2\right]^{-(\frac{1}{2}+\lambda)} = \sum_{m=0}^{\infty} \left(\frac{\varrho_\oplus}{\Delta_{\oplus\mathbb{D}}}\right)^m C_m^{(\lambda)}(\cos\psi), \qquad (3.166)$$

and of which the Legendre polynomials are a special case for $\lambda = 0$. In this particular case $-\frac{1}{2}(n+1) = -(\frac{1}{2}+\lambda)$ or $\lambda = n/2$ and, therefore,

$$\frac{1}{r} = \frac{1}{\Delta_{\oplus\mathbb{D}}} \sum_{n=0}^{\infty} \left(\frac{\varrho'_{\mathbb{D}}}{\Delta_{\oplus\mathbb{D}}}\right)^n P_n^{(0)}(\cos\psi') \sum_{m=0}^{\infty} \left(\frac{\varrho_\oplus}{\Delta_{\oplus\mathbb{D}}}\right)^m C_m^{(n/2)}(\cos\psi); \qquad (3.167)$$

from Gegenbauer's polynomials it is necessary to go back to Legendre's polynomials, because it is they that contain the definition formula of geopotential coefficients. The general relations are known, however, for the purposes of geodynamics and it is sufficient to restrict the solution to degrees $n = m = 4$. The necessary relations will be given below.

For integration (3.160) to be conducive to the geopotential coefficients of both bodies, it is necessary to modify function $P_n^{(k)}(\sin\delta'_\oplus)\cos k(\Lambda'_{\mathbb{D}} + T'_\oplus)$ so that it does not contain arguments $\delta'_\oplus$, $T'_\oplus$, the selenocentric equatorial coordinates of the general mass element $dm_\oplus$ of the Earth, which preclude direct transition to the geopotential coefficients. This can be done by means of transformation

$$P_n^{(k)}(\sin \delta_\oplus') \frac{\cos kT_\oplus'}{\sin kT_\oplus'} = P_n^{(k)}(\sin \delta_O') \frac{\cos kT_O'}{\sin kT_O'} + \frac{Q_n^{(k)}}{R_n^{(k)}}. \qquad (3.168)$$

In view of the definition

$$P_n^{(k)}(\sin \delta) \frac{\cos kT}{\sin kT} = \left\{ \sum_{m=0}^{\frac{1}{2}(n-k)} \frac{(-1)^m (2n-2m)!}{2^n m!(n-m)!(n-k-2m)!} \sin^{n-k-2m} \delta \right\}$$

$$\mathrm{Re} \sum_{t=0}^{k} \binom{k}{t} \frac{\mathrm{i}^t}{\mathrm{i}^{t-1}} (\cos \delta \cos T)^{k-t} (\cos \delta \sin T)^t, \quad \mathrm{i} = \sqrt{-1}, \qquad (3.169)$$

correction terms Q and R are

$$\frac{Q_n^{(k)}}{R_n^{(k)}} = \sum_{m=0}^{\frac{1}{2}(n-k)} \frac{(-1)^m (2n-2m)!}{2^n m!(n-m)!(n-k-2m)!}$$

$$\times \left\{ \sum_{s=1}^{n-k-2m} \binom{n-k-2m}{s} (\sin \delta_O')^{n-k-2m-s} L^s \right.$$

$$\times \mathrm{Re} \sum_{t=0}^{k} \binom{k}{t} \frac{\mathrm{i}^t}{\mathrm{i}^{t-1}} \sum_{j=0}^{k-t} \binom{k-t}{j} (\cos \delta_O' \cos T_O')^{k-t-j} M^j$$

$$\times \sum_{p=0}^{t} \binom{t}{p} (\cos \delta_O' \sin T_O')^{t-p} N^p$$

$$+ \sin^{n-k-2m} \delta_O' \, \mathrm{Re} \sum_{t=0}^{k} \binom{k}{t} \frac{\mathrm{i}^t}{\mathrm{i}^{t-1}} \sum_{j=1}^{k-t} \binom{k-t}{j}$$

$$\times (\cos \delta_O' \cos T_O')^{k-t-j} M^j \sum_{p=1}^{t} \binom{t}{p} (\cos \delta_O' \sin T_O')^{t-p} N^p$$

$$+ \sin^{n-k-2m} \delta_O' \, \mathrm{Re} \sum_{t=0}^{k} \binom{k}{t} \frac{\mathrm{i}^t}{\mathrm{i}^{t-1}} \left[(\cos \delta_O' \cos T_O')^{k-t} \right.$$

$$\times \sum_{p=1}^{t} \binom{t}{p} (\cos \delta_O' \sin T_O')^{t-p} N^p$$

$$\left. \left. + (\cos \delta_O' \sin T_O')^t \sum_{j=1}^{k-t} \binom{k-t}{j} (\cos \delta_O' \cos T_O')^{k-t-j} M^j \right] \right\}, \qquad (3.170)$$

with (we can now omit the symbols $\oplus$, $\mathbb{C}$ for coordinates ϱ, ϕ, Λ)

$$L = \sin \delta_O' \sum_{n=1}^{\infty} \left(\frac{\varrho}{\Delta}\right)^n P_n^{(0)}(\cos \psi)$$

$$+ [\cos \phi \cos \Lambda \cos(x_1, x_3') + \cos \phi \sin \Lambda \cos(x_2, x_3')$$

$$+ \sin \phi \cos(x_3, x_3')] \sum_{n=0}^{\infty} \left(\frac{\varrho}{\Delta}\right)^{n+1} P_n^{(0)}(\cos \psi), \qquad (3.171)$$

$$\frac{M}{N} = \cos \delta'_0 \frac{\cos T'_0}{\sin T'_0} \sum_{n=1} \left(\frac{\varrho}{\Delta}\right)^n P_n^{(0)}(\cos \psi)$$

$$+ \left[\cos \phi \cos \Lambda \frac{\cos(x_1, x'_1)}{\cos(x_1, x'_2)} + \cos \phi \sin \Lambda \frac{\cos(x_2, x'_1)}{\cos(x_2, x'_2)} \right.$$

$$\left. + \sin \phi \frac{\cos(x_3, x'_1)}{\cos(x_3, x'_2)}\right] \sum_{n=0} \left(\frac{\varrho}{\Delta}\right)^{n+1} P_n^{(0)}(\cos \psi) . \tag{3.172}$$

Expressions $\cos(x_1, x'_1), \ldots, \cos(x_3, x'_3)$ can be derived as functions of Euler's angles which directionally link systems x_j and x'_j of the Earth's and Moon's ellipsoids of inertia. General relations (3.13) hold true for them; to distinguish them from those in (3.13) they have here been marked φ', ϑ', ψ'. Putting $n \leqq 4$, $m \leqq 4$, which is sufficient for solving all current geodynamic and astrodynamic problems, the relation sought for (3.167) reads

$$\frac{1}{r} = \frac{1}{\Delta_{\oplus\mathbb{D}}} \left\{1 + \sum_{m=1}^{4} \Pi^m P_m^{(0)}(\cos \psi) + \sum_{n=1}^{4} \Pi'^n P_n^{(0)}(\cos \psi') + R \right\}, \tag{3.173}$$

with

$$R = \Pi' P_1^{(0)}(\cos \psi') \sum_{m=1}^{3} \Pi^m C_m^{(1/2)}(\cos \psi)$$

$$+ \Pi'^2 P_2^{(0)}(\cos \psi') \sum_{m=1}^{2} \Pi^m C_m^{(1)}(\cos \psi) + \Pi'^3 \Pi C_1^{(3/2)}(\cos \psi) , \tag{3.174}$$

$$\Pi = \varrho_\oplus / \Lambda_{\oplus\mathbb{D}}, \quad \Pi' = \varrho'_\mathbb{D} / \Delta_{\oplus\mathbb{D}} . \tag{3.175}$$

Since

$$C_m^{(0)}(\cos \psi) = P_m^{(0)}(\cos \psi) ,$$

$$C_1^{(1/2)}(\cos \psi) = 2P_1^{(0)}(\cos \psi), \quad C_2^{(1/2)} = \tfrac{1}{3} + \tfrac{8}{3} P_2^{(0)}(\cos \psi) ,$$

$$C_3^{(1/2)}(\cos \psi) = \tfrac{4}{5} [P_1^{(0)}(\cos \psi) + 4P_3^{(0)}(\cos \psi)] ,$$

$$C_1^{(1)}(\cos \psi) = 3 P_1^{(0)}(\cos \psi), \quad C_2^{(1)}(\cos \psi) = 1 + 5 P_2^{(0)}(\cos \psi),$$

$$C_1^{(3/2)}(\cos \psi) = 4P_1^{(0)}(\cos \psi) , \tag{3.176}$$

finally,

$$R = 2\Pi\Pi' P_1^{(0)}(\cos \psi) P_1^{(0)}(\cos \psi') + 3\Pi\Pi'^2 P_1^{(0)}(\cos \psi) P_2^{(0)}(\cos \psi')$$

$$+ \tfrac{1}{3} \Pi^2 \Pi' [1 + 8 P_2^{(0)}(\cos \psi)] P_1^{(0)}(\cos \psi')$$

$$+ \Pi^2 \Pi'^2 [1 + 5P_2^{(0)}(\cos \psi)] P_2^{(0)}(\cos \psi')$$

$$+ \tfrac{4}{5} \Pi^3 \Pi' [P_1^{(0)}(\cos \psi) + 4 P_3^{(0)}(\cos \psi)] P_1^{(0)}(\cos \psi')$$

$$+ 4 \Pi\Pi'^3 P_1^{(0)}(\cos \psi) P_3^{(0)}(\cos \psi') . \tag{3.177}$$

Force function (3.160) then reads

$$
V_{\oplus\mathbb{D}} = G \frac{M_{\oplus} M_{\mathbb{D}}}{\varDelta_{\oplus\mathbb{D}}} \left\{ 1 + \sum_{n=2}^{\infty} \sum_{k=0}^{n} \left(\frac{a_0}{\varDelta_{\oplus\mathbb{D}}} \right)^n (J_n^{(k)} \cos kT_{O'} \right.
$$

$$
- S_n^{(k)} \sin kT_{O'}) P_n^{(k)} (\sin \delta_{O'}) + \sum_{n=2}^{\infty} \sum_{k=0}^{n} \left(\frac{a_0'}{\varDelta_{\oplus\mathbb{D}}} \right)^n [(J_n^{(k)})' \cos kT_O'
$$

$$
\left. - (S_n^{(k)})' \sin kT_O'] P_n^{(k)} \sin \delta_O' + \varDelta R_{\oplus\mathbb{D}} \right\} , \tag{3.178}
$$

and with an accuracy of up to the fourth order inclusive,

$$
\varDelta R_{\oplus\mathbb{D}} = \left(\frac{a_0}{\varDelta_{\oplus\mathbb{D}}} \right)^2 \left(\frac{a_0'}{\varDelta_{\oplus\mathbb{D}}} \right)^2 [F_{22}^{00} J_2^{(0)} (J_2^{(0)})' + F_{22}^{22} J_2^{(2)} (J_2^{(2)})'
$$

$$
+ G_{22}^{22} S_2^{(2)} (S_2^{(2)})' + F_{22}^{02} J_2^{(0)} (J_2^{(2)})' + F_{22}^{20} J_2^{(2)} (J_2^{(0)})'
$$

$$
+ G_{22}^{02} J_2^{(0)} (S_2^{(2)})' + G_{22}^{20} S_2^{(2)} (J_2^{(0)})'
$$

$$
+ H_{22}^{22} J_2^{(2)} (S_2^{(2)})' + K_{22}^{22} S_2^{(2)} (J_2^{(2)})'] . \tag{3.179}
$$

For brevity the expressions for coefficients $F_{22}^{00}, \ldots, K_{22}^{22}$ are not given; these can be found in Burša (1979a). A quite general solution of the problem was presented by Šidlichovský (1978). Further on we shall use in concrete form only

$$
F_{22}^{00} = \tfrac{3}{4} - \tfrac{15}{4} (\sin^2 \delta_{O'} + \sin^2 \delta_O') + \tfrac{105}{4} \sin^2 \delta_O' \sin^2 \delta_{O'}
$$

$$
+ \tfrac{3}{2} \cos^2 (x_3, x_3') + 15 \sin \delta_O' \sin \delta_{O'} \cos (x_3, x_3')
$$

$$
= \tfrac{3}{4} [1 - 5 (\sin^2 \delta_{O'} + \sin^2 \delta_O') + 35 \sin^2 \delta_{O'} \sin^2 \delta_O'
$$

$$
+ 2 \cos^2 \vartheta' + 20 \sin \delta_{O'} \sin \delta_O' \cos \vartheta'] . \tag{3.180}
$$

In (3.179) $J_n^{(k)}$, $S_n^{(k)}$ are the geopotential coefficients of the Earth, $(J_n^{(k)})'$, $(S_n^{(k)})'$ are the potential coefficients of the Moon and a_0, a_0' are auxiliary factors of length which are optional (see Sect. 1.3.2).

The resultant force function, $V_{\oplus\mathbb{D}O}$, besides containing $V_{\oplus\mathbb{D}}$, also contains $V_{\oplus O}$, i.e. the partial force function of the Earth–Sun system. We know only very little about the Sun's gravitational field; however, we can assume that it is rotationally symmetrical, i.e. that all non-zonal potential coefficients are equal to zero. Here we shall retain only the second zonal potential coefficient which is the result of the Sun's polar flattening, of which we may assume that it is of the order of $(5 \pm 0.7) \times 10^{-5}$ (Dicke and Goldenberg 1967). The axes of the heliocentric coordinate system are marked x_j'', the Sun's centre of mass is denoted O'' and its geocentric equatorial coordinates are $\delta_{O''}$, $T_{O''}$; Euler's angles, defining the mutual directions of axes x_j'' and x_j, are similarly φ'', ϑ'', ψ'' with general definition (3.13), the second zonal potential coefficient of the Sun is $(J_2^{(0)})''$, the auxiliary optional factor of lengths for describing the solar gravitational field is

a_0'' and $\Delta_{\oplus\odot} = \overline{OO''}$. Then

$$V_{\oplus\odot} = G\,\frac{M_\oplus M_\odot}{\Delta_{\oplus\odot}}\left[1 + \left(\frac{a_0''}{\Delta_{\oplus\odot}}\right)^2 (J_2^{(0)})''\, P_2^{(0)}(\sin\delta_{O''}) + \Delta R_{\oplus\odot}\right], \qquad (3.181)$$

where, as in (3.179),

$$\Delta R_{\oplus\odot} = \left(\frac{a_0}{\Delta_{\oplus\odot}}\right)^2 \left(\frac{a_0''}{\Delta_{\oplus\odot}}\right)^2 (F_{22}^{00})_{\oplus\odot}\, J_2^{(0)}(J_2^{(0)})'' , \qquad (3.182)$$

$$(F_{22}^{00})_{\oplus\odot} = \tfrac{3}{4}\,[1 - 5\,(\sin^2\delta_{O''} + \sin^2\delta_O'')$$
$$+\; 35\sin^2\delta_{O''}\sin^2\delta_O'' + 2\cos^2\vartheta'' + 20\sin\delta_{O''}\sin\delta_O''\cos\vartheta''] .$$
$$(3.183)$$

3.9.2 Right-Hand Sides of Euler's Dynamic Equations as Functions of the Gravitational Perturbations due to the Moon and the Sun

The expressions for components G_j of the resultant moment of external forces due to the Moon and Sun, which form the rhs of Euler's dynamic equations, have been expressed as functions of $V_{\oplus\mathbb{)}\odot}$ (3.157). It is now necessary to calculate derivatives $\partial V_{\oplus\mathbb{)}\odot}/\partial\psi$, $\partial V_{\oplus\mathbb{)}\odot}/\partial\vartheta$, $\partial V_{\oplus\mathbb{)}\odot}/\partial\varphi$, and substitute them into (3.157). The result can be checked by calculating the components G_j of vector (Burša 1979b) (see Fig. 3.19)

$$\mathbf{G} = G \int\limits_{M_\oplus}\int\limits_{M_\mathbb{)}} \left[\varrho_\oplus \times \operatorname{grad}\left(\frac{1}{r}\right)\right] dm_\oplus\, dm_\mathbb{)}$$
$$+\; G \int\limits_{M_\oplus}\int\limits_{M_\odot} \left[\varrho_\mathbb{)}' \times \operatorname{grad}\left(\frac{1}{r}\right)\right] dm_\oplus\, dm_\odot . \qquad (3.184)$$

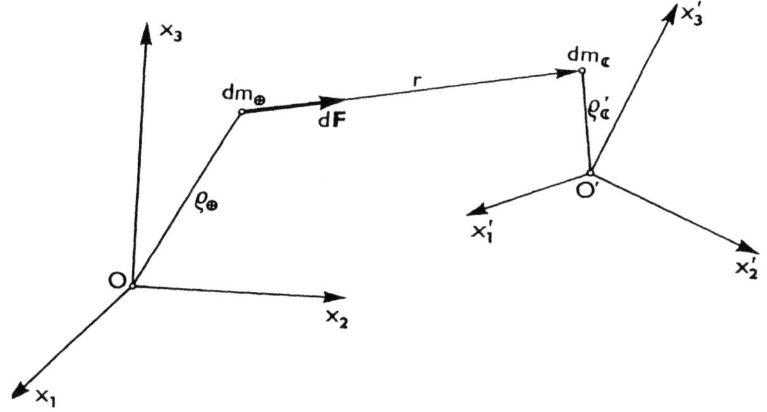

Fig. 3.19. The moment of external forces with which the Moon acts on the Earth

We shall indicate the solution up to the fourth degree inclusive, considering only the harmonic terms, which are clearly dominant, in the gravitational fields of all three bodies: in the Earth's field $J_2^{(0)}$, $J_3^{(0)}$, $J_4^{(0)}$, $J_2^{(2)}$, $S_2^{(2)}$, $J_3^{(1)}$, $S_3^{(3)}$, $J_4^{(3)}$; in the Moon's field $(J_2^{(0)})'$, $(J_2^{(2)})'$; and in the Sun's field $(J_2^{(0)})''$. As regards the terms containing products of the potential coefficients, we shall only consider $J_2^{(0)}(J_2^{(0)})'$ and $J_2^{(0)}(J_2^{(0)})''$, because all the others are at least three orders of magnitude smaller and are not as large as a whole series of linear terms which we are neglecting anyway.

In addition to these, we must also consider terms of the zeroth order which describe spherically symmetrical fields, and which become zero when differentiated with respect to Euler's angles ψ, ϑ, φ. Therefore, instead of $V_{\oplus\mathbb{D}\odot}$, we shall be working with $R_{\oplus\mathbb{D}\odot}$, which does not contain these terms, and which in concrete form reads

$$
\begin{aligned}
R_{\oplus\mathbb{D}\odot} = G\,\frac{M_\oplus M_\mathbb{D}}{\varDelta_{\oplus\mathbb{D}}} &\left\{ \left(\frac{a_0}{\varDelta_{\oplus\mathbb{D}}}\right)^2 \left[J_2^{(0)}\, P_2^{(0)}(\sin\delta_{O'}) + \right. \right. \\
&\quad \left. + (J_2^{(2)}\cos 2T_{O'} - S_2^{(2)}\sin 2T_{O'})\, P_2^{(2)}(\sin\delta_{O'}) \right] \\
&\quad + \left(\frac{a_0}{\varDelta_{\oplus\mathbb{D}}}\right)^3 \left[J_3^{(0)}\, P_3^{(0)}(\sin\delta_{O'}) \right. \\
&\quad \left. + J_3^{(1)}\, P_3^{(1)}(\sin\delta_{O'})\cos T_{O'} - S_3^{(3)}\, P_3^{(3)}(\sin\delta_{O'})\sin 3T_{O'} \right] \\
&\quad + \left(\frac{a_0}{\varDelta_{\oplus\mathbb{D}}}\right)^4 \left[J_4^{(0)}\, P_4^{(0)}(\sin\delta_{O'}) + J_4^{(3)}\, P_4^{(3)}(\sin\delta_{O'})\cos 3T_{O'} \right] \\
&\quad \left. + \left(\frac{a_0}{\varDelta_{\oplus\mathbb{D}}}\right)^2 \left(\frac{a_0'}{\varDelta_{\oplus\mathbb{D}}}\right)^2 F_{22}^{00}\, J_2^{(0)}(J_2^{(0)})' \right\} \\
+ G\,\frac{M_\oplus M_\odot}{\varDelta_{\oplus\odot}} &\left\{ \left(\frac{a_0}{\varDelta_{\oplus\mathbb{D}}}\right)^2 \left[J_2^{(0)}\, P_2^{(0)}(\sin\delta_{O''}) \right. \right. \\
&\quad \left. + (J_2^{(2)}\cos 2T_{O''} - S_2^{(2)}\sin 2T_{O''})\, P_2^{(2)}(\sin\delta_{O''}) \right] \\
&\quad + \left(\frac{a_0}{\varDelta_{\oplus\odot}}\right)^3 \left[J_3^{(0)}\, P_3^{(0)}(\sin\delta_{O''}) \right. \\
&\quad \left. + J_3^{(1)}\, P_3^{(1)}(\sin\delta_{O''})\cos T_{O''} - S_3^{(3)}\, P_3^{(3)}(\sin\delta_{O''})\sin 3T_{O''} \right] \\
&\quad + \left(\frac{a_0}{\varDelta_{\oplus\odot}}\right)^4 \left[J_4^{(0)}\, P_4^{(0)}(\sin\delta_{O''}) + J_4^{(3)}\, P_4^{(3)}(\sin\delta_{O''})\cos 3T_{O''} \right] \\
&\quad \left. + \left(\frac{a_0}{\varDelta_{\oplus\odot}}\right)^2 \left(\frac{a_0''}{\varDelta_{\oplus\odot}}\right)^2 (F_{22}^{00})_{\oplus\odot}\, J_2^{(0)}(J_2^{(0)})'' \right\}.
\end{aligned}
\tag{3.185}
$$

F_{22}^{00} is given by Eq. (3.180) and $(F_{22}^{00})_{\oplus\odot}$ by (3.183).

To obtain derivatives $\partial R_{\oplus\mathbb{D}\odot}/\partial\psi$, $\partial R_{\oplus\mathbb{D}\odot}/\partial\vartheta$, $\partial R_{\oplus\mathbb{D}\odot}/\partial\varphi$ we shall use the transformation equations for the direction cosines of the geocentric direction to the perturbing body (Moon, Sun) which follow from matrix $\varphi\,\vartheta\,\psi$ (3.15).

Once we have the derivatives of the perturbing function, $\partial R_{\oplus \mathbb{D} \odot}/\partial \psi$, $\partial R_{\oplus \mathbb{D} \odot}/\partial \vartheta$, $\partial R_{\oplus \mathbb{D} \odot}/\partial \varphi$, we shall be able to express the components of the resultant moment of external forces, G_j. The reader is reminded that function $R_{\oplus \mathbb{D} \odot}$ is identical with tidal force function (4.121), and that the precession and nutation moment is generated by tidal forces (Melchior and Georis 1968), provided that the Earth is considered to be a perfectly rigid body.

For the sake of brevity, we shall now restrict ourselves to degree $n = 2$ in the perturbing (tidal force) functions, and we shall, therefore, consider the gravitational fields of the Moon and Sun to be spherically symmetrical. Then

$$
\begin{aligned}
R_{\oplus \mathbb{D} \odot} = {} & G \frac{M_\oplus M_\mathbb{D}}{\Delta_{\oplus \mathbb{D}}} \sum_{k=0}^{2} \left(\frac{a_0}{\Delta_{\oplus \mathbb{D}}}\right)^2 (J_2^{(k)} \cos k T_{O'} \\
& - S_2^{(k)} \sin k T_{O'}) \, P_2^{(k)} (\sin \delta_{O'}) \\
& + G \frac{M_\oplus M_\odot}{\Delta_{\oplus \odot}} \sum_{k=0}^{2} \left(\frac{a_0}{\Delta_{\oplus \odot}}\right)^2 (J_2^{(k)} \cos k T_{O''} \\
& - S_2^{(k)} \sin k T_{O''}) \, P_2^{(k)} (\sin \delta_{O''})
\end{aligned}
\tag{3.186}
$$

and, since

$$
\frac{\partial}{\partial \psi} P_2^{(0)} (\sin \delta_{O'}) = 3 \sin \delta_{O'} \sin \vartheta \cos \beta_{O'} \cos (\psi - \lambda_{O'}) \,,
$$

$$
\frac{\partial}{\partial \vartheta} P_2^{(0)} (\sin \delta_{O'}) = 3 \sin \delta_{O'} [\cos \vartheta \cos \beta_{O'} \sin (\psi - \lambda_{O'}) - \sin \vartheta \sin \beta_{O'}] \,,
$$

$$
\frac{\partial}{\partial \varphi} P_2^{(0)} (\sin \delta_{O'}) = 0 \,,
$$

$$
\frac{\partial}{\partial \psi} [P_2^{(2)} (\sin \delta_{O'}) \cos 2T_{O'}] = 6 \cos \delta_{O'} \cos \beta_{O'} [\sin (\lambda_{O'} - \psi) \cos (\varphi + T_{O'})
$$
$$
- \cos \vartheta \cos (\lambda_{O'} - \psi) \sin (\varphi + T_{O'})] \,,
$$

$$
\frac{\partial}{\partial \vartheta} [P_2^{(2)} (\sin \delta_{O'}) \cos 2T_{O'}] = - 6 \cos \delta_{O'} [\sin \vartheta \cos \beta_{O'} \sin (\lambda_{O'} - \psi)
$$
$$
- \cos \vartheta \sin \beta_{O'}] \sin (\varphi + T_{O'}) \,,
$$

$$
\frac{\partial}{\partial \varphi} [P_2^{(2)} (\sin \delta_{O'}) \cos 2T_{O'}] = - 6 \cos^2 \delta_{O'} \sin 2T_{O'}
$$
$$
= - 2P_2^{(2)} (\sin \delta_{O'}) \sin 2T_{O'} \,,
$$

$$
\frac{\partial}{\partial \psi} P_2^{(0)} (\sin \delta_{O''}) = 3 \sin \delta_{O''} \sin \vartheta \cos (\psi - \lambda_{O''}) \,,
$$

$$
\frac{\partial}{\partial \vartheta} P_2^{(0)} (\sin \delta_{O''}) = 3 \sin \delta_{O''} \cos \vartheta \sin (\psi - \lambda_{O''}) \,,
$$

$$\frac{\partial}{\partial \varphi}\, P_2^{(0)}(\sin \delta_{O''}) = 0\,,$$

$$\frac{\partial}{\partial \psi}\left[P_2^{(2)}(\sin \delta_{O''}) \cos 2 T_{O''}\right] = 6 \cos \delta_{O''}\left[\sin(\lambda_{O''} - \psi)\cos(\varphi + T_{O''})\right.$$

$$\left. - \cos \vartheta \cos(\lambda_{O''} - \psi)\sin(\varphi + T_{O''})\right]\,,$$

$$\frac{\partial}{\partial \vartheta}\left[P_2^{(2)}(\sin \delta_{O''}) \cos 2 T_{O''}\right] = -6 \cos \delta_{O''}\sin \vartheta \sin(\lambda_{O''} - \psi)\sin(\varphi + T_{O''})\,,$$

$$\frac{\partial}{\partial \varphi}\left[P_2^{(2)}(\sin \delta_{O''}) \cos 2 T_{O''}\right] = -6 \cos^2 \delta_{O''}\sin 2 T_{O''}$$

$$= -2 P_2^{(2)}(\sin \delta_{O''})\sin 2 T_{O''}\,,$$

the G_j-components come out as

$$G_1 = \tfrac{3}{2}\, G(J_2^{(0)} + 2J_2^{(2)})\left[\frac{M_\oplus M_{\mathbb{D}}}{\Delta_{\oplus\mathbb{D}}}\left(\frac{a_0}{\Delta_{\oplus\mathbb{D}}}\right)^2 \sin 2\delta_{O'} \sin T_{O'}\right.$$

$$\left. + \frac{M_\oplus M_\odot}{\Delta_{\oplus\odot}}\left(\frac{a_0}{\Delta_{\oplus\odot}}\right)^2 \sin 2\,\delta_{O''} \sin T_{O''}\right]$$

$$= \tfrac{3}{2}(B - C)\left(\frac{GM_{\mathbb{D}}}{\Delta_{\oplus\mathbb{D}}^3}\sin 2\,\delta_{O'}\sin T_{O'} + \frac{GM_\odot}{\Delta_{\oplus\odot}^3}\sin 2\,\delta_{O''}\sin T_{O''}\right)\,,$$

$$G_2 = \tfrac{3}{2}\, G(J_2^{(0)} - 2J_2^{(2)})\left[\frac{M_\oplus M_{\mathbb{D}}}{\Delta_{\oplus\mathbb{D}}}\left(\frac{a_0}{\Delta_{\oplus\mathbb{D}}}\right)^2 \sin 2\delta_{O'} \cos T_{O'}\right.$$

$$\left. + \frac{M_\oplus M_\odot}{\Delta_{\oplus\odot}}\left(\frac{a_0}{\Delta_{\oplus\odot}}\right)^2 \sin 2\delta_{O''} \cos T_{O''}\right]$$

$$= \tfrac{3}{2}(A - C)\left(\frac{GM_{\mathbb{D}}}{\Delta_{\oplus\mathbb{D}}^3}\sin 2\,\delta_{O'}\cos T_{O'} + \frac{GM_\odot}{\Delta_{\oplus\odot}^3}\sin 2\,\delta_{O''}\cos T_{O''}\right)\,,$$

$$G_3 = -6G\,\frac{M_\oplus M_{\mathbb{D}}}{\Delta_{\oplus\mathbb{D}}}\left(\frac{a_0}{\Delta_{\oplus\mathbb{D}}}\right)^2 J_2^{(2)} \cos^2 \delta_{O'}\sin 2 T_{O'}$$

$$- 6G\,\frac{M_\oplus M_\odot}{\Delta_{\oplus\odot}}\left(\frac{a_0}{\Delta_{\oplus\mathbb{D}}}\right)^2 J_2^{(2)} \cos^2 \delta_{O''}\sin 2 T_{O''}$$

$$= -\tfrac{3}{2}(B - A)\left(\frac{GM_{\mathbb{D}}}{\Delta_{\oplus\mathbb{D}}^3}\cos^2 \delta_{O'}\sin 2 T_{O'} + \frac{GM_\odot}{\Delta_{\oplus\odot}^3}\cos^2 \delta_{O''}\sin 2 T_{O''}\right)\,.$$

$$(3.187)$$

The components of the moment for $n = 2$ indicate that the effects of the Earth's polar flattening (coefficient $J_2^{(0)}$) and equatorial flattening (coefficient $J_2^{(2)}$) can be considered jointly in the theory of precession, if we insert $J_2^{(0)} + 2J_2^{(2)}$ (in G_1) and $J_2^{(0)} - 2J_2^{(2)}$ (in G_2) in the equatorial components instead of $J_2^{(0)}$. The effect of the equatorial flattening thus does not have to be treated separately (in G_1 and G_2).

3.10 Approximate Solution for the Precession-Nutation Motion Under Equal Equatorial Moments of Inertia

The exact solution of Euler's complete dynamic equations (3.156), i.e. with the rhs (3.157) $G_j \neq 0$, has not been found yet. However, there are approximate solutions in which the densities of the perturbing bodies are considered to be spherically symmetrical, and in the Earth's gravitational field only the second zonal ($n = 2$, $k = 0$) harmonic term with geopotential coefficient $J_2^{(0)}$ is retained, i.e. the Earth's gravitational field is considered to be completely rotationally symmetrical, symmetrical with respect to the equatorial plane, and all even zonal terms with $n > 2$ are neglected. The effect of some of the neglected terms is then included in corrections (e.g. Woolard 1953). In this section we shall only describe the approximate solution after Woolard (1953). A more detailed treatment of the problem would require much more time and space.

In this extremely simplified case

$$\frac{\partial R_{\oplus \mathbb{D} \odot}^{(0)}}{\partial \varphi} = 0, \tag{3.188}$$

and Euler's dynamic equations take the form

$$A\dot{\omega}_1 + (C - A)\omega_0 \omega_2 = \frac{3}{2} G \frac{M_\oplus M_\mathbb{D}}{\Delta_{\oplus \mathbb{D}}} \left(\frac{a_0}{\Delta_{\oplus \mathbb{D}}}\right)^2 J_2^{(0)} \sin 2\delta_{O'} \sin T_{O'}$$

$$+ \frac{3}{2} G \frac{M_\oplus M_\odot}{\Delta_{\oplus \odot}} \left(\frac{a_0}{\Delta_{\oplus \odot}}\right)^2 J_2^{(0)} \sin 2\delta_{O''} \sin T_{O''},$$

$$A\dot{\omega}_2 - (C - A)\omega_0 \omega_1 = \frac{3}{2} G \frac{M_\oplus M_\mathbb{D}}{\Delta_{\oplus \mathbb{D}}} \left(\frac{a_0}{\Delta_{\oplus \mathbb{D}}}\right)^2 J_2^{(0)} \sin 2\delta_{O'} \cos T_{O'}$$

$$+ \frac{3}{2} G \frac{M_\oplus M_\odot}{\Delta_{\oplus \odot}} \left(\frac{a_0}{\Delta_{\oplus \odot}}\right)^2 J_2^{(0)} \sin 2\delta_{O''} \cos T_{O''},$$

$$C\dot{\omega}_3 = 0. \tag{3.189}$$

The solution of the system of homogeneous equations, without the rhs ($G_1 = 0$, $G_2 = 0$), describes free nutation and, in view of (3.49), reads

$$\omega_1 = \left[\frac{TC - L^2}{A(C - A)}\right]^{1/2} \cos\left(\omega_0 \frac{C - A}{A} t - \beta_0\right),$$

$$\omega_2 = \left[\frac{TC - L^2}{A(C - A)}\right]^{1/2} \sin\left(\omega_0 \frac{C - A}{A} t - \beta_0\right). \tag{3.190}$$

The first integrals T, L and the third integration constant β_0 are no longer constants, but functions of time:

$$T = T(t), \quad L = L(t), \quad \beta_0 = \beta_0(t). \tag{3.191}$$

If $A = B$, T and L are not independent quantities as, in view of (3.47),

$$L^2 - TA = \omega_0^2 C(C - A) = \text{const}, \tag{3.192}$$

assuming that moments of inertia A and C are considered constant, i.e. that the Earth is a perfectly rigid body.

Instead of (3.191) we now have only two independent functions of time, either $\beta_0(t)$ and $L(T)$, or $\beta_0(t)$ and $T(t)$, or rather

$$g_1(t) = \left[\frac{CT(t) - L^2(t)}{A(C - A)} \right]^{1/2} \cos \beta_0(t),$$

$$g_2(t) = \left[\frac{CT(t) - L^2(t)}{A(C - A)} \right]^{1/2} \sin \beta_0(t), \tag{3.193}$$

with

$$\omega_1 = g_1(t) \cos k_0 t + g_2(t) \sin k_0 t,$$

$$\omega_2 = g_1(t) \sin k_0 t - g_2(t) \cos k_0 t,$$

$$\omega_3 = \omega_0;$$

$$k_0 = \omega_0 \frac{C - A}{A}. \tag{3.194}$$

Equations (3.194) describe the motion of vector $\boldsymbol{\omega}$ in the system of axes x_j of the ellipsoid of inertia. Euler's kinematic equations (3.25) describe this motion in the inertial system (of the 'fixed' ecliptic) $\bar{x}_j$:

$$\dot{\psi} = [g_1(t) \sin (k_0 t + \varphi) - g_2(t) \cos (k_0 t + \varphi)] \csc \vartheta,$$

$$\dot{\vartheta} = g_1(t) \cos (k_0 t + \varphi) + g_2(t) \sin (k_0 t + \varphi),$$

$$\dot{\varphi} = \omega_0 - \dot{\psi} \cos \vartheta$$

$$= \omega_0 - [g_1(t) \sin (k_0 t + \varphi) - g_2(t) \cos (k_0 t + \varphi)] \cot \vartheta. \tag{3.195}$$

We thus seek a solution identical in form with that of the solution of the homogeneous equations.

Functions $g_1(t)$ and $g_2(t)$ are related to the first integrals and phase angle β_0 as

$$\frac{CT(t) - L^2(t)}{A(C - A)} = g_1^2(t) + g_2^2(t),$$

$$\tan \beta_0(t) = \frac{g_2(t)}{g_1(t)}. \tag{3.196}$$

This yields

$$A[\dot{g}_1(t)\cos k_0 t + \dot{g}_2(t)\sin k_0 t] = G_1^{(0)} ,$$

$$A[\dot{g}_1(t)\sin k_0 t - \dot{g}_2(t)\cos k_0 t] = G_2^{(0)} , \qquad (3.197)$$

$$\dot{g}_1(t) = \frac{1}{A}(G_1^{(0)}\cos k_0 t + G_2^{(0)}\sin k_0 t) ,$$

$$\dot{g}_2(t) = \frac{1}{A}(G_1^{(0)}\sin k_0 t - G_2^{(0)}\cos k_0 t) \qquad (3.198)$$

and, after integration,

$$g_1(t) = g_1(0) + \frac{1}{A}\int_{t=0}^{t}(G_1^{(0)}\cos k_0 t + G_2^{(0)}\sin k_0 t)\,dt ,$$

$$g_2(t) = g_2(0) + \frac{1}{A}\int_{t=0}^{t}(G_1^{(0)}\sin k_0 t - G_2^{(0)}\cos k_0 t)\,dt ; \qquad (3.199)$$

$g_1(0)$ and $g_2(0)$ are the values of functions $g_1(t)$ and $g_2(t)$ for $t = 0$.

After substituting (3.199) into (3.194) and (3.195) we obtain equations for $\omega_1, \omega_2, \dot{\psi}, \dot{\vartheta}, \dot{\varphi}$. The unknowns also occur on the rhs, but this can be dealt with by applying the method of successive approximations. Six integration constants $g_1(0), g_2(0), \omega_0, \psi(0), \vartheta(0)$ and $\varphi(0)$ have to be determined from the initial conditions (for $t = 0$).

3.11 Numerical Results

The numerical integration of the equation for precession and nutation is tedious, and it would be difficult to fit its description into this chapter. The final solution contains:

(a) Terms of type $at + bt^2 + ct^3 + \ldots$ ($a, b, c, \ldots$ are constants) increasing with time. Strictly speaking, these terms are also of a periodic nature; however, their periods are very long in comparison to b.

(b) Periodic terms of type $a_1\sin(\alpha_1 t + \beta_1) + a_2\sin(\alpha_2 t + \beta_2) + \ldots$ ($\alpha_1, \alpha_2, \ldots, \beta_1, \beta_2, \ldots, a_1, a_2, \ldots$ are constants). Terms a are called precession terms and terms b are nutation terms. A total of 70 terms have so far been derived, the most important being:

1. Precession due to the Moon and Sun (lunisolar precession; precession cone with apex angle $\sim 47°$) with a period of $\sim 25\,770$ years (precession constant $-50.41''$/year) and amplitude $\sim 23.5°$.

2. Lunar nutation with a period of 18.6 years and amplitude $9.2''$ in ϑ (nutation constant).

3. Semi-annual solar nutation with an amplitude of $0.55''$ in angle ϑ.

4. A 14-day lunar nutation with an amplitude of $0.098''$ in angle ϑ.

The integral mean value of the variation of the precession angle

$$\bar{\dot{\psi}} = -\frac{3}{2}\frac{1}{\omega_0}\frac{C-A}{C}\left(\frac{GM_{\mathrm{)}}}{\Delta_{\oplus\mathrm{)}}^3} + \frac{GM_{\odot}}{\Delta_{\oplus\odot}^3}\right)\cos\vartheta\,. \tag{3.200}$$

Since $GM_{\mathrm{)}}/\Delta_{\oplus\mathrm{)}}^3 = 9\times10^{-14}\,\mathrm{s}^{-2}$, $GM_{\odot}/\Delta_{\oplus\odot}^3 = 4\times10^{-14}\,\mathrm{s}^{-2}$, the effect of the Sun on precession is about half the effect of the Moon. The aggregate value of the secular term, including planetary precession (0.10″ in the sense opposite to that of the lunisolar precession) and geodetic or relativistic precessions (0.02″, also in the opposite sense), is $-50.29''$/year.

A new value of the precession constant for epoch $T_0 = 2000$ was adopted at the 16th General Assembly of the IAU: $50.290\,966'' + 0.000\,222\,2''\,T$; T is expressed in Julian centuries beginning with epoch T_0.

The period of precession motion

$$T_\psi = \frac{2\pi}{\bar{\dot{\psi}}} \sim 25\,770 \text{ tropical years.} \tag{3.201}$$

The precession of the vernal equinox amounts to about $1°$ in 72 years. In about 1200 years the Earth's axis of rotation will be pointing to αLyrae (Vega).

The precession of the Earth's axis of rotation was discovered by Hipparchus roughly 140 years B.C. from observations of star positions; the physical essence of the phenomenon was explained by Newton. As a result of precession the symbols of the Zodiac no longer agree with the appropriate constellations; the shift in ecliptical longitude is now nearly $30°$.

The retrograde motion of the nodal line of the Moon's orbit with a period of ~ 18.6 years, caused by perturbations of the Moon's motion due to the Sun, has the largest effect on nutation. It results in periodic variations of nutation angle ϑ with the same period (18.6 years). The order of magnitude of the amplitude of motion $\dot{\vartheta}$ is approximately

$$\dot{\vartheta} = -\frac{3}{2}\frac{1}{\omega_0}\frac{C-A}{C}\frac{GM_{\mathrm{)}}}{\Delta_{\oplus\mathrm{)}}^3}\sin 2(\bar{\lambda}_{o'} - \psi)\,, \tag{3.202}$$

the effect of the Sun on $\dot{\vartheta}$ being completely neglected. The principal term (nutation constant), discovered and explained by Bradley in 1746 (unpubl), has an amplitude of $9.206''$. To be more precise, the motion is elliptical, and the nutation constant represents the semimajor axis (pointing towards the pole of the ecliptic); the semiminor axis is $\sim 6.84''$.

The effect of the planets on the precession of the Earth's axis of rotation is considerably smaller. Table 3.5 shows the numerical values of coefficient

$$A_{p\oplus} = \frac{3}{2}\frac{C-A}{\omega_0 C}\frac{GM_p}{\Delta_{\oplus p}^3}$$

for each planet (p), where $\Delta_{\oplus p}$ is the minimum geocentric distance of the planet and GM_p the planetocentric gravitational constant. One can see that the precession-nutation effect, with which the planets act on the dynamics of the Earth's

Table 3.5. The effect of planets on the precession of the Earth's axis of rotation

Planet (p)	GM_p $(10^9 \, \mathrm{m}^3 \, \mathrm{s}^{-2})$	$\Delta_{\oplus p}$ $(10^{11} \, \mathrm{m})$	$GM_p/\Delta_{\oplus p}^3$ (s^{-2})	$A_{p\oplus}$ $(\mathrm{arcsec} \, \mathrm{cy}^{-1})$
Mercury	22 031.8	0.92	2.8×10^{-20}	1.2×10^{-3}
Venus	324 858.8	0.41	4.7×10^{-18}	0.21
Mars	42 828.44	0.78	9.0×10^{-20}	4.0×10^{-3}
Jupiter	$126 686.9 \times 10^3$	6.3	5.1×10^{-19}	2.2×10^{-2}
Saturn	$37 940.6 \times 10^3$	12.0	2.2×10^{-20}	9.6×10^{-4}
Uranus	$\sim 585 \times 10^4$	27.0	3.0×10^{-22}	1.3×10^{-5}
Neptune	$\sim 686 \times 10^4$	44.0	8.1×10^{-23}	3.5×10^{-6}

rotation, is several orders of magnitude smaller than that of the lunisolar precession and nutation. Vondrák (1982) has calculated the planetary nutation terms.

The planets also affect the position of the ecliptic because they cause perturbations in the Earth's orbital elements, i.e. in its motion around the Sun. This effect on the precession angle ($\dot\psi$ – motion of the vernal equinox) amounts to $\sim + 0.10''$/year and is referred to as 'planetary precession', which is not very apt since it does not represent the precession described by Euler's dynamic equations. In the narrower sense of the term it has very little to do with the Earth's rotation.

The planets have an even smaller effect on the Sun's precession, because their orbital planes are deflected very little from the plane of the solar equator. The latter plane makes an angle of $\sim 7.25°$ with the plane of the ecliptic. Table 3.6 gives the amplitudes of the precession of the Sun's axis of rotation

$$A_{p\odot} = \frac{3}{2} \frac{C_\odot - A_\odot}{C_\odot \omega_0} \frac{GM_p}{[a(1 - e)]^3} ,$$

where a is the semimajor axis, e the eccentricity of the planet's orbit (see Table 6.6), $A_\odot$, $C_\odot$ the Sun's principal moments of inertia, ω_0 the angular velocity of its rotation and $\Delta_{\odot p} = a(1 - e)$ the heliocentric distance in the perihelion.

Table 3.6. The effect of planets on the precession of the Sun's axis of rotation

Planet (p)	$\Delta_{\odot p}$ $(10^{11} \, \mathrm{m})$	$GM_p/\Delta_{\odot p}^3$ (s^{-2})	$A_{p\odot}$ $(\mathrm{arcsec} \, \mathrm{cy}^{-1})$
Mercury	0.46	2.2×10^{-19}	5.2×10^{-2}
Venus	1.07	2.6×10^{-19}	5.8×10^{-2}
Earth plus Moon	1.47	1.2×10^{-19}	2.7×10^{-2}
Mars	2.07	3.7×10^{-21}	8.5×10^{-4}
Jupiter	7.41	2.7×10^{-19}	6.2×10^{-2}
Saturn	13.0	1.3×10^{-20}	3.0×10^{-3}
Uranus	27.0	2.5×10^{-22}	5.6×10^{-5}
Neptune	45.0	7.5×10^{-23}	1.7×10^{-5}

4 The Earth's Tides. Tidal Deformation of the Earth's Crust

4.1 Introduction

The Earth's tides are related to viscoelastic deformations of the Earth generated by the gravitational effects of the Moon and Sun.

A force, resulting from the gravitational effects of the Earth, Moon and Sun, and from the centrifugal force due to the Earth's rotation, acts at each point of the Earth's surface. All these components vary with time, and the predominant component due to the Earth's gravitation varies least. Its time variations are due to two effects: to seasonal transfers of atmospheric masses with a prevalent period of 1 year, on the one hand, and to convective flow in the Earth's core and mantle, which is a very slow process, on the other hand. These time variations are not considered in the theory of tides. The gravitational forces due to the Moon and Sun vary relatively rapidly with time as a result of the rapidly varying positions of these bodies relative to the observer on the Earth. The effect of these forces of lunisolar origin on the rotation dynamics of the Earth was discussed in Chapter 3. On the Earth's surface tidal forces are reflected in observable time variations of the vertical component of the acceleration of gravity and in variations of the direction of the vertical.

The history of tides goes back to Newton's *Principia* (1687; Chapter 3). In 1769, Euler discussed tides at the end of his treatise *Section prima de statu aequilibrii fluidorum*. He then developed the static theory of tides based on expressing the difference between the gravitational potential due to the Moon at the Earth's mass centre and at a general point on its surface in 1783. Both Newton and Euler attempted to explain the motion of the water masses in oceans and seas. Laplace devoted considerable attention to tides particularly in the second volume of four of his *Celestial Mechanics* (1805).

The next milestone in the history of the theory of tides is the year 1845 when Peters explained exactly the time variations of the vertical. In the subsequent period the static theory of tides was developed further; the Earth was considered a rigid body and only the variations of the gravity field on the Earth's surface due to the perturbing forces of the Moon and Sun were investigated. Thomson (Lord Kelvin, about 1876) was first to consider the deformation of the rigid Earth, whose surface changed due to tidal forces very much like the ocean surface but to a lesser extent. Further progress was made in the theory of tides when Love (1909) introduced parameters now referred to as Love numbers;

these provide an integral description of the elastic response of the Earth to tidal force, and can be calculated from observations of gravity variations and variations of the vertical. Numerous researchers (Poincaré, Jeffreys, Vicente and Molodenskii) proved that the Earth's liquid core had a considerable effect on tidal phenomena. The theory of tides in a spherically symmetrical Earth was developed by Pekeris, Takeuchi, Saito and others. Takeuchi was the first to integrate numerically the differential equations for displacement and he was able to estimate the magnitude of the shear modulus (rigidity) in the Earth's liquid core. Molodenskii (1953) studied the field of tidal displacements in the Earth taking into account viscosity. The general solution of the tidal problem under lateral variations of density and of the elastic parameters was presented by Molodenskii (1980). A complete bibliography of papers on Earth tides, including a brief history of the development of theoretical methods and observations of tides, can be found in Melchior (1983).

The problem of Earth tides can be divided into several more or less independent partial problems:

1. Determination and analysis of lunisolar gravitational forces acting on the Earth's surface.

2. Determination of the field of elastic displacements within the Earth due to the effect of tidal forces acting on the surface, the distribution of density and elastic moduli, or viscosity, within the Earth being considered known. The displacement field for a particular model of the Earth can then be used to derive the values of the Love numbers; this is referred to as the direct problem.

3. From a geophysical point of view the inverse problem is of particular importance. The observed values of the Love numbers are, in this case, used to estimate the distribution of elastic parameters and density within the Earth.

In this chapter we shall concentrate mainly on the first of these problems, i.e. the analysis of tidal forces due to the Moon and Sun, which conforms to the concept of this book. The solution of this problem provides the boundary conditions necessary to solve the second problem, i.e. to determine the displacement pattern within the Earth and the corresponding values of the Love numbers, the figure of the Earth and the distribution of density and elastic parameters within it being given. This problem is solved by integrating the differential equations which express the equilibrium of all forces acting on the Earth. This problem differs but little from the problem of the Earth's free oscillations. In studying Earth tides and free oscillations the same equilibrium equations and models of the Earth are employed. The difference is that in the case of the free oscillations one investigates the displacements within the Earth, which has been made to oscillate in some manner and then left to oscillate without any external forces acting, whereas in the case of tides the same differential equations for displacement are solved with time variable tidal forces acting on the Earth's surface. This problem exceeds the scope of this book and we shall only deal with it marginally in the simple case of static tides of a rigid Earth, and in the case in which the Earth's response to tidal forces is expressed integrally in terms of Love numbers without solving the differential equations for displacement. We shall not deal with the other aspects of Earth tides such as the inverse problem and observations and interpretation of tidal effects.

4.2 Tide-Generating Potential of a Perfectly Rigid Earth

The basis for solving the problem is the tide-generating (tidal) potential V_s which, at general point $M(\varrho, \phi, \Lambda)$ on the Earth's surface (Fig. 4.1), is

$$V_s(M) = V_{s\,\mathbb{D}}(M) + V_{s\,\odot}(M), \tag{4.1}$$

where

$$V_{s\,\mathbb{D}}(M) = V_{\mathbb{D}}(M) + Q_{\oplus\mathbb{D}}(M) \tag{4.2}$$

is its part due to the Moon and

$$V_{s\,\odot} = V_{\odot}(M) + Q_{\oplus\odot}(M) \tag{4.3}$$

is its part due to the Sun; $V_{\mathbb{D}}(M)$ and $V_{\odot}(M)$ are the gravitational potentials of the Moon and Sun at point M; $Q_{\oplus\mathbb{D}}(M)$ and $Q_{\oplus\odot}(M)$ are the potentials of centrifugal forces at point M due to the translational motion of the Earth–Moon and Earth–Sun systems, respectively, in each case about a common barycentre. Here we shall only derive the part due to the Moon, $V_{s\,\mathbb{D}}$; the part due to the Sun, $V_{s\,\odot}$, can be derived quite analogously.

We shall start with two natural conditions:

$$V_{s\,\mathbb{D}}(O) = 0, \tag{4.4}$$

i.e. the tidal potential is zero at the Earth's centre of mass O, and

$$\mathbf{w}_{\oplus\mathbb{D}}(O) = -\mathbf{g}_{\mathbb{D}}(O), \tag{4.5}$$

i.e. at the centre of mass centrifugal acceleration $\mathbf{w}_{\oplus\mathbb{D}}$ equals gravitational

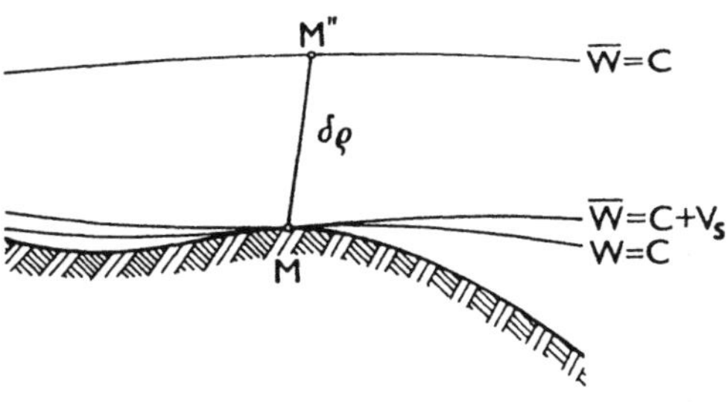

Fig. 4.1. Tide-generating potential V_s and radial tidal displacement $\delta\varrho$ of the equipotential surface of the geopotential

acceleration $\mathbf{g}_{\!\!\;\raisebox{0pt}{\tiny ☽}}(O)$ due to the Moon. In view of (4.2), (4.4) yields

$$Q_{\oplus\!\!\;☽}(O) = - V_{☽}(O) . \tag{4.6}$$

Condition (4.5) is not satisfied exactly because of the considerable deviation of the Moon's gravitational field from a spherically symmetrical field.

Acceleration field $\mathbf{w}_{\oplus\!\!\;☽}$ is homogeneous (Fig. 4.2); hence

$$\mathbf{w}_{\oplus\!\!\;☽}(M) = \mathbf{w}_{\oplus\!\!\;☽}(O) = \mathbf{w}_{\oplus\!\!\;☽} \tag{4.7}$$

and equipotential surfaces $Q_{\oplus\!\!\;☽} = \text{const}$ (Fig. 4.2) are perpendicular to $\mathbf{w}_{\oplus\!\!\;☽}$. The difference between potentials $Q_{\oplus\!\!\;☽}$ at points M and O is thus (Fig. 4.2)

$$Q_{\oplus\!\!\;☽}(M) - Q_{\oplus\!\!\;☽}(O) = w_{\oplus\!\!\;☽}\varrho\cos\psi \tag{4.8}$$

or, in view of (4.5),

$$Q_{\oplus\!\!\;☽}(M) - Q_{\oplus\!\!\;☽}(O) = - g_{☽}(O)\varrho\cos\psi ; \tag{4.9}$$

ψ is the angle between the geocentric radius-vector of point M and the geocentric radius-vector of the Moon's centre of mass.

Taking into account (4.9) and (4.6), tidal potential (4.2) becomes

$$V_{s☽}(M) = V_{☽}(M) - V_{☽}(O) - g_{☽}(O)\varrho\cos\psi , \tag{4.10}$$

and

$$V_{☽}(M) = G \int\limits_{M_{☽}} \frac{dm_{☽}}{r} , \tag{4.11}$$

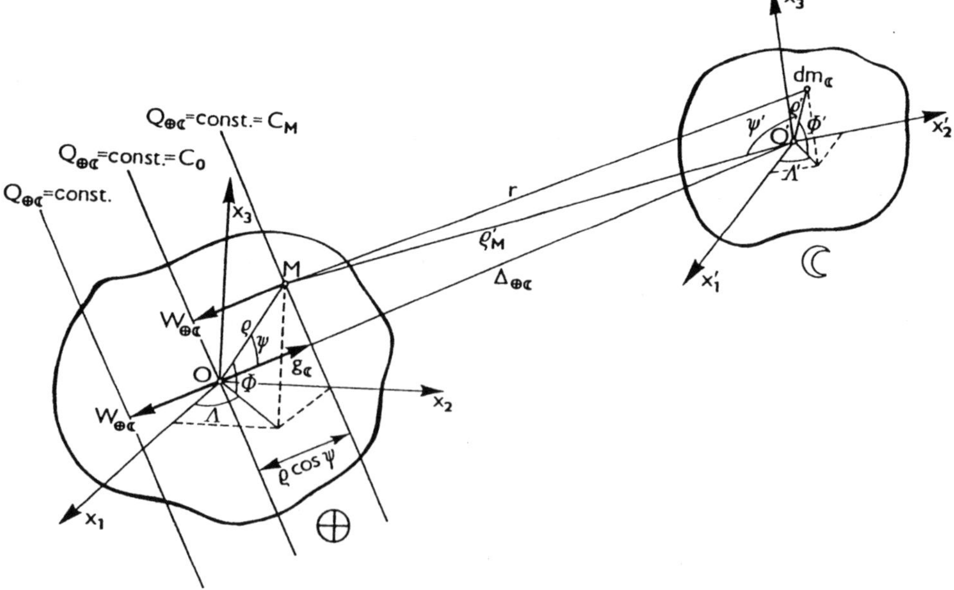

Fig. 4.2. Diagram used in deriving the tide-generating potential

$\mathrm{d}m_{\mathrm{D}}$ being a general mass element of the Moon, and $r = \overline{M\,\mathrm{d}m_{\mathrm{D}}}$ (Fig. 4.2). Since, similarly to (1.68),

$$\frac{1}{r} = \frac{1}{\varrho'_M} \sum_{m=0}^{\infty} \left(\frac{\varrho'}{\varrho'_M}\right)^m \mathrm{P}_m^{(0)}(\cos\psi')\,, \tag{4.12}$$

$$\frac{1}{\varrho'_M} = \frac{1}{\varDelta_{\oplus\mathrm{D}}} \sum_{n=0}^{\infty} \left(\frac{\varrho}{\varDelta_{\oplus\mathrm{D}}}\right)^n \mathrm{P}_n^{(0)}(\cos\psi)\,, \tag{4.13}$$

gravitational potential $V_{\mathrm{D}}(M)$ may now be expressed as

$$\begin{aligned}
V_{\mathrm{D}}(M) = \frac{G}{\varDelta_{\oplus\mathrm{D}}} \int_{M_{\mathrm{D}}} \sum_{m=0}^{\infty} &\left\{\left(\frac{\varrho'}{\varDelta_{\oplus\mathrm{D}}}\right)^m \mathrm{P}_m^{(0)}(\cos\psi')\right.\\
&\times \left[\sum_{n=0}^{\infty}\left(\frac{\varrho}{\varDelta_{\oplus\mathrm{D}}}\right)^n \mathrm{P}_n^{(0)}(\cos\psi)\right]^{m+1}\Bigg\}\,\mathrm{d}m_{\mathrm{D}}\\
= \frac{G}{\varDelta_{\oplus\mathrm{D}}} \int_{M_{\mathrm{D}}} \sum_{m=0}^{\infty} &\left[\left(\frac{\varrho'}{\varDelta_{\oplus\mathrm{D}}}\right)^m \mathrm{P}_m^{(0)}(\cos\psi')\right.\\
&\times \sum_{n=0}^{\infty}\left(\frac{\varrho}{\varDelta_{\oplus\mathrm{D}}}\right)^n \mathrm{C}_n^{(m/2)}(\cos\psi)\bigg]\mathrm{d}m_{\mathrm{D}}
\end{aligned} \tag{4.14}$$

or

$$\begin{aligned}
V_{\mathrm{D}}(M) = \frac{G}{\varDelta_{\oplus\mathrm{D}}} \sum_{m=0}^{\infty} &\left\{\left[\sum_{n=0}^{\infty}\left(\frac{\varrho}{\varDelta_{\oplus\mathrm{D}}}\right)^n \mathrm{C}^{(m/2)}(\cos\psi)\right]\right.\\
&\times \int_{M_{\mathrm{D}}}\left(\frac{\varrho'}{\varDelta_{\oplus\mathrm{D}}}\right)^m \mathrm{P}_m^{(0)}(\cos\psi')\,\mathrm{d}m_{\mathrm{D}}\bigg\}\,,
\end{aligned} \tag{4.15}$$

where $\mathrm{C}_n^{m/2}$ are the ultraspheric (Gegenbauer) polynomials defined by generating function (3.166). As in Section 1.3.2 we shall introduce the potential coefficients of the Moon $(J_m^{(k)})'$, $(S_m^{(k)})'$, defined generally by Eq. (1.72):

$$\begin{aligned}
V_{\mathrm{D}}(M) = \frac{GM_{\mathrm{D}}}{\varDelta_{\oplus\mathrm{D}}} \sum_{m=0}^{\infty} &\left\{\left[\sum_{n=0}^{\infty}\left(\frac{\varrho}{\varDelta_{\oplus\mathrm{D}}}\right)^n \mathrm{C}_n^{(m/2)}(\cos\psi)\right]\sum_{k=0}^{m} v_{\mathrm{D}}^m\right.\\
&\times \left[(J_m^{(k)})'\cos kT'_M - (S_m^{(k)})'\sin kT'_M\right]\mathrm{P}_m^{(k)}(\sin\delta'_M)\bigg\}\,;
\end{aligned} \tag{4.16}$$

δ'_M and T'_M are the selenocentric equatorial coordinates of point M on the Earth's surface; $v_{\mathrm{D}} = (a_0)_{\mathrm{D}}/\varDelta_{\oplus\mathrm{D}}$, where $(a_0)_{\mathrm{D}}$ is an optional quantity of length, e.g. the semimajor axis of the Moon's ellipsoid. The Moon's potential coefficients do not exceed 3×10^{-3} and, therefore, considering the present accuracy of tidal observations, we can put $m = 0$ in (4.16) and use the approximate relation

$$V_{\mathrm{D}}(M) = \frac{GM_{\mathrm{D}}}{\varDelta_{\oplus\mathrm{D}}} \sum_{n=0}^{\infty} \left(\frac{\varrho}{\varDelta_{\oplus\mathrm{D}}}\right)^n \mathrm{C}_n^{(0)}(\cos\psi)\,. \tag{4.17}$$

Similarly,

$$V_{\text{☽}}(O) = \frac{GM_{\text{☽}}}{\Delta_{\text{⊕☽}}} \sum_{m=0}^{\infty} \sum_{k=0}^{m} v_{\text{☽}}^{m}(J_m^{(k)})' \cos kT'_0$$

$$- (S_m^{(k)})' \sin kT'_0] P_m^{(k)}(\sin \delta'_0), \tag{4.18}$$

where δ'_0 and T'_0 are the selenocentric coordinates of the Earth's centre of mass. If we put $m = 0$ in (4.18)

$$V_{\text{☽}}(O) = \frac{GM_{\text{☽}}}{\Delta_{\text{⊕☽}}}. \tag{4.19}$$

Finally, we shall express the gravitational acceleration due to the Moon at the Earth's centre of mass. In view of (2.161),

$$g_{\text{☽}}^{(0)} = \frac{GM_{\text{☽}}}{\Delta_{\text{⊕☽}}^2} \left\{ 1 + (C_0^{(0)})' + \sum_{m=2}^{\infty} \sum_{k=0}^{m} [(C_m^{(k)})' \cos kT'_0 \right.$$

$$\left. - (D_m^{(k)})' \sin kT'_0] P_m^{(k)}(\sin \delta'_0) \right\}; \tag{4.20}$$

coefficients $(C_m^{(k)})'$ and $(D_m^{(k)})'$ are defined as in the Earth's gravity field. However, this case only involves the gravitational component and, therefore, it is necessary to put $\omega = 0$ $(q = 0)$. With $m = 0$ we once again arrive at the approximate relation

$$g_{\text{☽}}(O) = \frac{GM_{\text{☽}}}{\Delta_{\text{⊕☽}}^2}. \tag{4.21}$$

In this approximation, i.e. with $(J_m^{(k)})'$ and $(S_m^{(k)})' = 0$ for all $n > 0$, in other words if we consider the Moon's gravitational field to be spherically symmetrical in view of Eqs. (4.17), (4.19) and (4.21), tidal potential (4.10) can be expressed simply as

$$V_{s\text{☽}} = \frac{GM_{\text{☽}}}{\Delta_{\text{⊕☽}}} \sum_{n=0}^{\infty} \left(\frac{\varrho}{\Delta_{\text{⊕☽}}}\right)^n P_n^{(0)}(\cos \psi) - \frac{GM_{\text{☽}}}{\Delta_{\text{⊕☽}}} - \frac{GM_{\text{☽}}}{\Delta_{\text{⊕☽}}^2} \varrho \cos \psi, \tag{4.22}$$

having taken Eqs. (3.176) into account. Of course

$$P_0^{(0)}(\cos \psi) = 1, \quad P_1^{(0)}(\cos \psi) = \cos \psi, \tag{4.23}$$

and, therefore,

$$V_{s\text{☽}} = \frac{GM_{\text{☽}}}{\Delta_{\text{⊕☽}}} \sum_{n=2}^{\infty} \left(\frac{\varrho}{\Delta_{\text{⊕☽}}}\right)^n P_n^{(0)}(\cos \psi); \tag{4.24}$$

$$P_n^{(0)}(\cos \psi) = P_n^{(0)}(\sin \phi) P_n^{(0)}(\sin \delta_{O'})$$

$$+ 2 \sum_{k=1}^{n} \frac{(n-k)!}{(n+k)!} P_n^{(k)}(\sin \phi) P_n^{(k)}(\sin \delta_{O'}) \cos k(\Lambda + T_{O'}), \tag{4.25}$$

where $\delta_{O'}$ and $T_{O'}$ are the geocentric equatorial coordinates [declination and hour angle relative to the prime meridian $(x_1\,x_3)$] of the Moon's centre of mass O'. The selenocentric gravitational constant $GM_{\mathbb{D}} = 4902.8 \times 10^9\,\mathrm{m}^3\,\mathrm{s}^{-2}$.

At any point M on the surface of an ideally rigid regularized Earth, the resultant gravity potential at the instant of tidal action

$$\bar{W}(M) = W(M) + V_{s\mathbb{D}}(M) = \frac{GM_{\oplus}}{\varrho}\left\{1 + \sum_{n=2}^{\infty}\sum_{k=0}^{n}\left(\frac{a_0}{\varrho}\right)^n (J_n^{(k)}\cos k\Lambda \right.$$

$$+ S_n^{(k)}\sin k\Lambda)\,\mathrm{P}_n^{(k)}(\sin\phi) + \frac{1}{3}q\left(\frac{a_0}{\varrho}\right)^{-3}[1 - \mathrm{P}_2^{(0)}(\sin\phi)]$$

$$\left. + \frac{M_{\mathbb{D}}}{M_{\oplus}}\sum_{n=2}^{\infty}\left(\frac{\varrho}{\varDelta_{\oplus\mathbb{D}}}\right)^{n+1}\mathrm{P}_n^{(0)}(\cos\psi)\right\}; \tag{4.26}$$

the tidal part (for $\psi = 0$) amounts to $\sim 6 \times 10^{-8}$ of the total value of the geopotential. This is of the same order as the geopotential coefficients of the Earth's gravitational field of degrees ~ 10–15. Neither point M nor any other mass element of the Earth change their position in space.

Tidal potential (4.24) defines the tidal force field. However, this can also be defined directly without involving potential $Q_{\oplus\mathbb{D}}$ (or $Q_{\oplus\odot}$) of the centrifugal forces of translational motion about the common barycentre. This situation is illustrated schematically in Fig. 4.3. Gravitational force $\mathbf{F}$ due to the Moon (Sun) acts at point M; at the Earth's centre of mass this force is $\mathbf{F}_O$. The homogeneous force field generates the Earth's translational motion about the common

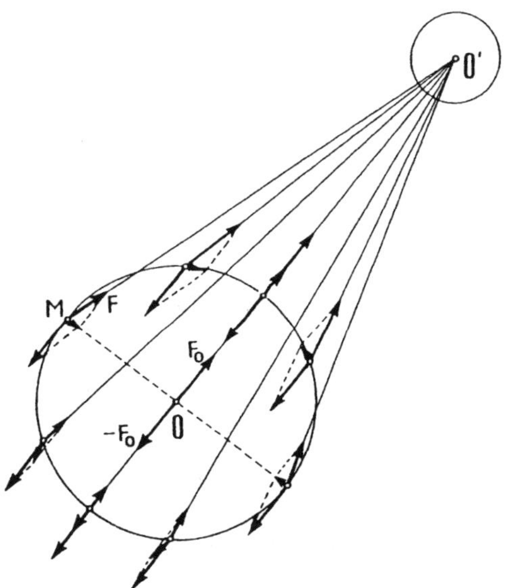

Fig. 4.3. Field of tidal forces

barycentre of the Earth–Moon (Sun) system. The tidal force field **S** is then

$$\mathbf{S} = \mathbf{F} - \mathbf{F}_O \, .$$

If the Moon's gravitational field were spherically symmetrical, both definitions of the tidal force field would be equivalent. However, this is in fact not so since force $\mathbf{F}_O$, acting at the Earth's centre of mass O, does not pass through the Moon's centre of mass O' and, therefore, the potential of forces $\mathbf{F}_O$ is not identical with potential $Q_{\oplus\mathbb{D}}$.

These subtleties of a theoretical nature in the actual definition of the tidal potential only surfaced in recent years when space methods disclosed that the Moon's gravitational field differed from the spherical quite substantially. So far, however, they are still of a rather theoretical significance.

The tidal potential gets its fundamental features from term $n = 2$ with the largest amplitude. The amplitude of the harmonic term of the third degree only amounts to about 1.5% of the principal term $n = 2$, which reads

$$(V_{s\mathbb{D}})_2 = \frac{GM_{\mathbb{D}}}{\varDelta_{\oplus\mathbb{D}}} \left(\frac{\varrho}{\varDelta_{\oplus\mathbb{D}}} \right)^2 \left[(\tfrac{3}{2}\sin^2\phi - \tfrac{1}{2})(\tfrac{3}{2}\sin^2\delta_{O'} - \tfrac{1}{2}) \right.$$
$$+ \tfrac{3}{4}\sin 2\phi \sin 2\delta_{O'} \cos(\varLambda + T_{O'})$$
$$\left. + \tfrac{3}{4}\cos^2\phi \cos^2\delta_{O'} \cos 2(\varLambda + T_{O'}) \right] \, . \tag{4.27}$$

It contains the zonal part $(n = 2, k = 0)$, which is independent of hour angle $T_{O'}$,

$$(V_{s\mathbb{D}})_2^{(0)} = \frac{GM_{\mathbb{D}}}{\varDelta_{\oplus\mathbb{D}}} \left(\frac{\varrho}{\varDelta_{\oplus\mathbb{D}}} \right)^2 (\tfrac{3}{2}\sin^2\phi - \tfrac{1}{2})(\tfrac{3}{2}\sin^2\delta_{O'} - \tfrac{1}{2}) \, , \tag{4.28}$$

the tesseral part $(n = 2, k = 1)$

$$(V_{s\mathbb{D}})_2^{(1)} = \frac{3}{4}\frac{GM_{\mathbb{D}}}{\varDelta_{\oplus\mathbb{D}}} \left(\frac{\varrho}{\varDelta_{\oplus\mathbb{D}}} \right)^2 \sin 2\phi \sin 2\delta_{O'} \cos(\varLambda + T_{O'}) \tag{4.29}$$

and the sectorial part $(n = 2, k = 2)$

$$(V_{s\mathbb{D}})_2^{(2)} = \frac{3}{4}\frac{GM_{\mathbb{D}}}{\varDelta_{\oplus\mathbb{D}}} \left(\frac{\varrho}{\varDelta_{\oplus\mathbb{D}}} \right)^2 \cos^2\phi \cos^2\delta_{O'} \cos 2(\varLambda + T_{O'}) \, . \tag{4.30}$$

Were we to put $\varrho = \mathrm{const}$, i.e. substitute the Earth with a sphere, the geometric locus of points with the same values of the principal term of tidal-potential $n = 2$,

$$(V_{s\mathbb{D}})_2^{(0)} + (V_{s\mathbb{D}})_2^{(1)} + (V_{s\mathbb{D}})_2^{(2)} = (V_{s\mathbb{D}})_2 = \mathrm{const} \, ,$$

relative to the Earth's sphere, would be (in the first approximation) the surface of a rotational ellipsoid whose major axis, lying in meridional plane $\varLambda = -T_{O'}$, would point to the Moon's centre of mass, O' $(\mathbb{D})$ in Fig. 4.4).

The zonal part, which can be expressed as

$$(V_{s\mathbb{D}})_2^{(0)} = \frac{1}{16}\frac{GM_{\mathbb{D}}}{\varDelta_{\oplus\mathbb{D}}} \left(\frac{\varrho}{\varDelta_{\oplus\mathbb{D}}} \right)^2 (1 - 3\cos 2\phi)(1 - 3\cos 2\delta_{O'}) \, , \tag{4.31}$$

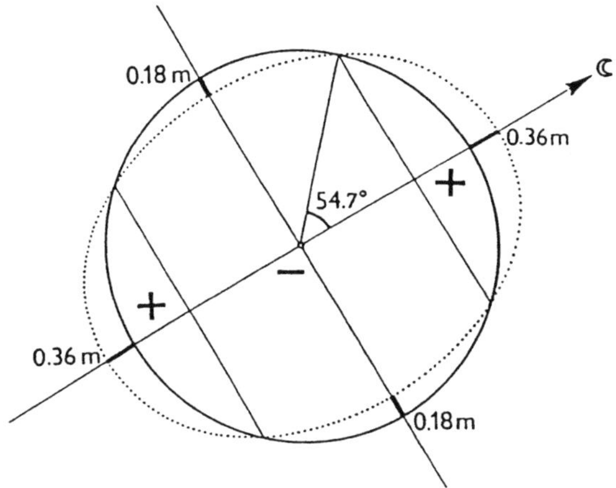

Fig. 4.4. Tidal perturbations of the idealized boundary equipotential surface of the geopotential

contains a term which is constant for the Earth as a whole, regardless of the position of potential point M,

$$\frac{1}{16}\frac{GM_{\mathbb{D}}}{\Delta_{\oplus\mathbb{D}}}\left(\frac{\varrho}{\Delta_{\oplus\mathbb{D}}}\right)^2 ,$$

a term which is constant along a particular terrestrial latitude, $\phi = \text{const}$,

$$-\frac{1}{16}\frac{GM_{\mathbb{D}}}{\Delta_{\oplus\mathbb{D}}}\left(\frac{\varrho}{\Delta_{\oplus\mathbb{D}}}\right)^2 \cos 2\phi ,$$

and, finally, a term which varies periodically with the Moon's declination ($\sim \pm 29°$) and a period of ~ 14 days,

$$-\frac{3}{16}\frac{GM_{\mathbb{D}}}{\Delta_{\oplus\mathbb{D}}}\left(\frac{\varrho}{\Delta_{\oplus\mathbb{D}}}\right)^2 \cos 2\delta_{0'}(1 - 3\cos 2\phi)$$

and whose integral mean value is

$$\sim -\frac{9\sqrt{3}}{32\pi}\frac{GM_{\mathbb{D}}}{\Delta_{\oplus\mathbb{D}}}\left(\frac{\varrho}{\Delta_{\oplus\mathbb{D}}}\right)^2 (1 - 3\cos 2\phi) .$$

The tesseral part (4.29) is short-period with a period of ~ 24 h, the sectorial part (4.30) is also short-period with a period of ~ 12 h. Distance $\Delta_{\oplus\mathbb{D}}$ of the centres of mass of both bodies, however, is not constant but varies periodically within the interval of $\sim (364 - 407) \times 10^3$ km, and also contributes to the variability of $V_{s\mathbb{D}}$.

The lines of zero values of the separate tidal parts (4.23), (4.29) and (4.30) are shown in Fig. 4.5. Since $P_2^{(0)}(\sin\phi) = 0$ for $\phi = \pm 35.264°$ and $P_2^{(0)}(\sin\delta_{0'}) < 0$ always, the sign of the zonal part,

$$(V_{s\mathbb{D}})_2^{(0)} = \frac{GM_{\mathbb{D}}}{\Delta_{\oplus\mathbb{D}}}\left(\frac{\varrho}{\Delta_{\oplus\mathbb{D}}}\right)^2 P_2^{(0)}(\sin\phi)\,P_2^{(0)}(\sin\delta_{0'}) \tag{4.32}$$

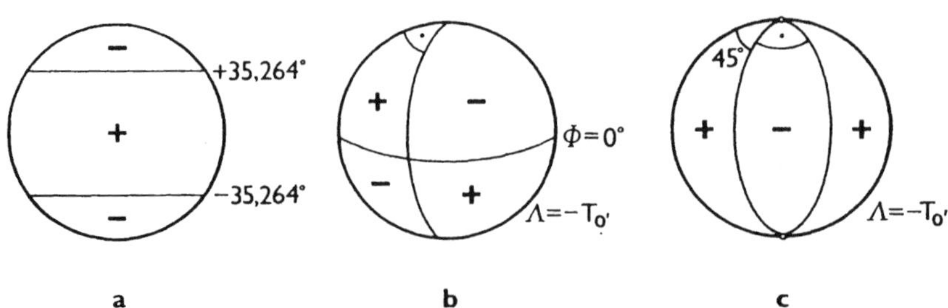

Fig. 4.5. Schematic global diagram of **a** zonal, **b** tesseral and **c** sectorial components

is always opposite to that of the second zonal Legendre polynomial $P_2^{(0)}$ $(\sin \phi)$. This part then generates the long-period variations in the tidal and gravity potential and, since the variations depend on latitude, the shapes of the equipotential surfaces change too.

Zonal part $(V_{s})^{(0)}_2$ of the tidal potential is responsible for the radial displacement $\delta\varrho$ of the equipotential surfaces of gravity potential $W = C = W_M$ after tidal deformation. It represents the work of the tidal forces along the path (Fig. 4.1) from the position before (M) to the position after the deformation (M'').

If the angle between the normal to surface $W = \text{const} = C$ and the geocentric radius-vector is neglected, then according to Bruns theorem (Pick et al. 1973) the radial displacement $\delta\varrho = \varrho(M'') - \varrho(M)$ becomes

$$\delta\varrho = -\frac{1}{\dfrac{\partial W}{\partial \varrho}} V_{s}.$$ (4.33)

Quantity $\delta\varrho$ is referred to as the height of the static tidal wave of an equipotential surface of the perfectly rigid Earth.

If $\psi = 0°, 180°$ (Fig. 4.4)

$$\delta\varrho = \frac{M_{\mathbb{D}}}{M_\oplus}\left(\frac{\varrho}{\varDelta_{\oplus\mathbb{D}}}\right)^3 \varrho \sim 0.36 \text{ m},$$ (4.34)

and if $\psi = 90°, 270°$

$$\delta\varrho = -\frac{1}{2}\frac{M_{\mathbb{D}}}{M_\oplus}\left(\frac{\varrho}{\varDelta_{\oplus\mathbb{D}}}\right)^3 \varrho \sim -0.18 \text{ m},$$ (4.35)

where $\dfrac{M_\oplus}{M_{\mathbb{D}}} = 81.302$. This means that if the Earth's gravitational field were ideally spherically symmetrical (and the effect of rotation were disregarded), spherical equipotential surfaces would flatten into roughly ellipsoidal shape. The major axes of these ellipsoids would copy the line connecting the mass centres of the Earth and the Moon, and the flattenings of the equipotential surfaces close to the Earth's surface would amount to $\sim 8 \times 10^{-8}$.

The zonal deformation is obtained by substituting the zonal part of tidal potential $(V_{s\mathrm{D}})_2^{(0)}$ on the rhs of (4.33). This then yields a subsidence of -0.18 m at the terrestrial poles and a maximum elevation of $+0.09$ m at the equator, given $\delta_{O'} = 0°$ in both cases. This means that the zonal tidal term causes a flattening of the equipotential surface of gravity

$$\delta\alpha \sim 4.2 \times 10^{-8}, \tag{4.36}$$

and changes the denominator of the Earth's polar flattening, which is $\sim 1/298.257$ for an Earth unperturbed by tides, by as much as 0.0038 (for details see Sect. 5.4).

The zonal tidal term is thus responsible for an additional flattening of the Earth's equipotential surfaces and varies periodically with the Moon's declination. Its periodicity is given by the Moon's declination varying in the interval of $\sim \pm 29°$. Maximum zonal deformations occur if $\delta_{O'} = 0°$. For $\delta_{O'} = \pm 30°$ function $P_2^{(0)} (\sin \delta_{O'})$ displays the least possible value and the deformations are minimal (subsidence of ~ -0.04 m at the poles, elevation of ~ 0.02 m at the equator).

The tesseral part (4.29), i.e.

$$(V_{s\mathrm{D}})_2^{(1)} = \frac{1}{3} \frac{GM_{\mathrm{D}}}{\Delta_{\oplus\mathrm{D}}} \left(\frac{\varrho}{\Delta_{\oplus\mathrm{D}}}\right)^2 P_2^{(1)}(\sin \phi)\, P_2^{(1)}(\sin \delta_{O'})\cos(\Lambda + T_{O'}), \tag{4.37}$$

is zero if $P_2^{(1)} (\sin \phi) = 0$, or $\cos(\Lambda + T_{O'}) = 0$, or $P_2^{(1)} (\sin \delta_{O'}) = 0$, i.e. for $\phi = 0°$, or $\phi = \pm 90°$, or $\Lambda + T_{O'} = 90°, 270°$, or $\delta_{O'} = 0$. Figure 4.5b shows the signs and lines of zero value of the tesseral tidal part for $\delta_{O'} > 0$; if $\delta_{O'} < 0$ the signs are reversed. Tesseral perturbations are thus not symmetrical with respect to axis x_3; they are short-period with a period of 1 day, and the maximum absolute value of their amplitude occurs at the points of intersection of parallels $\phi = \pm 45°$ with meridian $\Lambda = -T_{O'}$, passing through the Moon's centre of mass. For $\delta_{O'} = 30°$, $\phi = 45°$ and $\Lambda = -T_{O'}$, the elevation is $+0.23$ m.

The tidal sectorial part (4.30), i.e.

$$(V_{s\mathrm{D}})_2^{(2)} = \frac{1}{12} \frac{GM_{\mathrm{D}}}{\Delta_{\oplus\mathrm{D}}} \left(\frac{\varrho}{\Delta_{\oplus\mathrm{D}}}\right)^2 P_2^{(2)}(\sin \phi)\, P_2^{(2)}(\sin \delta_{O'})\cos 2(\Lambda + T_{O'}), \tag{4.38}$$

is zero if $P_2^{(2)} (\sin \phi) = 0$, or $\cos 2(\Lambda + T_{O'}) = 0$, i.e. for $\phi = \pm 90°$, or for meridians $\Lambda = -T_{O'} + 45°$ and $\Lambda = -T_{O'} + 135°$. Its period is thus semidiurnal. The lines of zero value of the sectorial harmonic term and the relevant signs are shown in Fig. 4.5c. Neither variations of the Moon's declination nor of the latitude, ϕ, of the potential point have an effect on the sign. Maximum elevation occurs for $\delta_{O'} = 0$ on the equator at a point with longitude $\Lambda = -T_{O'}$ and at a point with longitude $\Lambda = -T_{O'} + 180°$ (in the meridional plane containing the Moon's centre of mass), and amounts to $+0.26$ m. Maximum subsidence occurs for $\delta_{O'} = 0$ at equatorial points with longitudes $\Lambda = -T_{O'} + 90°$ and $\Lambda = -T_{O'} + 270°$, and amounts to -0.26 m.

We have been considering a single body causing the tidal perturbations of the Earth's equipotential surfaces, namely, the Moon. However, the tidal potential is also generated by the Sun. In analogy with (4.24) the tidal potential due to the Sun is

$$V_{s\odot} = \frac{GM_\odot}{\varDelta_{\oplus\odot}} \sum_{n=2}^{\infty} \left(\frac{\varrho}{\varDelta_{\oplus\odot}}\right)^n P_n^{(0)}(\cos\psi_\odot),$$
(4.39)

where $\psi_\odot$ is the angle between the geocentric radius-vector of point M and the geocentric radius-vector of the Sun's centre of mass. The heliocentric gravitational constant

$$GM_\odot = 132\,712\,440 \times 10^{12}\,\mathrm{m}^3\,\mathrm{s}^{-2}.$$

Retaining the principal terms $n = 2$, the ratio oi the two tidal potentials

$$\frac{V_{s\mathbb{D}}}{V_{s\odot}} = \frac{GM_\mathbb{D}}{GM_\odot} \left(\frac{\varDelta_{\oplus\odot}}{\varDelta_{\oplus\mathbb{D}}}\right)^3 \frac{P_2^{(0)}(\cos\psi_\mathbb{D})}{P_2^{(0)}(\cos\psi_\odot)},$$
(4.40)

where

$$\frac{GM_\mathbb{D}}{GM_\odot} \left(\frac{\varDelta_{\oplus\odot}}{\varDelta_{\oplus\mathbb{D}}}\right)^3 \sim 2.$$
(4.41)

The lunar and solar tidal terms are in this approximate ratio ($\sim 2:1$) and their sum represents the aggregate effect. Its maximum occurs if the mass centres of all three bodies are located in the plane of a single meridian (at full or new moon) and the minimum occurs at quarter moon (the Moon is at the beginning of the second or last quarter when the tidal potential of the Sun is subtracted from that of the Moon).

Of the other bodies of the Solar System, the largest tidal perturbations are caused by Venus and Jupiter. Their planetocentric gravitational constants are

$$GM_\varphi = 324\,858.8 \times 10^9\,\mathrm{m}^3\,\mathrm{s}^{-2},$$
$$GM_4 = 126\,686\,900 \times 10^9\,\mathrm{m}^3\,\mathrm{s}^{-2}.$$
(4.42)

Once again, retaining only the principal terms $n = 2$, the ratio of the lunar tidal potential to that of Venus and Jupiter is

$$\frac{GM_\mathbb{D}}{GM_\varphi} \left(\frac{\varDelta_{\oplus\varphi}}{\varDelta_{\oplus\mathbb{D}}}\right)^3 \frac{P_2^{(0)}(\cos\psi_\mathbb{D})}{P_2^{(0)}(\cos\psi_\varphi)},$$
$$\frac{GM_\mathbb{D}}{GM_4} \left(\frac{\varDelta_{\oplus 2}}{\varDelta_{\oplus\mathbb{D}}}\right)^3 \frac{P_2^{(0)}(\cos\psi_\mathbb{D})}{P_2^{(0)}(\cos\psi_2)},$$
(4.43)

with

$$\frac{GM_\mathbb{D}}{GM_\varphi} \left(\frac{\varDelta_{\oplus\varphi}}{\varDelta_{\oplus\mathbb{D}}}\right)^3 \sim 5 \times 10^6 - 2 \times 10^4$$
(4.44)

for the maximum and minimum geocentric distance of Venus, $(259 - 40) \times 10^6$ km, and

$$\frac{GM_{\mathbb{D}}}{GM_4}\left(\frac{\Delta_{\oplus 4}}{\Delta_{\oplus \mathbb{D}}}\right)^3 \sim (6.1 - 1.4) \times 10^5$$

for the maximum and minimum geocentric distance of Jupiter $(965 - 591) \times 10^6$ km. With regard to the instruments currently available, the tidal forces due to Venus, Jupiter and other bodies of the Solar System cannot be measured. The maximum tidal effects of the planets on the Earth as compared with lunisolar effects are given in Table 4.1 (M is the mass of the perturbing, i.e. tide-generating, body).

In defining the fundamental equipotential surface (the geoid) it is necessary to decide whether to take into account the permanent (zero-frequency) part of the zonal term (4.32) in the tidal potential (see Appendix), because the integral mean value of function $P_2^{(0)}(\sin \delta)$ is the same if the tide-generating body is the Moon ($\delta = \delta_{0'}$) or the Sun ($\delta - \delta_{0''}$), specifically (Honkasalo 1964; Zadro and Marussi 1973)

$$[P_2^{(0)}(\sin \delta)]_{\text{mean}} = \tfrac{3}{4}\sin^2 \varepsilon_0 - \tfrac{1}{2} = \tfrac{1}{2}[P_2^{(0)}(\sin \varepsilon_0) - \tfrac{1}{2}],$$

where ε_0 is the angle between the Earth's equatorial plane and the plane of the ecliptic; for epoch J2000

$$\varepsilon_0 = 23°26'21.4119''.$$

Table 4.1. Maximum tidal effects of the Moon, Sun and planets on the Earth: radial tidal displacements of equipotential surfaces of a perfectly rigid Earth ($GM_\oplus = 398\,600.440 \times 10^9$ m^3 s^{-2})

Body	GM $[10^9$ m^3 s$^{-2}]$	$M/M_\oplus$	Minimum geocentric distance $[10^9$ m$]$	$\delta\varrho$ [m]
Moon	4 902.799	0.012 300	0.3565	0.36
Sun	13 271 244.0 $\times 10^4$	332 946.0	147	0.17
Venus	324 858.60	0.814 999	40	2×10^{-5}
Jupiter	126 686 537.0	317.828	591	2×10^{-6}
Mars	42 828.3	0.107 447	56	10^{-6}
Mercury	22 032.09	0.055 274	82	10^{-7}
Saturn	37 931 272.0	95.161 1	1199	6×10^{-8}
Uranus	5 793 939.0	14.535 7	2586	10^{-9}
Neptune	6 835 096.0	17.148	4309	4×10^{-10}

The permanent (or zero-frequency) part of the zonal tidal term (marked with a bar), generated by both these bodies, is thus (Rapp et al. 1991)

$$(\bar{V}_{s\,\rm D})_2^{(0)} + (\bar{V}_{s\,\odot})_2^{(0)}$$

$$= \frac{1}{2}\left[P_2^{(0)}(\sin\varepsilon_0) - \frac{1}{2}\right] P_2^{(0)}(\sin\phi)\left[\frac{GM_{\rm D}}{\varDelta_{\oplus\rm D}}\left(\frac{\varrho}{\varDelta_{\oplus\rm D}}\right)^2\right.$$

$$\left. + \frac{GM_\odot}{\varDelta_{\oplus\odot}}\left(\frac{\varrho}{\varDelta_{\oplus\odot}}\right)^2\right] = -[1.9538\ \rm m^2\,s^{-2}]P_2^{(0)}(\sin\phi)$$

$$= -[3.126\times10^{-8}\,W_0]P_2^{(0)}(\sin\phi)$$

(see Appendix).

The appropriate radial displacement of the equipotential surfaces (in metres) is $-0.199\ P_2^{(0)}(\sin\phi)$.

4.3 Tide-Generating Potential of a Perfectly Elastic Earth

We have so far considered the Earth to be perfectly rigid and undeformable. The tidal perturbations we dealt with concerned only the equipotential surfaces of the gravity potential. However, the Earth is not perfectly rigid but is capable of deformation. We shall now consider it to be perfectly elastic in shape. Mass displacements then occur within the body itself, potential point M on the Earth's surface moves to position M' (Fig. 4.6) and additional potential $\delta V_{s\rm D}$ is generated as a result of the mass transfer within the body.

Assume once again that W is the gravity potential prior to tidal action, V_S the tidal potential (we shall omit the symbol $\rm D$ now to save space) and $\bar{W}(M')$ the

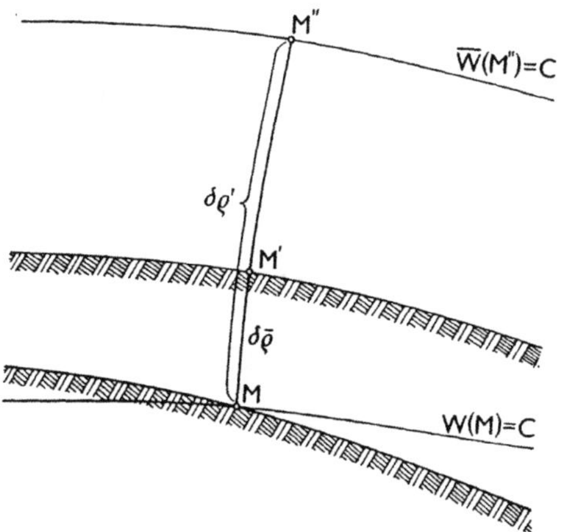

Fig. 4.6. Radial tidal displacement of the Earth's crust

resultant potential at M' after deformation; hence

$$\bar{W}(M') = W(M') + V_s(M') + \delta V_s(M')$$

$$= W(M) + \frac{\partial W}{\partial \varrho}\delta\bar{\varrho} + V_s(M') + \delta V_s(M');$$

$$\delta\bar{\varrho} = [\varrho(M') - \varrho(M)] \,. \tag{4.45}$$

As regards the radial displacement of the original equipotential surface $W(M) = C$, this will be larger due to δV_s. The surface will move to position M'' (Fig. 4.6) and

$$\bar{W}(M'') = C = W(M'') + V_s + \delta V_s = W(M) + \frac{\partial W}{\partial \varrho}\delta\varrho' + V_s + \delta V_s, \tag{4.46}$$

with $\delta\varrho' = \varrho(M'') - \varrho(M) > \delta\varrho$, i.e. the radial displacement of the equipotential surface in an elastic Earth is larger than in a perfectly rigid Earth (Fig. 4.1).

According to Hooke's law,

$$\frac{\delta V_s}{V_s} = k \,, \tag{4.47}$$

$$\frac{\delta\bar{\varrho}}{\delta\varrho} = h \,. \tag{4.48}$$

We have again neglected the deflection of geocentric radius-vector ϱ from the normal to surface $W = \text{const}$; the constants of proportionality h and k are called Love numbers. The term k expresses the ratio of the additional potential, generated by mass transfer, to the tidal potential; h is the ratio of the radial displacement (tidal amplitude) of the Earth's crust to the radial displacement $\delta\varrho$ of the equipotential surface of the tidal potential for a perfectly rigid Earth (4.33). For a perfectly rigid body $k = 0$, $h = 0$; for a body of homogeneous fluid $k = 3/2$, $h = 5/2$, as indicated by (4.50). For a homogeneous spherical body of radius R, density σ, gravity g on its surface and shear modulus μ (see Table 2.13) $k = \frac{3}{2}[1 + 19\mu/(2g\sigma R)]^{-1}$.

With sufficient accuracy

$$h = \frac{\dfrac{\partial W}{\partial \varrho}\delta\bar{\varrho}}{\dfrac{\partial W}{\partial \varrho}\delta\varrho} = \frac{W(M') - C}{\bar{W}(M) - C} = \frac{\bar{W}(M) - \bar{W}(M')}{V_s} = -\frac{\dfrac{\partial W}{\partial \varrho}\delta\bar{\varrho}}{V_s} \,. \tag{4.49}$$

The relation between the radial displacements of equipotential surfaces of a perfectly elastic Earth, $\delta\varrho'$, and of a perfectly rigid Earth, $\delta\varrho$, reads

$$\frac{\delta\varrho'}{\delta\varrho} = \frac{\dfrac{\partial W}{\partial \varrho}\delta\varrho'}{\dfrac{\partial W}{\partial \varrho}\delta\varrho} = \frac{V_s + \delta V_s}{V_s} = 1 + \frac{\delta V_s}{V_s} = 1 + k \,. \tag{4.50}$$

Taking into account (4.47)–(4.50), (4.45) can be expressed as follows:

$$\bar{W}(M') - W(M) = V_s(1 + k - h),$$

(4.51)

having put $\partial V_s/\partial \varrho = 0$ and $\partial \delta V_s/\partial \varrho = 0$ in interval MM''; similarly,

$$\bar{W}(M') - W(M') = V_s + \delta V_s = V_s(1 + k)$$

(4.52)

and (4.45) and (4.49) also yield

$$W(M') - W(M) = -hV_s.$$

(4.53)

Constants k and h can be determined from observations of tidal force components (Melchior 1983). However, their values, calculated using different data, differ quite considerably. Here we shall adopt (for $n = 2$)

$$k = 0.30, \quad h = 0.62$$

(4.54)

$$1 + k - h = 0.68.$$

(4.55)

The radial displacements of equipotential surfaces of the elastic Earth are larger than those of a perfectly rigid Earth as illustrated by the overview in Table 4.2. The zonal perturbations of equipotential surfaces of the perfectly elastic Earth also produce a larger flattening than in a perfectly rigid Earth, $\sim 5.5 \times 10^{-8}$ (as much as 0.0049 in the flattening denominator), and the constant part of the zonal deformation is 1.3 times larger than for the perfectly rigid Earth.

In view of (4.49) the radial tidal perturbations of the Earth's crust, $\delta \bar{\varrho} = \varrho(M') - \varrho(M)$,

$$\delta \bar{\varrho} = -h \frac{V_s}{\dfrac{\partial W}{\partial \varrho}};$$

(4.56)

the figures for the separate tidal terms are given in Table 4.2.

Hence, tidal deformations take place in the perfectly elastic Earth and, as a consequence, the Earth's ellipsoid of inertia changes its orientation and shape, because the tensor of inertia changes (Sect. 5.4). The zonal tidal terms (14-day due to the Moon, 6-month due to the Sun; the periodicity is due to the square of the sine of the declinations of the Moon and the Sun) are responsible for the long-period variations of its flattening and for the variations of the angular velocity of the Earth's rotation (with periods of 14 days and 6 months): if $C\omega = \text{const}$, (Sect. 3.5),

$$\frac{\delta \omega}{\omega} = -\frac{\delta C}{C},$$

(4.57)

where δC is the tidal variation of the largest principal Earth's moment of inertia, which practically does not change the direction of the smallest axis of the ellipsoid of inertia. The initial estimates of the variations of angular velocity of the Earth's rotation, generated by tidal deformations of the Earth, were presented by Jeffreys (1928). As a result of the displacement of terrestrial masses the

Table 4.2. Radial lunar tidal perturbations of equipotential surfaces of the gravity potential and of the Earth's crust

Type of perturbation	Position	$\delta\varrho$ $(10^{-2}\,\text{m})$	$\delta\varrho'$ $(10^{-2}\,\text{m})$	$\delta\bar{\varrho}$ $(10^{-2}\,\text{m})$
Total	$\psi = 0°, 180°$	36	46	21
	$\psi = 90°, 270°$	-18	-23	-11
Zonal	$\delta_{O'} = 0°, \phi = \pm 90°$	-18	-23	-11
	$\delta_{O'} = 0°, \phi = 0°$	9	12	5
Tesseral	$\delta_{O'} = 29°, \phi = 45°,$ $\Lambda = -T_{O'}$	15	19	9
Sectorial	$\delta_{O'} = 0°, \phi = 0°,$ $\Lambda = -T_{O'}$	26	34	15

zonal term again has a permanent (zero-frequency) part (sc. indirect effect),

$$[-3.126k \times 10^{-8} W_0]\,\text{P}_2^{(0)}(\sin\phi)\,;$$

the appropriate radial deformation of the equipotential surfaces (in metres) is $-0.199k\text{P}_2^{(0)}(\sin\phi)$.

The tesseral tidal terms cause precession and nutation of the smallest axis of the ellipsoid of inertia, but have no practical effect on the magnitude of the largest moment of inertia. These diurnal tidal waves perturb the Earth's ellipsoid of inertia in that its smallest axis moves periodically about the instantaneous axis of rotation. Since tidal deformations do not take place instantaneously, but lag behind rotation, the sectorial tidal terms cause secular deceleration of the Earth's rotation (Sect. 3.5). Sectorial perturbations have practically no effect on either the periodic changes in rotation or variations in the direction of the axis of rotation. They do not perturb the largest moment C, but they do cause variations in A and B.

The tidal changes of the tensor of inertia are discussed in detail in Section 5.4.

4.4 Additional Potential in Outer Space due to the Earth's Tidal Deformation

Tidal forces are the cause of additional potential δV_s (4.47), generated by the tidal deformation of the Earth. In general, at a point on the Earth's surface

$$\delta V_s = \frac{GM_{\mathbb{D}}}{\Delta_{\oplus\mathbb{D}}} \sum_{n=2}^{\infty} k_n \left(\frac{\varrho}{\Delta_{\oplus\mathbb{D}}}\right)^n \text{P}_n^{(0)}(\cos\psi_{\mathbb{D}})$$

$$+ \frac{GM_{\odot}}{\Delta_{\oplus\odot}} \sum_{n=2}^{\infty} k_n \left(\frac{\varrho}{\Delta_{\oplus\odot}}\right)^n \text{P}_n^{(0)}(\cos\psi_{\odot})\,. \tag{4.58}$$

Since this is a value on a boundary surface, we shall put $\delta V_s = \overline{\delta V_s}$. We shall seek to determine δV_s in outer space, i.e. to solve Dirichlet's first boundary-value problem for this particular case. For conciseness we shall only consider a single tide-generating body, the Moon.

As per (1.71), at general external point $N(\Delta, \delta, T)$ (Fig. 4.7)

$$\delta V_s = \frac{GM_\oplus}{\Delta} \sum_{n=2}^{\infty} \sum_{j=0}^{n} \left(\frac{a_0}{\Delta}\right)^n (\delta J_n^{(j)} \cos jT - \delta S_n^{(j)} \sin jT) P_n^{(j)}(\sin \delta), \qquad (4.59)$$

where $\delta J_n^{(j)}$ and $\delta S_n^{(j)}$ are the variations of the geopotential coefficients due to tidal deformations; these are expressed as dimensionless parameters (normalized to $M_\oplus a_0^n$). The order of the spherical harmonics is now denoted j to distinguish it from the Love number k.

Constants $\delta J_n^{(j)}$ and $\delta S_N^{(j)}$ must be determined so that (4.59) yields boundary value $\overline{\delta V_s}$ on the boundary surface, i.e. for $\Delta = \varrho, \delta = \phi, T = -\Lambda$.

For the sake of simplicity we shall consider the boundary surface to be a sphere $\bar{S}$, i.e. we shall put $\varrho = \text{const} = R$, and develop $\overline{\delta V_s}$ into a series of spherical harmonics on this sphere. On the sphere the spherical harmonics satisfy conditions of orthogonality, and for coefficients $p_n^{(j)}$ and $q_n^{(j)}$ in the expansion of $\overline{\delta V_s}$ on boundary sphere $\bar{S}$,

$$\overline{\delta V_s}(\phi, \Lambda) = \sum_{n=0}^{\infty} \sum_{j=0}^{n} (p_n^{(j)} \cos j\Lambda + q_n^{(j)} \sin j\Lambda) P_n^{(j)}(\sin \phi); \qquad (4.60)$$

evidently

$$\begin{aligned}
\frac{p_n^{(j)}}{q_n^{(j)}} &= \frac{\int_{\bar{S}} \overline{\delta V_s} P_n^{(j)}(\sin \phi) \genfrac{}{}{0pt}{}{\cos j\Lambda}{\sin j\Lambda}\, d\bar{S}}{\int_{\bar{S}} \left[P_n^{(j)}(\sin \phi) \genfrac{}{}{0pt}{}{\cos j\Lambda}{\sin j\Lambda} \right]^2 d\bar{S}} \\[2mm]
&= \frac{(2 - \delta_{j,0})(2n + 1)(n - j)!}{4\pi(n + j)!} \int_{S} \overline{\delta V_s} P_n^{(j)}(\sin \phi) \genfrac{}{}{0pt}{}{\cos j\Lambda}{\sin j\Lambda}\, d\bar{S}, \qquad (4.61)
\end{aligned}$$

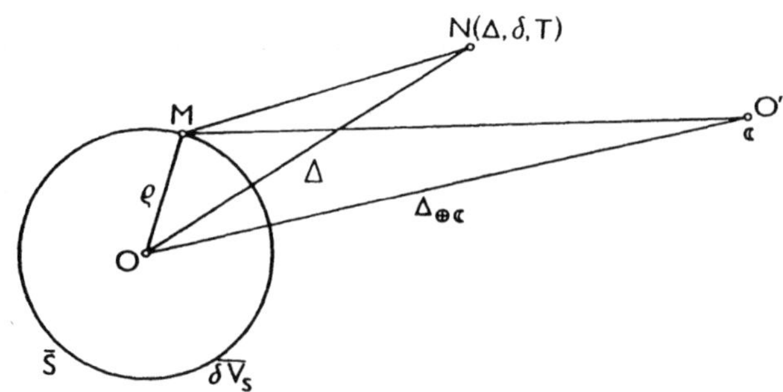

Fig. 4.7. Diagram used to derive the additional potential in outer space

where $\delta_{j,0}$ is the Kronecker symbol. In a more general solution, where the boundary surface is, for example, the surface of the geoid, $W = W_0$, the expansion would be more complicated. The tidal perturbation theory as a whole has so far been developed for the case of a sphere, so that a more accurate solution at this point would be inappropriate.

Since

$$\overline{\delta V_s} = \sum_{n=2}^{\infty} k_n \sum_{j=0}^{n} (V_s)_n^{(j)}, \tag{4.62}$$

and if $(V_s)_n^{(j)}$ are the separate harmonic terms of n-th degree and j-th order of tide-generating potential (4.24) (the symbol $\mathbb{D}$ has been omitted),

$$(V_s)_n^{(j)} = \frac{GM_{\mathbb{D}}}{\Delta_{\oplus\mathbb{D}}} \left(\frac{R}{\Delta_{\oplus\mathbb{D}}}\right)^n (2 - \delta_{j,0}) \frac{(n-j)!}{(n+j)!}$$
$$\times [\,P_n^{(j)}(\sin\phi)\,P_n^{(j)}(\sin\delta_{O'})\cos j(\Lambda + T_{O'})]\,, \tag{4.63}$$

then clearly also

$$\begin{matrix} p_n^{(j)} \\ q_n^{(j)} \end{matrix} = (2 - \delta_{j,0})\frac{(n-j)!}{(n+j)!}\frac{GM_{\mathbb{D}}}{\Delta_{\oplus\mathbb{D}}} k_n \left(\frac{R}{\Delta_{\oplus\mathbb{D}}}\right)^n P_n^{(j)}(\sin\delta_{O'}) \begin{matrix} \cos jT_{O'} \\ -\sin jT_{O'} \end{matrix}. \tag{4.64}$$

Expansion (4.60) must be identical with (4.59) if $\Delta = R, \delta = \phi, T = -\Lambda$, and the following conditions must be satisfied:

$$\frac{GM_{\oplus}}{R}\left(\frac{a_0}{R}\right)^n \begin{matrix} \delta J_n^{(j)} \\ \delta S_n^{(j)} \end{matrix} = \begin{matrix} p_n^{(j)} \\ q_n^{(j)} \end{matrix}. \tag{4.65}$$

Consequently,

$$\begin{matrix} \delta J_n^{(j)} \\ \delta S_n^{(j)} \end{matrix} = \frac{R^{n+1}}{GM_{\oplus} a_0^n} \begin{matrix} p_n^{(j)} \\ q_n^{(j)} \end{matrix}$$
$$= (2 - \delta_{j,0})\frac{(n-j)!}{(n+j)!} k_n \frac{M_{\mathbb{D}}}{M_{\oplus}}\left(\frac{R}{a_0}\right)^n \left(\frac{R}{\Delta_{\oplus\mathbb{D}}}\right)^{n+1} P_n^{(j)}(\sin\delta_{O'}) \begin{matrix} \cos jT_{O'} \\ -\sin jT_{O'} \end{matrix}, \tag{4.66}$$

which follows by direct comparison of (4.59) with the lunar part in (4.58), anyway. After substituting into (4.59) we arrive at the potential, generated by the Earth's tidal deformations, in outer space:

$$\delta V_s = \frac{GM_{\mathbb{D}}}{\Delta} \sum_{n=2}^{\infty} k_n \left(\frac{R}{\Delta}\right)^n \left(\frac{R}{\Delta_{\oplus\mathbb{D}}}\right)^{n+1} \sum_{j=0}^{n} (2 - \delta_{j,0})\frac{(n-j)!}{(n+j)!}$$
$$\times P_n^{(j)}(\sin\delta)\,P_n^{(j)}(\sin\delta_{O'})\cos j(T - T_{O'})\,. \tag{4.67}$$

If we restrict ourselves to harmonic spherical terms of the second degree in the expansion of the tidal potential, the variations of geopotential coefficients (4.66) due to tidal deformations will also be harmonic coefficients of the second degree (for details refer to Sect. 5.4):

$$\delta J_2^{(0)} = \frac{M_{\mathbb{D}}}{M_{\oplus}} k_2 \left(\frac{R}{a_0}\right)^2 \left(\frac{R}{\Delta_{\oplus\mathbb{D}}}\right)^3 \mathrm{P}_2^{(0)}(\sin \delta_{O'}),$$

$$\begin{matrix} \delta J_2^{(1)} \\ \delta S_2^{(1)} \end{matrix} = \frac{1}{3} \frac{M_{\mathbb{D}}}{M_{\oplus}} k_2 \left(\frac{R}{a_0}\right)^2 \left(\frac{R}{\Delta_{\oplus\mathbb{D}}}\right)^3 \mathrm{P}_2^{(1)}(\sin \delta_{O'}) \quad \begin{matrix} \cos T_{O'} \\ -\sin T_{O'} \end{matrix},$$

$$\begin{matrix} \delta J_2^{(2)} \\ \delta S_2^{(2)} \end{matrix} = \frac{1}{12} \frac{M_{\mathbb{D}}}{M_{\oplus}} k_2 \left(\frac{R}{a_0}\right)^2 \left(\frac{R}{\Delta_{\oplus\mathbb{D}}}\right)^3 \mathrm{P}_2^{(2)}(\sin \delta_{O'}) \quad \begin{matrix} \cos 2T_{O'} \\ -\sin 2T_{O'} \end{matrix}. \tag{4.68}$$

Coefficients $\delta J_2^{(0)}$ and $\delta J_2^{(2)}$ can be used to express the deformation variations, $\delta A, \delta B, \delta C$, of the principal moments of the Earth's inertia assuming that (Munk and MacDonald 1960)

$$\delta A + \delta B + \delta C = 0. \tag{4.69}$$

Equations (2.247) immediately yield

$$\delta C - \delta A = M_{\oplus} a_0^2 (-\delta J_2^{(0)} + 2\delta J_2^{(2)}),$$

$$\delta C - \delta B = M_{\oplus} a_0^2 (-\delta J_2^{(0)} - 2\delta J_2^{(2)}),$$

$$\delta B - \delta A = 4M_{\oplus} a_0^2 \delta J_2^{(2)},$$

and, in view of the given condition,

$$\delta C = -\tfrac{2}{3} M_{\oplus} a_0^2 \delta J_2^{(0)},$$

$$\delta A = \tfrac{1}{3} M_{\oplus} a_0^2 \delta J_2^{(0)} - 2M_{\oplus} a_0^2 \delta J_2^{(2)},$$

$$\delta B = \tfrac{1}{3} M_{\oplus} a_0^2 \delta J_2^{(0)} + 2M_{\oplus} a_0^2 \delta J_2^{(2)}. \tag{4.70}$$

Perturbations (4.68) are of the order of 10^{-8} or smaller. They can be used to calculate the variations along the axes of the Earth's ellipsoid of inertia in terms of variations $\delta v, \delta \varepsilon, \delta \mu$ (Sect. 5.4; Burša 1983) of Cardan's angles defined by Eqs. (5.131):

$$\delta v = -1.03'' \mathrm{P}_2^{(1)}(\sin \delta_{O'}) \cos T_{O'},$$

$$\delta \varepsilon = 1.03'' \mathrm{P}_2^{(1)}(\sin \delta_{O'}) \sin T_{O'},$$

$$\delta \mu = 67.1'' \mathrm{P}_2^{(2)}(\sin \delta_{O'} \sin 2T_{O'} - 38.7'' \mathrm{P}_2^{(2)}(\sin \delta_{O'}) \cos 2T_{O'}. \tag{4.71}$$

The zonal deformation has no effect on the directions of the axes of the ellipsoid of inertia; however, it does affect its polar flattening. It has a constant part which increases the Earth's polar flattening (Groten 1970).

The dynamics of the Earth's ellipsoid of inertia is discussed in Section 5.4. The tidal variations of the second-degree geopotential coefficients play the most important part in it.

4.5 Effect of the Moon's Motion on the Tide-Generating Potential

Quantities $\Delta_{\oplus\mathrm{D}}$, $\delta_{O'}$, $T_{O'}$ are functions of time and of the orbital elements of the Moon's centre of mass. They can be expressed in the manner described in Section 1.5.1, and the tide-generating potential at general point (ϱ, ϕ, Λ) on the Earth's surface can be transformed to read similarly as (1.120), having taken into account (1.121) (Lambeck 1977):

$$V_{s\mathrm{D}} = \frac{GM_{\mathrm{D}}}{a_{\mathrm{D}}} \sum_{n=2}^{\infty} \left(\frac{\varrho}{a_{\mathrm{D}}}\right)^n \sum_{k=0}^{n} (2 - \delta_{k0}) \frac{(n-k)!}{(n+k)!} \, \mathrm{P}_n^{(k)}(\sin\phi)$$

$$\times \sum_{r=0}^{n} F_{nkr}(i_{\mathrm{D}}) \sum_{s=-\infty}^{\infty} G_{nrs}(e_{\mathrm{D}}) S_{nkrs}(\omega_{\mathrm{D}}, M_{\mathrm{D}}'', \Omega_{\mathrm{D}}, \Theta) ; \qquad (4.72)$$

$$S_{nkrs} = \begin{cases} \cos k\Lambda \\ -\sin k\Lambda \end{cases} \begin{matrix} (n-k) \text{ even} \\ (n-k) \text{ odd} \end{matrix} \cos[(n-2r)\omega_{\mathrm{D}}$$

$$+ (n-2r+s)M_{\mathrm{D}}'' + k(\Omega_{\mathrm{D}} - \Theta)]$$

$$+ \begin{cases} \sin k\Lambda \\ \cos k\Lambda \end{cases} \begin{matrix} (n-k) \text{ even} \\ (n-k) \text{ odd} \end{matrix} \sin[(n-2r)\omega_{\mathrm{D}}$$

$$+ (n-2r+s)M_{\mathrm{D}}'' + k(\Omega_{\mathrm{D}} - \Theta)]$$

$$= \frac{\cos[(n-2r)\omega_{\mathrm{D}} + (n-2r+s)M_{\mathrm{D}}'' + k(\Omega_{\mathrm{D}} - \Theta) - k\Lambda]_{(n-k)\text{even}}}{\sin[(n-2r)\omega_{\mathrm{D}} + (n-2r+s)M_{\mathrm{D}}'' + k(\Omega_{\mathrm{D}} - \Theta) - k\Lambda]_{(n-k)\text{odd}}} , \quad (4.73)$$

where Ω_{D}, i_{D}, ω_{D}, a_{D}, e_{D}, M_{D}'' are elements of the lunar orbit (the mean anomaly being denoted M_{D}'' to differentiate it from the Moon's mass M_{D}), referred to the Earth's equator and vernal equinox. Angle i_{D} between the plane of the Moon's osculating orbit and the plane of the equator varies periodically within the interval $23.5° \pm 5°$; for our purposes it is sufficient to take $i_{\mathrm{D}} = 23.5° = \varepsilon_0$. The geocentric declination of the Moon's centre of mass varies periodically in the interval from $+28°45'$ to $-28°45'$, and the period is 27.321 days (sidereal month).

Since the eccentricities of the Moon's and Earth's orbits are small, it is sufficient to consider $q = 0$, $q = \pm 1$, $q = \pm 2$ in the summation for $G_{nrs}(e)$. Besides, if we restrict ourselves to the first harmonic term in the tide-generating potential, i.e. to the second-degree term, we shall get

$$(V_{s\mathrm{D}})_2 = \frac{GM_{\mathrm{D}}}{a_{\mathrm{D}}} \left(\frac{\varrho}{a_{\mathrm{D}}}\right)^2 \sum_{k=0}^{2} (2 - \delta_{k0}) \frac{(2-k)!}{(2+k)!} \, \mathrm{P}_n^{(k)}(\sin\phi)$$

$$\times \sum_{r=0}^{2} F_{2kr}(i_{\mathrm{D}}) \sum_{s=-2}^{2} G_{2rs}(e_{\mathrm{D}})$$

$$\times \cos[(2-2r)\omega_{\mathrm{D}} + (2-2r+s)M_{\mathrm{D}}'' + k(\Omega_{\mathrm{D}} - \Theta) - k\Lambda]_{k=0,2}$$

$$\times \sin[(2-2r)\omega_{\mathrm{D}} + (2-2r+s)M_{\mathrm{D}}'' + k(\Omega_{\mathrm{D}} - \Theta) - k\Lambda]_{k=1} . \quad (4.74)$$

The expression for the tide-generating potential, (4.72), enables us to identify a whole series of other periodic terms (waves). Their general form is

$$(V_{s\mathbb{D}})^{(k)}_{n,rs} = A^{(k)}_{n,rs} P^{(k)}_n (\sin \phi) \cos (\omega^{(k)}_{n,rs} t + \beta^{(k)}_{n,rs}) \,.$$

(4.75)

As in (4.72) the additional potential generated by the tidal deformations of the Earth (4.67) in outer space, for example, at an artificial satellite, ϱ_s, δ_s, T_s (Kaula 1964; Lambeck 1977), can be expressed as

$$\delta R(\varrho_s, \delta_s, T_s) = \frac{GM_{\mathbb{D}}}{a_{\mathbb{D}}} \sum_{n=2}^{\infty} k_n \left(\frac{\varrho}{a_{\mathbb{D}}}\right)^n \left(\frac{\varrho}{a}\right)^{n+1} \sum_{k=0}^{n} (2 - \delta_{k0})$$

$$\times \frac{(n-k)!}{(n+k)!} \sum_{r=0}^{n} \sum_{s=-\infty}^{\infty} \sum_{p=0}^{n} \sum_{q=-\infty}^{\infty} F_{nkr}(i_{\mathbb{D}}) G_{nrs}(e_{\mathbb{D}}) F_{nkp}(i)$$

$$\times G_{npq}(e) \cos [(n-2r)\omega_{\mathbb{D}} + (n-2r+s)M''_{\mathbb{D}} + k(\Omega_{\mathbb{D}} - \Omega)$$

$$- (n-2p)\omega - (n-2p+q)M]\,,$$

(4.76)

where $\Omega, i, \omega, a, e, M$ are the elements of the satellite's orbit and k_n is the Love number of n-th degree (Sect. 4.7).

After substituting (4.76) into Lagrange's equations (1.105) we arrive at the perturbations of the satellite's orbital elements. For example, Lambeck (1977) derived the following expression for the variation of the inclination of the plane of the osculating orbit:

$$\frac{di}{dt} = \frac{GM_{\mathbb{D}}}{a_{\mathbb{D}}} \frac{1}{\{GM_{\oplus}[a(1-e^2)]\}^{1/2} \sin i} \sum_{n=2}^{\infty} \sum_{k=0}^{n} \sum_{r=0}^{n} \sum_{s=-\infty}^{\infty} \sum_{p=0}^{n} \sum_{q=-\infty}^{\infty}$$

$$\times [(n-2r)\cos i - k] k_n \left(\frac{\varrho}{a_{\mathbb{D}}}\right)^n \left(\frac{\varrho}{a}\right)^{n+1} (2 - \delta_{k0}) \frac{(n-k)!}{(n+k)!}$$

$$\times F_{nkr}(i_{\mathbb{D}}) F_{nkp}(i) G_{nrs}(e_{\mathbb{D}}) G_{npq}(e) \sin [(n-2r)\omega_{\mathbb{D}} + (n-2r+s)M''_{\mathbb{D}}$$

$$+ k(\Omega_{\mathbb{D}} - \Omega) - (n-2p)\omega - (n-2p+q)M]\,.$$

(4.77)

The tidal perturbations of orbital elements of artificial satellites have a periodic character. Apart from the Earth's rotation (Θ) the periodicity is also caused by the terms in the summations over r and s, the variations of elements $\omega_{\mathbb{D}}$ and $\Omega_{\mathbb{D}}$ due to the Earth's flattening, and the effect of the Sun.

The principal periodic terms (waves) in the tide-generating potential are the sectorial terms $(n = 2, k = 2, r = 0, s = 0)$. In Darwin's symbolics these are waves M_2 (due to the Moon) and S_2 (due to the Sun). The principal diurnal waves are the second-degree tesseral terms $(n = 2, k = 1, r = 1, s = 0)$, O_1 (due to the Moon), P_1 (due to the Sun) and K_1 (due to the combined effects of the Moon and Sun).

If in (4.77)

$$n - 2p + q = 0\,,$$

(4.78)

Table 4.3. Principal tidal waves

International wave designator	Period	Character	Origin
None	18.6 year	Zonal	Moon
S_a	1.0 year	Zonal	Sun
S_{sa}	0.5 year	Zonal	Sun
M_m	27.55 days	Zonal	Moon
M_f	13.66 days	Zonal	Moon
O_1	25.82 h	Tesseral	Moon
P_1	24.07 h	Tesseral	Sun
K	23.93 h	Tesseral	Moon + Sun
N_2	12.66 h	Sectorial	Moon
M_2	12.42 h	Sectorial	Moon
S_2	12.00 h	Sectorial	Sun
K_2	11.97 h	Sectorial	Moon + Sun

perturbations with a period in excess of 1 day may occur and these are particularly important with regard to the orbital analysis of artificial satellites. For all the other terms which do not satisfy condition (4.78), the amplitudes of the tidal perturbations are small and their frequencies high.

A list of the periods of the principal tidal waves and their international designators are given in Table 4.3.

4.6 Components of Tidal Forces

By differentiating tide-generating potential $V_{s\text{)}}$ we obtain the appropriate component of the tidal force $\mathbf{F}_{s\text{)}}$. We shall do this in the vertical and the two basic horizontal directions, i.e. in the planes of the meridian and prime vertical (Fig. 4.8). It is indeed in these directions that the components are measured. Once again the procedure will only be carried out for $n = 2$. Now, and also later (Table 4.4), we shall consider tidal forces per unit mass, i.e. in units of acceleration.

Vertical component $-\partial V_{s\text{)}}/\partial n$ is practically equal to the radial component in view of the small angle between normal n to the gravity equipotential surface and the geocentric radius-vector ϱ of the potential point:

$$-\partial V_{s\text{)}}/\partial\varrho = F_\varrho = -\frac{GM_\text{)}}{\Delta_{\oplus\text{)}}^2}\sum_{n=2}^{\infty} n\left(\frac{\varrho}{\Delta_{\oplus\text{)}}}\right)^{n-1} P_n^{(0)}(\cos\psi)\,. \tag{4.79}$$

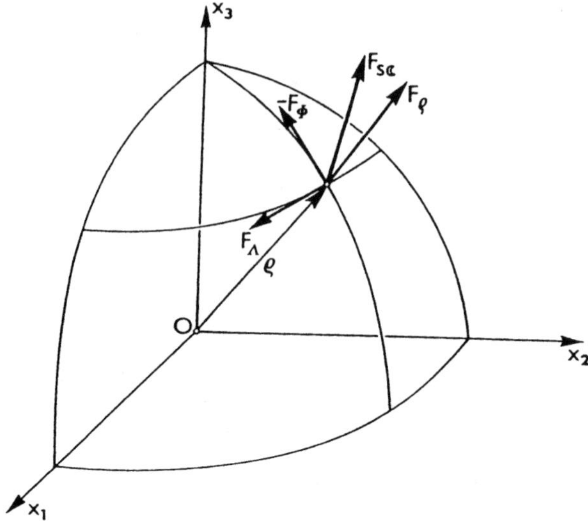

Fig. 4.8. Components of the tide-generating force

In view of (4.28) and assuming a perfectly rigid Earth (and restricting ourselves to $n = 2$), its zonal part is

$$(F_\varrho)_2^{(0)} = -\frac{\partial (V_{s)})_2^{(0)}}{\partial \varrho} = -\frac{1}{2}\frac{GM_{)}}{\Delta_{\oplus)}^2}\frac{\varrho}{\Delta_{\oplus)}}(3\sin^2\phi - 1)(3\sin^2\delta_{o'} - 1)$$

$$= -\frac{1}{8}\frac{GM_{)}}{\Delta_{\oplus)}^2}\frac{\varrho}{\Delta_{\oplus)}}(1 - 3\cos 2\phi)(1 - 3\cos 2\delta_{o'}) . \qquad (4.80)$$

The zero lines are the same as in Fig. 4.5a. Quantity $(F_\varrho)_2^{(0)}$ contains a term which is constant (assuming $\Delta_{\oplus)}$ and ϱ to be invariable) for the whole Earth regardless of the position of point M, where the vertical tidal component is being calculated:

$$-\frac{1}{8}\frac{GM_{)}}{\Delta_{\oplus)}^2}\frac{\varrho}{\Delta_{\oplus)}} \sim -6 \times 10^{-8}\,\mathrm{m\,s^{-2}}\,(0.006\;\mathrm{mgal}) . \qquad (4.81)$$

It also contains a term which is constant along the parallel, $\phi = \mathrm{const}$,

$$\frac{3}{8}\frac{GM_{)}}{\Delta_{\oplus)}^2}\frac{\varrho}{\Delta_{\oplus)}}\cos 2\phi ,$$

and, finally, a term which varies periodically with the Moon's declination

$$\frac{3}{8}\frac{GM_{)}}{\Delta_{\oplus)}^2}\frac{\varrho}{\Delta_{\oplus)}}\cos 2\delta_{o'}(1 - 3\cos 2\phi)$$

and whose integral mean value is

$$\sim \frac{9\sqrt{3}}{16\pi} \frac{GM_{\mathrm{D}}}{\Delta^2_{\oplus\mathrm{D}}} \frac{\varrho}{\Delta_{\oplus\mathrm{D}}} (1 - 3\cos 2\phi).$$

The zonal part of the tidal force thus causes an increase in gravity at the poles and its decrease at the equator.

The tesseral part of the radial (vertical) component of the tidal force,

$$(F_\varrho)_2^{(1)} = -\frac{\partial(V_{s\mathrm{D}})_2^{(1)}}{\partial\varrho} = -\frac{3}{2}\frac{GM_{\mathrm{D}}}{\Delta^2_{\oplus\mathrm{D}}}\frac{\varrho}{\Delta_{\oplus\mathrm{D}}}\sin 2\phi \sin 2\delta_{O'}\cos(\Lambda + T_{O'}), \qquad (4.82)$$

is short-periodic with a period of ~ 24 h, zero for $\phi = 0^0$ or $\phi = \pm 90°$, or $\Lambda + T_{O'} = 90°,\ 270°$, or $\delta_{O'} = 0$. For $\delta_{O'} > 0$ the zero line is the same as in Fig. 4.5b. Force (4.82) has its maximum absolute value at $\phi = 45°$, $|\delta_{O'}| = $ max., $\Lambda = -T_{O'}(+180°)$, and amounts to $0.070 \times 10^{-5}\ \mathrm{m\,s^{-2}}$.

The sectorial part,

$$(\Gamma_\varrho)_2^{(?)} = -\frac{\partial(V_{s\mathrm{D}})_2^{(2)}}{\partial\varrho} = -\frac{3}{2}\frac{GM_{\mathrm{D}}}{\Delta^2_{\oplus\mathrm{D}}}\frac{\varrho}{\Delta_{\oplus\mathrm{D}}}\cos^2\phi \cos^2\delta_{O'}\cos 2(\Lambda + T_{O'}), \qquad (4.83)$$

is short-periodic with a period of ~ 12 h, zero along meridians $\Lambda = -T_{O'} + 45°$, $\Lambda = -T_{O'} + 135°$, and the zero line is the same as in Fig. 4.5c. Its maximum absolute value is at $\phi = 0°$, $\delta = 0°$, and amounts to $0.083 \times 10^{-5}\ \mathrm{m\,s^{-2}}$.

If the Earth is perfectly elastic and Eq. (4.56) holds, the radial component of the tidal force

$$\bar{F}_\varrho = -\frac{\partial[\bar{W}(M') - W(M)]}{\partial\varrho} = -\frac{\partial}{\partial\varrho}\left(V_s + \delta V_s + \frac{\partial W}{\partial\varrho}\delta\bar\varrho\right). \qquad (4.84)$$

In view of (4.24)

$$\frac{\partial V_{s\mathrm{D}}}{\partial\varrho} = \frac{GM_{\mathrm{D}}}{\Delta^2_{\oplus\mathrm{D}}}\sum_{n=2}^{\infty} n\left(\frac{\varrho}{\Delta_{\oplus\mathrm{D}}}\right)^{n-1} \mathrm{P}_n^{(0)}(\cos\psi) \qquad (4.85)$$

and, if we restrict ourselves to a single term, $n = 2$,

$$\frac{\partial V_{s\mathrm{D}}}{\partial\varrho} = 2\frac{GM_{\mathrm{D}}}{\Delta^2_{\oplus\mathrm{D}}}\frac{\varrho}{\Delta_{\oplus\mathrm{D}}}\mathrm{P}_2^{(0)}(\cos\psi) = \frac{2V_{s\mathrm{D}}}{\varrho}. \qquad (4.86)$$

We have already expressed potential $\delta V_{s\mathrm{D}}$ in terms of series (4.59). Assuming that the tidal deformation of terrestrial masses will not change the position of the Earth's centre of mass, then also

$$\delta V_{s\mathrm{D}} = \frac{GM_\oplus}{\varrho}\sum_{n=2}^{\infty}\sum_{j=0}^{n}\left(\frac{a_0}{\varrho}\right)^n(\delta J_n^{(j)}\cos j\Lambda + \delta S_n^{(j)}\sin j\Lambda)\mathrm{P}_n^{(j)}(\sin\phi), \qquad (4.87)$$

$$\frac{\partial\delta V_{s\mathrm{D}}}{\partial\varrho} = -\frac{GM_\oplus}{\varrho^2}\sum_{n=2}^{\infty}\sum_{j=0}^{n}(n+1)\left(\frac{a_0}{\varrho}\right)^n(\delta J_n^{(j)}\cos j\Lambda$$
$$+ \delta S_n^{(j)}\sin j\Lambda)\mathrm{P}_n^{(j)}(\sin\phi) \qquad (4.88)$$

and, restricting ourselves once again to $n = 2$,

$$\frac{\partial \delta V_{s\mathrm{D}}}{\partial \varrho} = -3 \frac{GM_{\oplus}}{\varrho^2} \left(\frac{a_0}{\varrho}\right)^2 \sum_{j=0}^{2} (\delta J_2^{(j)} \cos j\Lambda$$

$$+ \delta S_2^{(j)} \sin j\Lambda) P_2^{(j)}(\sin\phi) = -3 \frac{\delta V_s}{\varrho}. \tag{4.89}$$

Also,

$$\frac{\partial^2 W}{\partial \varrho^2} = \frac{GM_{\oplus}}{\varrho^3} \left\{ 2 + \sum_{n=2}^{\infty} \sum_{j=0}^{n} (n+1)(n+2)\left(\frac{a_0}{\varrho}\right)^n (J_n^{(j)} \cos j\Lambda \right.$$

$$\left. + S_n^{(j)} \sin j\Lambda) P_n^{(j)}(\sin\phi) + \frac{2}{3} q \left(\frac{a_0}{\varrho}\right)^{-3} [1 - P_2^{(0)}(\sin\phi)] \right\}. \tag{4.90}$$

In view of the small radial perturbation of the Earth's crust, $\delta\bar{\varrho}$, only the spherical term need be considered in product $\dfrac{\partial^2 W}{\partial \varrho^2} \delta\bar{\varrho}$, so that

$$\frac{\partial^2 W}{\partial \varrho^2} \delta\bar{\varrho} = 2 \frac{GM_{\oplus}}{\varrho^3} \delta\bar{\varrho} = -2 \frac{\delta\bar{\varrho}}{\varrho} \frac{\partial W}{\partial \varrho}. \tag{4.91}$$

After substituting (4.86), (4.89) and (4.91) into (4.84), we arrive at

$$\bar{F}_{\varrho} = -2 \frac{V_{s\mathrm{D}}}{\varrho} + 3 \frac{\delta V_{s\mathrm{D}}}{\varrho} + 2 \frac{\delta\bar{\varrho}}{\varrho} \frac{\partial W}{\partial \varrho} = -\frac{V_{s\mathrm{D}}}{\varrho}(2 - 3k + 2h) \tag{4.92}$$

for the radial component of the tidal force, given a perfectly elastic Earth, in which case (4.47) and (4.48) apply. This means that the spectrum of the radial component of the tidal force (tidal variations of the gravity acceleration) is the same as that of the tide-generating potential.

Since

$$-2 \frac{V_{s\mathrm{D}}}{\varrho} = -\frac{\partial V_{s\mathrm{D}}}{\partial \varrho} = F_{\varrho}, \tag{4.93}$$

and if we again restrict ourselves to harmonic terms of the second degree, then

$$\bar{F}_{\varrho} = F_{\varrho}(1 - \tfrac{3}{2}k + h) \tag{4.94}$$

or, using the values of Love numbers (4.54),

$$1 - \tfrac{3}{2}k + h = 1.17, \tag{4.95}$$

$$\bar{F}_{\varrho} = 1.17 F_{\varrho}. \tag{4.96}$$

Some values of the zonal, tesseral and sectorial terms of the radial component of the tidal force due to the Moon for a perfectly rigid and perfectly elastic Earth are given in Table 4.4 for comparison. The pattern of the radial (vertical) component of the separate tidal terms is exactly the opposite to that in Fig. 4.5.

Table 4.4. Radial component of the tidal force due to the Moon

Type	Position	$F\varrho$ $(10^{-5}\,\mathrm{m\,s}^{-2})$	$\bar{F}\varrho$ $(10^{-5}\,\mathrm{m\,s}^{-2})$
Total	$\psi = 0°, 180°$	-0.111	-0.129
	$\psi = 90°, 270°$	0.056	0.064
Zonal	$\delta_{o'} = 0°, \phi = \pm 90°$	0.056	0.064
	$\delta_{o'} = 0°, \phi = 0°$	-0.028	-0.032
Tesseral	$\delta_{o'} = 29°, \phi = 45°,$ $\Lambda = -T_{o'}$	-0.070	-0.082
Sectorial	$\delta_{o'} = 0°, \phi = 0°,$ $\Lambda = -T_{o'}$	-0.083	-0.096

The horizontal component in the plane of the meridian (positive to the south),

$$-\frac{\partial V_{s\mathbb{D}}}{\varrho\,\partial\phi} = F_\phi \,, \tag{4.97}$$

again has a zonal part,

$$(F_\phi)_2^{(0)} = -\frac{GM_{\mathbb{D}}}{\Delta_{\oplus\mathbb{D}}^2}\frac{\varrho}{\Delta_{\oplus\mathbb{D}}}\,\mathrm{P}_2^{(1)}(\sin\phi)\,\mathrm{P}_2^{(0)}(\sin\delta_{o'}) \tag{4.98}$$

and two non-zonal terms. The first reads

$$(F_\phi)_2^{(1)} = -\frac{1}{3}\frac{GM_{\mathbb{D}}}{\Delta_{\oplus\mathbb{D}}^2}\frac{\varrho}{\Delta_{\oplus\mathbb{D}}}\,\mathrm{P}_2^{(1)}(\sin\delta_{o'})\cos(\Lambda + T_{o'})\frac{\mathrm{dP}_2^{(1)}(\sin\phi)}{\mathrm{d}\phi}\,;$$

$$\frac{\mathrm{dP}_2^{(1)}(\sin\phi)}{\mathrm{d}\phi} = -\tan\phi\,\mathrm{P}_2^{(1)}(\sin\phi) + \mathrm{P}_2^{(2)}(\sin\phi) = 3\cos 2\phi \tag{4.99}$$

and the second reads

$$(F_\phi)_2^{(2)} = -\frac{1}{12}\frac{GM_{\mathbb{D}}}{\Delta_{\oplus\mathbb{D}}^2}\frac{\varrho}{\Delta_{\oplus\mathbb{D}}}\,\mathrm{P}_2^{(2)}(\sin\delta_{o'})\cos 2(\Lambda + T_o)\frac{\mathrm{dP}_2^{(2)}(\sin\phi)}{\mathrm{d}\phi}\,;$$

$$\frac{\mathrm{dP}_2^{(2)}(\sin\phi)}{\mathrm{d}\phi} = -2\tan\phi\,\mathrm{P}_2^{(2)}(\sin\phi) = -3\sin 2\phi\,. \tag{4.100}$$

The zonal component (4.98) depends only on the Moon's declination, and not on its hour angle; it thus has a part which is a function of position on Earth only. It is schematically plotted in Fig. 4.9a and the force direction is shown in Fig. 4.10a. Figure 4.9b shows component (4.99) for $\delta_{\mathbb{D}} > 0$ which, however, depends on the Moon's hour angle; Fig. 4.10b shows the force directions. Figure 4.9c depicts the global pattern of component (4.100) which also depends on the

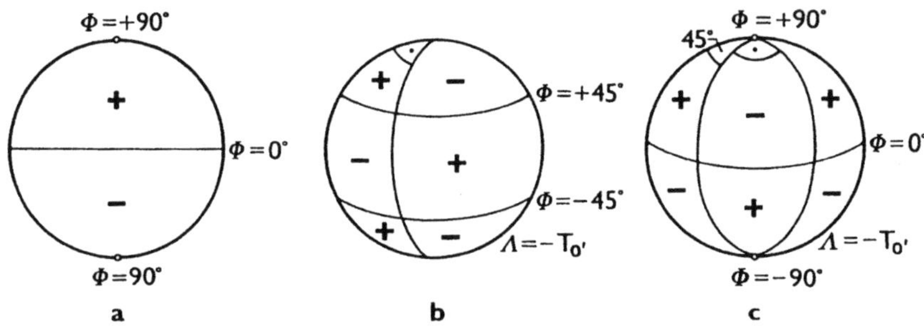

Fig. 4.9. Schematic global diagram of the **a** zonal, **b** tesseral and **c** sectorial horizontal component of the tide-generating force in the meridional plane

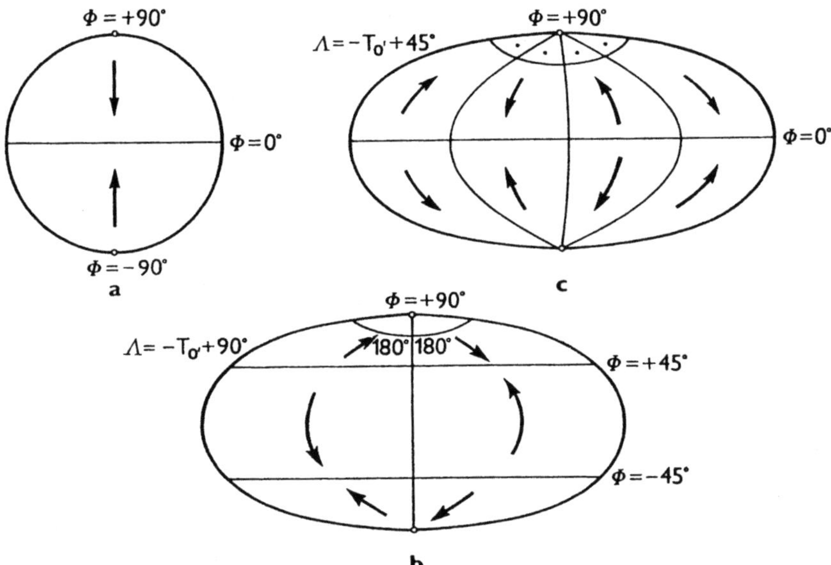

Fig. 4.10. Directions in which the horizontal component of the tide-generating force acts in the meridional plane. Zonal part **a** Eq. (4.98) and non-zonal parts **b** Eq. (4.99) and **c** Eq. (4.100)

Moon's hour angle; Fig. 4.10c shows the directions of acting forces. A force couple, which tends to deviate the ellipsoid of inertia so that its largest axis passes through the Moon's centre of mass, is created in the equatorial zone; a force couple is also created at the poles, acting in the opposite direction. If the Earth were spherically symmetrical these two couples would cancel each other; however, the real Earth is flattened at the poles and the equatorial couple predominates, and this is also responsible for precession and nutation (Sect. 3.9).

For a perfectly elastic Earth the horizontal component of the tidal force in the plane of the equator

$$\bar{F}_\phi = -\frac{\partial[\bar{W}(M') - W(M)]}{\varrho\,\partial\phi} = -\frac{\partial V_{s\mathbb{D}}}{\varrho\,\partial\phi}(1 + k - h) = F_\phi(1 + k - h),$$

$$(4.101)$$

having substituted (4.51) for $\bar{W}(M') - W(M)$. This means that for a perfectly elastic Earth $\bar{F}_\phi \sim 0.7\, F_\phi$.

Analogously, the horizontal component in the plane of the prime vertical (positive to the west)

$$-\frac{\partial V_s}{\varrho\cos\phi\,\partial\Lambda} = F_\Lambda\,.$$

$$(4.102)$$

Its zonal part is zero,

$$(F_\Lambda)_2^{(0)} = 0\,,$$

$$(4.103)$$

and its tesseral part (for $n = 2$; zero at the equator and along meridians $\Lambda = -T_{O'},\ \Lambda = -T_{O'} + \pi$),

$$(F_\Lambda)_2^{(1)} = \frac{3}{2}\frac{GM_\mathbb{D}}{\Delta_{\oplus\mathbb{D}}^2}\frac{\varrho}{\Delta_{\oplus\mathbb{D}}}\sin\phi\sin 2\delta_{O'}\sin(\Lambda + T_{O'})\,,$$

$$(4.104)$$

has the character of a tesseral spherical harmonic of the first degree and order; the sectorial part $\left(\text{zero along meridians } \Lambda = -T_{O'} + j\frac{\pi}{2}\right)$ reads

$$(F_\Lambda)_2^{(2)} = \frac{3}{2}\frac{GM_\mathbb{D}}{\Delta_{\oplus\mathbb{D}}^2}\frac{\varrho}{\Delta_{\oplus\mathbb{D}}}\cos\phi\cos^2\delta_{O'}\sin 2(\Lambda + T_{O'})\,.$$

$$(4.105)$$

For a perfectly elastic Earth, in analogy with (4.102),

$$\bar{F}_\Lambda = -\frac{\partial[\bar{W}(M') - W(M)]}{\varrho\cos\phi\,\partial\Lambda} = F_\Lambda(1 + k - h)\,.$$

$$(4.106)$$

The horizontal components of the tidal force determine the deflections of the vertical [relative to the Earth's crust (Melchior 1983)] in the plane of the meridian,

$$\delta\xi = \frac{1}{\dfrac{\partial W}{\partial n}}\bar{F}_\phi = \frac{1}{\dfrac{\partial W}{\partial\varrho}}\bar{F}_\phi$$

$$(4.107)$$

and in the plane of the prime vertical,

$$\delta\eta = \frac{1}{\dfrac{\partial W}{\partial\varrho}}\bar{F}_\Lambda\,.$$

$$(4.108)$$

The tangent to the vertical is deflected in the direction of the Moon (Sun). The constant part has a component along the meridian; it is derived from the zonal part (4.98) which can be expressed as

$$(F_\phi)_2^{(0)} = -\frac{3}{8}\frac{GM_{\mathbb{D}}}{\Delta_{\oplus\mathbb{D}}^2}\frac{\varrho}{\Delta_{\oplus\mathbb{D}}}\sin 2\phi\,(1 - 3\cos 2\delta_{0'})\,. \tag{4.109}$$

The corresponding deflection in seconds is

$$\delta\xi_2^{(0)} = -\frac{3}{8}\varrho''\frac{M_{\mathbb{D}}}{M_\oplus}\left(\frac{\varrho}{\Delta_{\oplus\mathbb{D}}}\right)^3\sin 2\,\phi\,(1 - 3\cos 2\delta_{0'})\,. \tag{4.110}$$

Its value is maximum for $\phi = 45°$ and amounts to $0.0016''$ for a perfectly elastic Earth and $0.0022''$ for a perfectly rigid Earth.

Component $\delta\eta$ of the sectorial part, derived from (4.105), displays the maximum amplitude overall. For $\delta_{0'} = 0$, $\phi = 0$, $\Lambda = -T_{0'} + 45°$, $\delta\eta_0^{(0)} \sim 0.018''$ given a perfectly rigid Earth.

In calculating the variations of verticals $\delta\xi$, $\delta\eta$ due to the horizontal component of the tidal force, F_ϕ, F_Λ, we did not take into account the displacement of the potential point as a result of the tidal deformation in the horizontal direction. In general, displacement occurs along the meridian, δs_ϕ, and along the prime vertical, δs_Λ, the amplitude being ~ 7 cm. Hence,

$$\delta s_\phi = -l\,\frac{1}{\dfrac{\partial W}{\partial \varrho}}\,\frac{\partial V_{s\mathbb{D}}}{\partial\phi}\,,$$

$$\delta s_\Lambda = -l\,\frac{1}{\dfrac{\partial W}{\partial \varrho}}\,\frac{\partial V_{s\mathbb{D}}}{\cos\phi\,\partial\Lambda}\,, \tag{4.111}$$

where (for $n = 2$) $l = 0.08$ is the Shida–Lambert constant (parameter, Shida number; for a perfectly rigid body $l = 0$ and for a body of homogeneous fluid mass $l = 0.23$). The change in the position of the vertical (astronomical zenith) due to this displacement of the point on the Earth's surface must be subtracted from

$$-(1 + k)\,\frac{1}{\dfrac{\partial W}{\partial \varrho}}\,\frac{\partial V_{s\mathbb{D}}}{\varrho\,\partial\phi}\,, \qquad -(1 + k)\,\frac{1}{\dfrac{\partial W}{\partial \varrho}}\,\frac{\partial V_{s\mathbb{D}}}{\varrho\cos\phi\,\partial\Lambda}\,,$$

if we are to get the variations of verticals, $\delta\xi'$ and $\delta\eta'$ relative to the Earth's axis (Melchior 1983):

$$\delta\xi' = -(1 + k - l)\,\frac{1}{\dfrac{\partial W}{\partial \varrho}}\,\frac{\partial V_{s\mathbb{D}}}{\varrho\,\partial\phi}\,,$$

$$\delta\eta' = -(1 + k - l)\,\frac{1}{\dfrac{\partial W}{\partial \varrho}}\,\frac{\partial V_{s\mathbb{D}}}{\varrho\cos\phi\,\partial\Lambda}\,. \tag{4.112}$$

Compared with (4.101) and (4.102), instead of coefficient $1 + k - h \sim 0.7$ these equations will contain coefficient $1 + k - l \sim 1.2$, and the actual amplitude may be as much as $0.025''$. Quantities (4.112) are recorded by astronomical instruments of the highest accuracy, quantities (4.107) and (4.108) by horizontal pendula, and quantities (4.92) by tidal gravity meters.

Horizontal (tangential) tidal forces are important with regard to the description of the dynamics of geotectonic processes. They tend to contract or expand parts of the Earth's crust horizontally. The horizontal tidal displacements of points of the Earth's surface are smaller than the vertical in the ratio of Love numbers, $h_2/l_2 \sim 7$, i.e. they amount to 5–7 cm at $\phi = \pm 45°$.

4.7 Love Numbers and Methods of Determining Them

Dimensionless parameters h, k and l, defined by Eqs (4.47), (4.48) and (4.111), belong to the set of Love numbers which describe the deformations of perfectly elastic bodies. They characterize the deviations of the Earth from perfect rigidity. Love (1909) introduced parameters h and k, while Shida and Matsuyama (1912) and Lambert (1943) introduced parameter l; strictly speaking, they are constants of proportionality of the appropriate deformations only with respect to the first term ($n = 2$) in the expansion of the tidal potential (4.24). Since all the deformations being considered can be expressed in terms of series of spherical harmonics, every harmonic term of degree n must have, as a result of their orthogonality, its own constant of proportionality with respect to the harmonic term in the tidal potential expansion of degree n. The condition is, of course, that the physical characteristics of the Earth's interior (density, Lamé constants λ, μ) are a function of the geocentric radius-vector only, i.e. that they are spherically symmetrical relative to the Earth's centre of mass.

In addition, the Love numbers are functions of geocentric radius-vector ϱ of the position of the potential point (assuming that the Earth is perfectly elastic and the density distribution is spherically symmetrical), i.e.

$$k = k(\varrho), \quad h = h(\varrho), \quad l = l(\varrho). \tag{4.113}$$

The additional potential due to the tidal deformations of the Earth's masses caused by the Moon is then

$$\delta V_{s\,\leftmoon} = \frac{GM_{\leftmoon}}{\varDelta_{\oplus\,\leftmoon}} \sum_{n=2}^{\infty} k_n(\varrho) \left(\frac{\varrho}{\varDelta_{\oplus\,\leftmoon}}\right)^n \sum_{j=0}^{n} (2 - \delta_{j0}) \frac{(n-j)!}{(n+j)!}$$
$$\times \mathrm{P}_n^{(j)}(\sin\phi)\, \mathrm{P}_n^{(j)}(\sin\delta_{O\cdot})\cos j(\varLambda + T_{O\cdot}), \tag{4.114}$$

and, analogously, the radial displacement, $\overline{MM'}$,

$$\overline{\delta\varrho} = -\frac{1}{\dfrac{\partial W}{\partial \varrho}} \frac{GM_{\leftmoon}}{\varDelta_{\oplus\,\leftmoon}} \sum_{n=2}^{\infty} h_n(\varrho) \left(\frac{\varrho}{\varDelta_{\oplus\,\leftmoon}}\right)^n \sum_{j=0}^{n} (2 - \delta_{j0}) \frac{(n-j)!}{(n+j)!}$$
$$\times \mathrm{P}_n^{(j)}(\sin\phi)\, \mathrm{P}_n^{(j)}(\sin\delta_{O\cdot})\cos j(\varLambda + T_{O\cdot}); \tag{4.115}$$

finally, the horizontal displacements

$$\delta s_\phi = -\frac{1}{\dfrac{\partial W}{\partial \varrho}} \frac{GM_{\mathbb{D}}}{\varDelta_{\oplus\mathbb{D}}} \sum_{n=2}^{\infty} l_n(\varrho) \left(\frac{\varrho}{\varDelta_{\oplus\mathbb{D}}}\right)^n$$

$$\times \left\{ P_n^{(1)}(\sin\phi)\, P_n^{(0)}(\sin\delta_{O'}) + 2 \sum_{j=1}^{n} \frac{(n-j)!}{(n+j)!} \left[P_n^{(j+1)}(\sin\phi) \right.\right.$$

$$\left.\left. - j\tan\phi\, P_n^{(j)}(\sin\phi) \right] P_n^{(j)}(\sin\delta_{O'}) \cos j(\varLambda + T_{O'}) \right\}, \qquad (4.116)$$

$$\delta s_\varLambda = -\frac{1}{\dfrac{\partial W}{\partial \varrho}} \frac{GM_{\mathbb{D}}}{\varDelta_{\oplus\mathbb{D}}} \sum_{n=2}^{\infty} l_n(\varrho) \left(\frac{\varrho}{\varDelta_{\oplus\mathbb{D}}}\right)^n \left[\sum_{j=0}^{n} (2 - \delta_{j0}) j \frac{(n-j)!}{(n+j)!} \right.$$

$$\left. \times P_n^{(j)}(\sin\phi)\, P_n^{(j)}(\sin\delta_{O'}) \sin j(\varLambda + T_{O'}) \right]. \qquad (4.117)$$

Thus, k_n, h_n, l_n are Love numbers of degree n. If $M(\varrho, \phi, \varLambda)$ is the potential point on the Earth's surface, then $k_n(\varrho) = k_n$, $h_n(\varrho) = h_n$, $l_n(\varrho) = l_n$ and, if $n = 2$, i.e. if we restrict ourselves to the first harmonic term in the expansion of the tidal potential which is what we usually did in the preceding section, $k_n = k_2 = k$, $h_n = h_2 = h$, $l_n = l_2 = l$. The Love numbers can be determined from gravity, tilt, satellite and astronomical observations.

The tidal perturbations of satellite orbits display amplitudes of as much as 50 m and periods of 10–300 days. The satellite method has so far been used to determine Love parameter k_2 (Lambeck 1977): $k_2 = 0.290$.

In principle, the high-precision laser observations can also be used to determine parameters h_2 and l_2. Displacements $\delta\bar\varrho$, δs_ϕ, $\delta s_\varLambda$, which are their functions, can indeed be calculated immediately from the variations of the spatial positions of satellite stations; this requires special satellite programmes and suitably positioned satellite stations on the Earth's surface.

The advantage of satellite methods of determining Love numbers is that the problem can be treated 'globally', i.e. the derived parameters really describe the global physical properties of the Earth, including the oceans. They are not affected by any local factors, which in general is the case with direct surface observations of tidal force components, $\dfrac{\partial V_{s\mathbb{D}}}{\partial\varrho}$, $\dfrac{\partial V_{s\mathbb{D}}}{\varrho\partial\phi}$, $\dfrac{\partial V_{s\mathbb{D}}}{\varrho\cos\phi\,\partial\varLambda}$.

Table 4.5 summarizes the values of some Love numbers;

$$\delta_2 = 1 - \tfrac{3}{2} k_2 + h_2, \quad \delta_3 = 1 + \tfrac{2}{3} h_3 - \tfrac{4}{3} k_3, \qquad (4.118)$$

the ratio of the observed vertical tidal component and the theoretical value for a perfectly rigid Earth;

$$\gamma_2 = 1 + k_2 - h_2, \quad \gamma_3 = 1 + k_3 - h_3, \qquad (4.119)$$

Table 4.5. Love numbers

n	k_n	h_n	l_n	δ_n	γ_n	Λ_n
2	0.30	0.62	0.08	1.17	0.68	1.22
3	0.07	0.26	0.015	1.08	0.81	

the ratio of the observed and theoretical horizontal component of the tidal deformation with respect to the Earth's crust;

$$\Lambda_2 = 1 + k_2 - l_2 , \tag{4.120}$$

and the ratio of the observed and theoretical tidal deflection of the vertical (Melchior 1983).

Molodenskii discovered an important property of parameters h_2 and k_2: the ratio h_2/k_2 remains constant (with a scatter of less than 1%) for various models of the structure of the Earth's interior.

Quantities δ_n, γ_n, Λ_n depend on the elasticity and density of the Earth. They can be computed theoretically if some dependence of elasticity and density on depth is introduced. By comparing the theoretical values of the computed parameters with the observed values, it is possible to estimate the probability with which a particular model of the Earth's crust structure will exist. Another option in this respect is to compute the theoretical value of Chandler's period (3.77), which should also agree with the observed value.

Strictly speaking, the Love numbers of the other bodies of the Solar System are not known, with the exception of the Moon for which $k_2 = 0.02$.

In all our deliberations we have considered Love numbers to be constants, independent of the periodicity of the phenomenon responsible for the deformations. However, these numbers do in fact depend on periods, and the problem becomes considerably more complicated (Zschau 1978).

4.8 The Precession-Nutation Torque of Tidal Forces

We shall prove that the moment of tidal forces acting on the Earth, **G**, can be determined precisely without knowing the density distribution within the Earth and using only the data on the external gravitational field of the Earth and the Moon. The best suited for this purpose are the geopotential coefficients known from satellite observations (Sect. 1.6.2).

We have two options: either to compute the components of **G**, which requires three integrations, or to calculate the tidal force function, $\bar{V}_s$, and calculate components G_j of moment **G** in terms of its derivatives with respect to the Euler angles defining the directions of axes x_j of the geocentric ellipsoid of inertia relative to the adopted axes of the fixed ecliptic for the given epoch. The solution via $\bar{V}_s$ is less tedious.

We shall thus introduce the concept of the tidal force function, and define it as the negative potential energy of tidal forces:

$$\bar{V}_{s\leftmoon} = \int_{M_\oplus} V_{s\leftmoon}\, dm_\oplus . \tag{4.121}$$

We shall assume that the axes of geocentric system x_j and of selenocentric system x'_j are the axes of the central ellipsoids of inertia of both bodies. With a view to (4.24), (4.25) and Section 1.3.2,

$$\bar{V}_{s\leftmoon} = G\,\frac{M_\oplus M_\leftmoon}{\Delta_{\oplus\leftmoon}} \sum_{n=2}^{\infty} \sum_{k=0}^{n} \left(\frac{a_0}{\Delta_{\oplus\leftmoon}}\right)^n (J_n^{(k)} \cos kT_{O'}$$

$$- S_n^{(k)} \sin kT_{O'})\, \mathrm{P}_n^{(k)}(\sin \delta_{O'}) + \int_{M_\oplus} \Delta V\, dm_\oplus ; \tag{4.122}$$

$J_n^{(k)}$ and $S_n^{(k)}$ are geopotential coefficients of degree n and order k, a_0 is the semi-major axis of the Earth's rotational ellipsoid used to define the geopotential coefficients, and $\mathrm{P}_n^{(k)}$ are associated Legendre functions of degree n and order k. Function ΔV expresses the effect of the deviations of the Moon's gravity field from the spherically symmetrical field (Burša 1974). This function was derived up to the fourth degree, powers $(\varrho_\oplus/\Delta_{\oplus\leftmoon})^n$ in Section 3.9.1 being taken into account. It contains products $J_2^{(0)}(J_2^{(0)})'$, $J_2^{(0)}(J_2^{(2)})'$, $J_2^{(2)}(J_2^{(0)})'$, $J_2^{(2)}(J_2^{(2)})'$, etc. It is a function of the parameters defining the mutual positions of the ellipsoids of inertia of both bodies, for example, of $\Delta_{\oplus\leftmoon}$, $\delta_{O'}$, T_O, and of Euler's angles ψ', ϑ', φ', defining axes x'_j relative to axes x_j.

We find that function $\bar{V}_s$ is part of force function (3.160). We are therefore able to use the relations we derived in Section 3.9 to determine the components of the precession and nutation moment of the tidal forces.

The following conclusions can thus be drawn:

1. The moments of tidal forces due to the Moon (Sun) can be determined exactly without knowing the density distribution within the Earth.
2. It is sufficient to know the external gravitational fields of both bodies to determine them exactly.
3. They can be expressed exactly as a function of only the potential coefficients of both bodies and of the parameters defining the mutual positions of their ellipsoids of inertia.
4. The precession and nutation moment of the tidal forces is identical with the moment of the force defined by the derivative of force function (3.160).

4.9 The Secular Love Number

This number was introduced by Munk and MacDonald (1960). Its definition is based on the model of a spherical, deformable body (radius R) which does not rotate in the initial epoch, i.e. its angular velocity of rotation $\omega = 0$, and the potential of the centrifugal force at all its points, therefore, is $Q = 0$.

If the body rotates at angular velocity ω, it will generate the centrifugal force potential

$$Q = \tfrac{1}{2}\varrho^2\omega^2\cos^2\phi = \tfrac{1}{3}\omega^2\varrho^2[1 - \mathrm{P}_2^{(0)}(\sin\phi)]$$

$$= \frac{1}{3}\frac{GM}{\varrho}\,q\left(\frac{R}{\varrho}\right)^{-3}[1 - \mathrm{P}_2^{(0)}(\sin\phi)]\,;$$

$$q = \omega^2 R^3/(GM) \tag{4.123}$$

and perturbing (deforming) potential

$$Q - \tfrac{1}{3}\omega^2\varrho^2 = \delta Q_s = -\tfrac{1}{3}\omega^2\varrho^2\,\mathrm{P}_2^{(0)}(\sin\phi)$$

$$= -\frac{1}{3}\frac{GM}{\varrho}\,q\left(\frac{R}{\varrho}\right)^{-3}\mathrm{P}_2^{(0)}(\sin\phi)\,, \tag{4.124}$$

which will cause the masses within the body to become displaced, generating geopotential variation δW_s. Assuming that Hooke's law may also be used in this case, at the boundary surface

$$\delta\bar{W}_s/\delta\bar{Q}_s = k_s\,, \tag{4.125}$$

where the constant of proportionality k_s is the 'secular' Love number. It reflects a measure of the body-yield-to-centrifugal deformation in the course of its development during the whole history of its evolution (Munk and MacDonald 1960). This means that at any point of the sphere the total deforming potential (we have omitted the constant part $-\tfrac{1}{3}\omega^2\varrho^2$ becuase we are only interested in the deformations dependent on ϕ)

$$\delta\bar{Q}_s + k_s\delta\bar{Q}_s = -\frac{1}{3}\frac{GM}{\varrho}\,q\left(\frac{R}{\varrho}\right)^{-3}\mathrm{P}_2^{(0)}(\sin\phi)(1 + k_s) \tag{4.126}$$

and the equipotential surface through this point will be radially deformed by

$$\delta\varrho_s = \frac{\delta\bar{Q}_s}{g}(1 + k_s) = -\frac{1}{3}\varrho q\left(\frac{R}{\varrho}\right)^{-3}\mathrm{P}_2^{(0)}(\sin\phi)(1 + k_s)\,. \tag{4.127}$$

We shall now consider the deformations of the boundary spherical surface, i.e. we shall put $\varrho = R$. The boundary value of the deforming potential at any point of the boundary surface

$$k_s\delta\bar{Q}_s = -\frac{1}{3}k_s\frac{GM}{R}\,q\,\mathrm{P}_2^{(0)}(\sin\phi)\,. \tag{4.128}$$

In outer space, at any point (ϱ_0, ϕ_0) without the body, deforming geopotential

$$\delta W_s = \frac{GM}{\varrho_0}\left(\frac{R}{\varrho_0}\right)^2\delta J_2^{(0)}\mathrm{P}_2^{(0)}(\sin\phi_0) \tag{4.129}$$

is then generated. Deformation $\delta J_2^{(0)}$ of the second zonal geopotential parameter, $J_2^{(0)}$, can then be calculated by solving the first (Dirichlet) boundary problem of the potential theory. Given $\varrho_0 = R$, $\phi_0 = \phi$, i.e. at the boundary surface,

$$\delta \bar{W}_s = k_s \delta \bar{Q}_s ; \tag{4.130}$$

hence

$$\frac{GM}{R} \delta J_2^{(0)} = -\frac{1}{3} k_s \frac{GM}{R} q , \tag{4.131}$$

and thus

$$\delta J_2^{(0)} = -\frac{1}{3} k_s \frac{\omega^2 R^3}{GM} ; \tag{4.132}$$

in outer space

$$\delta W_s = -\frac{1}{3} k_s \frac{\omega^2 R^5}{\varrho_0^3} P_2^{(0)} (\sin \phi_0) = k_s \left(\frac{R}{\varrho_0} \right)^5 \delta Q_s . \tag{4.133}$$

Since all geopotential parameters of the body, with the exception of $J_0^{(0)}$, were zero before rotation began,

$$\delta J_2^{(0)} = J_2^{(0)} = \frac{A - C}{MR^2} \tag{4.134}$$

and by comparing (4.134) and (4.132) we arrive at the relation for the secular Love number:

$$-\frac{1}{3} k_s \frac{\omega^2 R^3}{GM} = J_2^{(0)} \tag{4.135}$$

or

$$k_s = -3 J_2^{(0)} / q ; \tag{4.136}$$

for the real Earth (with $J_2^{(0)} = -1082.625 \times 10^{-6}$) $k_s = 0.938\,31$.

The rotation generates not only radial deformations of equipotential surfaces (4.127) but also radial deformations of the body's actual surface:

$$\delta \bar{\varrho} = h \frac{\delta \bar{Q}_s}{g} = -\frac{1}{3} h \varrho q \left(\frac{R}{\varrho} \right)^{-3} P_2^{(0)} (\sin \phi) , \tag{4.137}$$

where h is the Love number. If the body is of an ideal fluid, the deformation of the body's surface is identical in shape with the deformation of equipotential surfaces (4.127). The rotation is then responsible for the flattening of the equipotential surface,

$$\alpha_h = -\frac{1}{3} q (1 + k_s) [P_2^{(0)}(0) - P_2^{(0)}(1)] = \frac{1}{2} q (1 + k_s) . \tag{4.138}$$

By substituting (4.136) into (4.138) we arrive at the classical Clairaut equation, (2.189), which relates the flattening, rotation parameter q and the second zonal geopotential coefficient of the equipotential rotational ellipsoid in a linear approximation:

$$\alpha = \tfrac{1}{2}q(1 - 3J_2^{(0)}/q) = \tfrac{1}{2}q - \tfrac{3}{2}J_2^{(0)} . \tag{4.139}$$

The reader is reminded that this relation holds with an accuracy of the order of α, q. The more accurate relation reads (Burša 1970)

$$\alpha = \tfrac{1}{2}q - \tfrac{3}{2}J_2^{(0)} - \tfrac{11}{56}q^2 + \tfrac{3}{14}J_2^{(0)}q + \tfrac{9}{8}(J_2^{(0)})^2 + \tfrac{9}{98}q^3 - \tfrac{93}{784}J_2^{(0)}q^2 ; \tag{4.140}$$

it applies to any rotational spheroid regardless of the assumption of hydrostatic equilibrium. To be exact, if $J_2^{(0)}$ is the zonal geopotential coefficient of the real body, which is not in hydrostatic equilibrium, α is the flattening corresponding to its external real equipotential surface. If $J_2^{(0)} = (J_2^{(0)})_h$, i.e. the zonal geopotential coefficient of the real body is equal to parameter $(J_2^{(0)})_h$ for a body in hydrostatic equilibrium, we arrive at $\alpha = \alpha_h$, which is the hydrostatic flattening. Love number k_s in Eq. (4.138) corresponds to an Earth in ideal hydrostatic equilibrium. If Eqs. (4.136) and (4.138) are combined, constant $J_2^{(0)}$ should also reflect this condition, $(J_2^{(0)})_h = -1077 \times 10^{-6}$.

Parameter k_s reflects the radial mass structure of the Earth. If all the mass were concentrated in the Earth's centre, $J_2^{(0)}$ would be equal to zero and, consequently, $k_s = 0$. Equation (4.136) may also be changed by substituting $J_2^{(0)} = -HC/(M\bar{a}^2)$:

$$k_s = 3\,\frac{H}{q}\,\frac{C}{M\bar{a}^2} , \tag{4.141}$$

where H is constant (2.252) determined from the precession: $H = 0.003\,273\,965 \pm 2 \times 10^{-9}$ (Kinoshita and Souchay 1990). The value of the relative moment of inertia, $C/(M\bar{a}^2)$, is equal to 2/5 for any homogeneous rotational ellipsoid or sphere, and decreases below 2/5 as the density of the body increases towards its centre of mass. In contrast, if the body were formed by a boundary spherical layer this value would be equal to 2/3. This means that, for the real Earth,

$$k_s < 1.14 . \tag{4.142}$$

In view of (4.138), if the Earth is in ideal hydrostatic equilibrium, then

$$1 + k_s = 2\alpha_h/q , \tag{4.143}$$

α_h being the flattening of the boundary surface which bounds the body of the rotating ideal fluid. If $\alpha_h = 1/(299.0 \pm 1)$, which follows from the theory of hydrostatic equilibrium in linear approximation (Sect. 2.8),

$$\alpha_h = \frac{5}{2}q\left[1 + \left(\frac{5}{2} - \frac{15}{4}\frac{C}{M\bar{a}^2}\right)^2\right]^{-1} ; \tag{4.144}$$

hence

$$1 + k_s = 1.93 , \quad k_s = 0.93 . \tag{4.145}$$

If the current value for the real Earth, $\alpha = 1/298.257$, were taken as the flattening, we would get a very slightly higher value of $1 + k_s = 1.9373$, which is understandable since the difference, $\alpha_h - \alpha = 8.3 \times 10^{-6}$, is also a small value.

As regards Eq. (4.144) its real accuracy is of the order α, q. It might seem that the hydrostatic flattening can be determined with this accuracy only from rotation parameter q and from the relative moment of inertia $C/(M\bar{a}^2)$, independently of the radial distribution of density. Equation (4.144) was derived for the Radau–Darwin density model; if, for example, the mass is largely concentrated in the planet's centre (as in Jupiter), the formula is practically useless.

5 The Earth's Deformations and Variations in the Earth's Rotation

5.1 Introduction

In Chapters 3 and 4 we dealt separately with rotational and tidal dynamics. Both these geodynamic phenomena are responsible for the Earth's deformations and this is their common feature. The tidal deformations affect the Earth's rotation directly, and variations of the Earth's rotation generate additional deformations.

In this chapter we shall discuss some of the global phenomena which require a synthetic approach, i.e. those which involve rotational dynamics as well as dynamics of tidal and other deformations. We shall attempt to analyse the hypothesis of the Earth's expansion in the light of contemporary astronomical and geophysical data. We shall present the theory of the effect of the variations of the second zonal geopotential coefficient, detected by orbit analysis of the geodynamic satellite LAGEOS, on the Earth's polar motion.

5.2 Dynamics of the Tidal Deceleration of the Earth's Rotation

In the era of rotational time, the unit of the time scale was the sidereal period of the Earth's rotation,

$$T = 23 \text{ h } 56 \text{ min } 04.10 \text{ s} = 86\,164.10 \text{ s};\tag{5.1}$$

in other words, the time scale was based on the angular velocity of the Earth's rotation (3.103):

$$\omega = \frac{2\pi}{T} = 7.292\,115 \times 10^{-5} \text{ rad s}^{-1}.\tag{5.2}$$

The value

$$T = 23 \text{ h } 56 \text{ min } 04.098\,904 \text{ s}$$

was adopted for epoch 1900 (i.e. at 12 h 00 min 00 s ET on 31 December 1899), which was taken as the origin for ephemeris calculations. The time variations reflect milliseconds and, therefore, if the time to which quantities T and ω apply is not given, no more decimal places than in (5.1) and (5.2) can be given. For epoch 1900.0 the mean sidereal time 23 h 56 min 04.090 540 s, the mean solar day 24 h 00 min 00.000 000 s and tropical year 8765 h 48 min 45.9747 s were also adopted.

However, in the era of rotational time it was already known that the Earth's rotation was not uniform, i.e. that $\omega \neq$ const ($T \neq$ const). This was found by observing the positions of the Moon which, compared with the ephemerides calculated in the rotational time system, apparently accelerated its motion along its orbit.

Halley (1695) was the first to point out this in his treatise on the positions (longitudes) of ancient cities: 'And I could then pronounce in what proportion the Moon's motion does accelerate, which that it does, I think I can demonstrate, and shall (God willing) make it appear to the public'.

However, the quantification of the phenomenon was presented a half century later by Dunthorne (1750) on the basis of ancient, mediaeval and contemporaneous (telescopic as of about the year 1630) observations. He presented no interpretation or dynamics thereof. This was done 5 years later by Kant (1754), who first said that the origin should be sought in tidal friction, and who described the phenomenon's dynamics.

Darwin (1879, 1880) presented a detailed theory of the whole tidal mechanism of deceleration of the Earth's rotation and its effect on the Moon's motion. This theory is essentially used to this day to describe the phenomenon in its original form.

The real Earth is neither perfectly rigid nor perfectly elastic; it behaves like a viscoelastic body. This causes the tidal deformations (reactions) which continuously propagate along the Earth's surface to the west (in accord with the apparent motion of the tide-generating body), to lag in time; we shall denote the lag Δt.

The tidal deformations of the Earth cause periodic variations in the angular velocity of its rotation, $\omega_\oplus$, with a semi-annual, monthly and semi-monthly period (Jeffreys 1928), and secular deceleration, traditionally attributed to the energy dissipation of ocean tidal waves.

The maximum tidal deformation of the Earth due to the Moon will not occur at the given point of the Earth's surface with east geographical longitude Λ at the time the Moon's centre of mass O' is in the meridian of this point, i.e. when $T_{O'} = -\Lambda$, but will occur once the Earth has rotated through angle ε (Fig. 5.1), $\varepsilon = \omega_\oplus \Delta t$, where Δt is the time interval required to rotate through angle ε and $T_{O'}$ is the hour angle of the Moon's centre of mass O', reckoned from the prime meridian $\Lambda = 0°$.

Strictly speaking, it is also necessary to consider the Moon's own motion over time Δt, i.e.

$$\varepsilon = (\omega_\oplus - n_{\leftmoon}) \Delta t \,, \tag{5.3}$$

$n_{\leftmoon}$ being the mean motion of the Moon ($n_{\leftmoon} = 2\pi/T_{\leftmoon}$; $T_{\leftmoon}$ is the sidereal orbital period).

Due to the rotation of the Earth, therefore, the tides propagate in the direction of rotation, i.e. along $\overline{OA}$ (Fig. 5.1) and in front of the line joining the

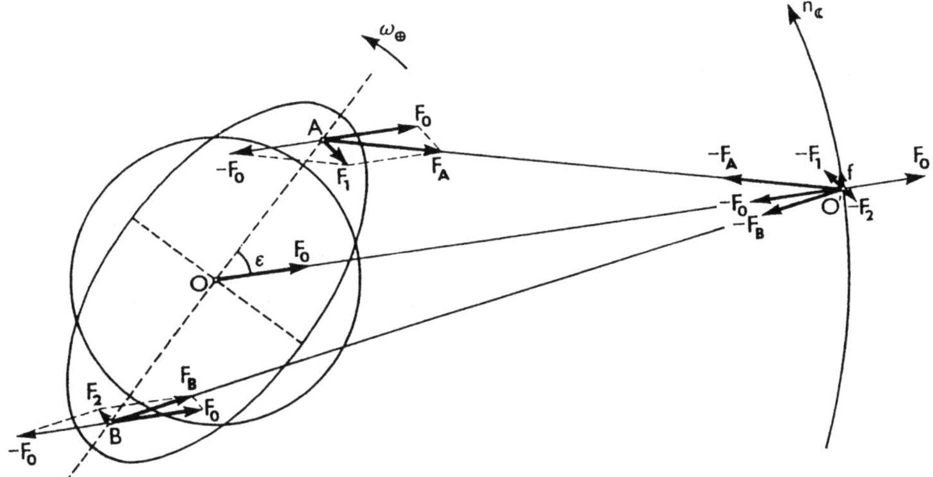

Fig. 5.1. Dynamics of the tidal deceleration of the Earth's rotation and of the Moon's mean motion

centres of mass of the Earth and the Moon, $\overline{OO'}$. In other words, the maximum tidal perturbation occurs at a given point only after the time of local culmination of the Moon. This generates two forces, $\mathbf{F}_1$, $\mathbf{F}_2$, as indicated in Fig. 5.1, which tend to turn the Earth against its rotation so that the line $\overline{AB}$, joining the centres of tidal bulges, passes through the Moon's mass centre O'. Forces $\mathbf{F}_A$, $\mathbf{F}_B$, $\mathbf{F}_O$ are the gravitational forces due to the Moon acting at the centres of tidal bulges A, B, at the Earth's centre of mass O, $|\mathbf{F}_A| > |\mathbf{F}_O| > |\mathbf{F}_B|$, because $\overline{AO'} < \overline{OO'} < \overline{BO'}$; $[(|\mathbf{F}_A| - |\mathbf{F}_B|)/|\mathbf{F}_A| \sim 5 \times 10^{-2}]$; $\mathbf{F}_1 = \mathbf{F}_A - \mathbf{F}_O$, $\mathbf{F}_2 = \mathbf{F}_B - \mathbf{F}_O$, $|\mathbf{F}_1| > |\mathbf{F}_2|$. The same forces act in the opposite direction at the Moon's centre of mass O'. Consequently, two forces, $-\mathbf{F}_1$, $-\mathbf{F}_2$, act at O', the difference $\mathbf{F}_2 - \mathbf{F}_1$ having a component along the normal to the orbit and component $\mathbf{f}$ acting along the tangent to it which, given the orientation of vectors $\omega_\oplus$ and $\mathbf{n}_{\mathbb{D}}$, acts in the direction of the Moon's orbital motion and, consequently, accelerates it.

The maximum tidal deformation thus occurs Δt later than the instant at which the Moon's (Sun's) mass centre was in the observer's meridian. This means that every tidal perturbation (wave) must be corrected by phase shift $\omega_\oplus \Delta t$, and that the arguments in (4.73), which depend on time, must be changed accordingly. If an approximate solution is sufficient, $k(\Theta + \omega_\oplus \Delta t)$ will replace $k\Theta$ and $(n - 2r + s)(M_{\mathbb{D}}'' + n_{\mathbb{D}})$ will replace $(n - 2r + s)M_{\mathbb{D}}''$ in (4.73) (Lambeck 1977). Then

$$S_{nkrs} = \frac{\cos}{\sin}\,[(n - 2r)\omega_{\mathbb{D}} + (n - 2r + s)(M_{\mathbb{D}}'' + n_{\mathbb{D}}\Delta t)$$

$$+ k(\Omega_{\mathbb{D}} - \Lambda - \Theta - \omega_\oplus \Delta t)]\,{\begin{matrix}(n-k)\ \text{even}\\(n-k)\ \text{odd}\end{matrix}}\,, \tag{5.4}$$

$$\varepsilon_{nkrs} = [(n - 2r + s)n_{\mathbb{D}} - k\omega_\oplus \Delta t] \tag{5.5}$$

is either positive or negative depending on the sign and magnitude of $(n - 2r + s)$.

We shall now only take the harmonic terms of the second degree into account. If $\beta_{2,rs}^{(k)}$ is the wave phase of the radial perturbation of the second degree, order k and groups r, s, and if the Earth is perfectly elastic, then the phase shift, in view of the viscosity, is

$$\varepsilon_{2,rs}^{(k)} = \beta_{2,rs}^{(k)}(1 - \tfrac{3}{2}k_2 + h_2)/(-\tfrac{3}{2}k_2 + h_2);\tag{5.6}$$

the phase shift of the components of the analogous wave of the horizontal perturbations is similarly

$$\varepsilon_{2,rs}^{(k)} = \beta_{2,rs}^{(k)}(1 + k_2 - h_2)/(k_2 - h_2).\tag{5.7}$$

Phase shifts $\varepsilon_{2,rs}^{(k)}$ usually amount to several tenths of a degree and do not exceed $0.5°$.

Angles $\varepsilon_{2,rs}^{(k)}$ are the measure of internal friction, or of the dissipation of tidal energy Q within the Earth (Lambeck 1977):

$$\varepsilon_{2,rs}^{(k)} = \frac{1}{Q_{2,rs}^{(k)}} = \frac{1}{2\pi}\frac{\Delta E_{2,rs}^{(k)}}{E_{2,rs}^{(k)}},\tag{5.8}$$

where ΔE is the tidal energy dissipated during a single cycle of the appropriate component of the tidal force and E is the maximum energy during the cycle. If, for example, $\varepsilon_2^{(k)} = 0.2°$, $Q_2^{(k)} = 300$ (Lambeck 1977).

Force component $\mathbf{f}$ can be calculated by differentiating the additional potential (4.58), generated at the Moon's mass centre O', by the changes in the mass distribution within the Earth after the tidal deformation.

In this particular case $\psi = \varepsilon$, i.e. (approximately)

$$\delta V(O') = \frac{1}{2}\frac{GM_{\mathrm{D}}}{\Delta_{\oplus\mathrm{D}}}\left(\frac{R}{\Delta_{\oplus\mathrm{D}}}\right)^5 k_2(3\cos^2\varepsilon - 1).\tag{5.9}$$

This immediately yields the component of the force acting along the tangent to the Moon's orbit:

$$-\frac{\partial\delta V_s}{\Delta_{\oplus\mathrm{D}}\,\partial\varepsilon}M_{\mathrm{D}} = \frac{3}{2}\frac{GM_{\mathrm{D}}^2}{\Delta_{\oplus\mathrm{D}}^2}\left(\frac{R}{\Delta_{\oplus\mathrm{D}}}\right)^5 k_2\sin 2\varepsilon.\tag{5.10}$$

Its moment relative to the Earth's mass centre O must be equal to the negative time derivative of the Earth's spin angular momentum $\mathbf{I}\cdot\boldsymbol{\omega}$ because the total angular momentum of the Earth–Moon system must be preserved, provided that the system is considered to be isolated. As regards the component acting in the direction of vector $\boldsymbol{\omega}_\oplus$,

$$-C\frac{d\omega_\oplus}{dt} = \frac{3}{2}\frac{GM_{\mathrm{D}}^2}{\Delta_{\oplus\mathrm{D}}}\left(\frac{R}{\Delta_{\oplus\mathrm{D}}}\right)^5 k_2\cos i_{\mathrm{D}}\sin 2\varepsilon,\tag{5.11}$$

i_{D} being the angle between the plane of the Moon's orbit and the Earth's equatorial plane.

The Earth–Moon system is in fact not isolated; the Sun contributes between one-third and one-seventh to the phenomenon in question. Assuming angle ε is the same under the effect of both bodies, the simplified expression for this component reads

$$-C\frac{d\omega_\oplus}{dt} = \frac{3}{2}\left[\frac{GM_{\leftmoon}^2}{\Delta_{\oplus\leftmoon}}\left(\frac{R}{\Delta_{\oplus\leftmoon}}\right)^5\cos i_{\leftmoon} + \frac{GM_\odot^2}{\Delta_{\oplus\odot}}\left(\frac{R}{\Delta_{\oplus\odot}}\right)^5\cos\tilde\varepsilon_0\right]k_2\sin 2\varepsilon$$

$$= \frac{3}{2}k_2\sin 2\varepsilon\,\frac{GM_{\leftmoon}^2}{\Delta_{\oplus\leftmoon}}\left(\frac{R}{\Delta_{\oplus\leftmoon}}\right)^5\cos i_{\leftmoon}\left[1 + \left(\frac{M_\odot}{M_{\leftmoon}}\right)^2\left(\frac{\Delta_{\oplus\leftmoon}}{\Delta_{\oplus\odot}}\right)^6\frac{\sin\tilde\varepsilon_0}{\cos i_{\leftmoon}}\right],$$

$$(5.12)$$

$\tilde\varepsilon_0$ being the mean value of the angle between the plane of the Earth's equator and the plane of the ecliptic. The ratio of the Moon's and Sun's tidal potentials and of the radial components of the tidal forces when the perturbing bodies are in the plane of the local meridian is indeed

$$\frac{GM_{\leftmoon}}{\Delta_{\oplus\leftmoon}}\left(\frac{\varrho}{\Delta_{\oplus\leftmoon}}\right)^2 : \frac{GM_\odot}{\Delta_{\oplus\odot}}\left(\frac{\varrho}{\Delta_{\oplus\odot}}\right)^2 = \frac{M_{\leftmoon}}{M_\odot}\left(\frac{\Delta_{\oplus\odot}}{\Delta_{\oplus\leftmoon}}\right)^3 \sim 1.95\,; \qquad (5.13)$$

however, the ratio of the time variations generated in the angular velocity of the Earth's rotation ($\Delta_{\oplus\odot} \sim A_\odot = 1.495\,978\,70 \times 10^{11}$ m, heliocentric gravitational constant $GM_\odot = 132\,712\,496.5 \times 10^{12}$ m^{3}s^{-2}),

$$\left(\frac{d\omega_\oplus}{dt}\right)_{\mathrm{Sun}} : \left(\frac{d\omega_\oplus}{dt}\right)_{\mathrm{Moon}} = \left[\frac{GM_\odot}{GM_{\leftmoon}}\left(\frac{\Delta_{\oplus\leftmoon}}{\Delta_{\oplus\odot}}\right)^3\right]^2, \qquad (5.14)$$

varies from $1:3.1$ [$(\Delta_{\oplus\odot})_{\min} = 147.1 \times 10^9$ m, $(\Delta_{\oplus\leftmoon})_{\max} = 406\,730 \times 10^3$ m] to $1:7.2$ [$(\Delta_{\oplus\odot})_{\max} = 152.1 \times 10^9$ m, $(\Delta_{\oplus\leftmoon})_{\min} = 364\,400 \times 10^3$ m].

If product $k_2\sin 2\varepsilon$ were known, $d\omega_\oplus/dt$ could be calculated and vice versa, because all the other quantities in (5.12) are known. The problem is, however, that we are not certain whether quantity

$$d\omega_\oplus/dt = -(5.4 \pm 0.5) \times 10^{-22}\ \mathrm{rad\,s}^{-2}, \qquad (5.15)$$

determined from observations, is indeed, and only, of tidal origin, i.e. we do not know whether mechanisms of non-tidal origin are also involved. Strictly speaking, product $k_2\sin\varepsilon$ is not known either. On the one hand, we are not certain whether we can use Love number $k_2 \sim 0.30$ at all in this case and, on the other hand, angle ε has not been determined reliably yet from observations: gravimetric tide data indicate a scatter larger than the value itself. If quantity (5.15) is attributed an exclusively tidal origin and if $GM_{\leftmoon} = 4902 \times 10^9$ m^{3}s^{-2}, $GM_\odot = 13\,271\,244.0 \times 10^{13}$ m^{3}s^{-2}, $\Delta_{\oplus\leftmoon} = 3.844 \times 10^8$ m, $\Delta_{\oplus\odot} = 1.495\,9787 \times 10^{11}$ m, $G = 6673 \times 10^{-14}$ m^{3}s^{-2}kg^{-1}, $C = 8.036 \times 10^{37}$ kg m^2, $\tilde\varepsilon_0 = 23°26'$

and $i_{\mathrm{D}} = \tilde{\varepsilon}_0 \pm 5°09' \sim \tilde{\varepsilon}_0$,

$$C\frac{d\omega_{\oplus}}{dt} = -4.34 \times 10^{16} \, \mathrm{kg\,m^2\,rad\,s^{-2}},$$

$$\frac{3}{2}\frac{GM_{\mathrm{D}}^2}{\Delta_{\oplus\mathrm{D}}}\left(\frac{R}{\Delta_{\oplus\mathrm{D}}}\right)^5 \cos i_{\mathrm{D}} = 1.612 \times 10^{18} \, \mathrm{kg\,m^2\,s^{-2}},$$

$$\frac{3}{2}\frac{GM_{\odot}^2}{\Delta_{\oplus\odot}}\left(\frac{R}{\Delta_{\oplus\odot}}\right)^5 \cos\tilde{\varepsilon}_0 = 3.40 \times 10^{17} \, \mathrm{kg\,m^2\,s^{-2}}, \tag{5.16}$$

we arrive at $k_2 \sin 2\varepsilon = 0.022\,23$ and, given $k_2 = 0.30$, $\varepsilon = 2.16°$.

The same mechanism that decelerates the Earth in its rotation accelerates the Moon tangentially to its orbit. This acceleration, according to Kepler's third law $n_{\mathrm{D}}^2 a_{\mathrm{D}}^3 = G(M_{\oplus} + M_{\mathrm{D}})$, results in the semimajor axis a_{D} of the Moon's orbit increasing in length, i.e. in the Moon moving away from the Earth, and in its mean motion decelerating. The secular increase da_{D}/dt of the semimajor axis of the Moon's orbit was proved, by laser ranging of lunar reflectors in the interval from 1969 to 1983, to be (Dickey et al. 1991)

$$\frac{da_{\mathrm{D}}}{dt} = (3.77 \pm 0.15)\,\mathrm{cm\,y^{-1}}, \tag{5.17}$$

(hereinafter y will be used for year) which, translated into the deceleration of the Moon's mean motion n_{D}, is

$$\frac{dn_{\mathrm{D}}}{dt} = -\frac{3}{2}\frac{n_{\mathrm{D}}}{a_{\mathrm{D}}}\frac{da_{\mathrm{D}}}{dt} = -(25.5'' \pm 1.0'')\,\mathrm{cy^{-2}}. \tag{5.18}$$

Quantities (5.15), (5.17) and/or (5.18) thus have the same cause and must be mutually related. This relation can be deduced from the condition that the total angular momentum of rotational and orbital motion $\mathbf{L}_{\oplus\mathrm{D}\odot}$ of the Earth–Moon–Sun system must be conserved:

$$\mathbf{L}_{\oplus\mathrm{D}\odot} = \mathbf{L}_{\oplus\mathrm{D}} + \mathbf{L}_{\odot B} + \mathbf{I}_{\oplus}\cdot\boldsymbol{\omega}_{\oplus} + \mathbf{I}_{\mathrm{D}}\cdot\boldsymbol{\omega}_{\mathrm{D}} + \mathbf{I}_{\odot}\cdot\boldsymbol{\omega}_{\odot} = \mathbf{C},$$

$$d\mathbf{L}_{\oplus\mathrm{D}\odot}/dt = 0, \tag{5.19}$$

where $\mathbf{L}_{\oplus\mathrm{D}}$ is the orbital angular momentum of the Earth–Moon system, $\mathbf{L}_{\odot B}$ is the orbital angular momentum of the system Sun-barycentre B of the Earth–Moon system, $\mathbf{I}_{\oplus}, \mathbf{I}_{\mathrm{D}}, \mathbf{I}_{\odot}$ are the tensors of inertia of the respective bodies, and $\boldsymbol{\omega}_{\oplus}, \boldsymbol{\omega}_{\mathrm{D}}, \boldsymbol{\omega}_{\odot}$ are the vectors of their instantaneous rotation.

The orbital angular momentum

$$\mathbf{L}_{\oplus\mathrm{D}} = \left[\mathbf{r}_{O'} \times M_{\mathrm{D}}\frac{d\mathbf{r}_{O'}}{dt}\right] + \left[\mathbf{r}_O \times M_{\oplus}\frac{d\mathbf{r}_O}{dt}\right], \tag{5.20}$$

where O and O' are the mass centres of the Earth and Moon, and $\mathbf{r}_O$, $\mathbf{r}_{O'}$ are the radius-vectors referred to barycentre B. Since, by definition of the barycentre,

$$\mathbf{r}_{O'} = M_{\oplus}(M_{\oplus} + M_{\mathrm{D}})^{-1}\mathbf{r}_{OO'}, \qquad \mathbf{r}_O = M_{\mathrm{D}}(M_{\oplus} + M_{\mathrm{D}})^{-1}\mathbf{r}_{OO'}; \tag{5.21}$$

hence

$$\mathbf{L}_{\oplus\rotatebox{0}{)}} = \left[\mathbf{r}_{oo'} \times \frac{\mathrm{d}\mathbf{r}_{oo'}}{\mathrm{d}t}\right] M_\oplus M_{\rotatebox{0}{)}} (M_\oplus + M_{\rotatebox{0}{)}})^{-1} \tag{5.22}$$

or

$$\mathbf{L}_{\oplus\rotatebox{0}{)}} = \mathbf{K}_{\oplus\rotatebox{0}{)}} M_\oplus M_{\rotatebox{0}{)}} (M_\oplus + M_{\rotatebox{0}{)}})^{-1}, \tag{5.23}$$

where $\mathbf{K}_{\oplus\rotatebox{0}{)}}$ is the vectorial integral of areas [see (1.29)]:

$$K_{\oplus\rotatebox{0}{)}}^2 = G(M_\oplus + M_{\rotatebox{0}{)}})a_{\rotatebox{0}{)}}(1 - e_{\rotatebox{0}{)}}^2) = n_{\rotatebox{0}{)}}^2 a_{\rotatebox{0}{)}}^4 (1 - e_{\rotatebox{0}{)}}^2); \tag{5.24}$$

$a_{\rotatebox{0}{)}}$ and $e_{\rotatebox{0}{)}}$ are the semi-axis and eccentricity of the Moon's orbit, respectively, and G is the gravitational constant.

Orbital angular momentum $\mathbf{L}_{\odot B}$ is analogously

$$\mathbf{L}_{\odot B} = \mathbf{K}_{\odot B} M_\odot (M_\oplus + M_{\rotatebox{0}{)}})(M_\odot + M_\oplus + M_{\rotatebox{0}{)}})^{-1}, \tag{5.25}$$

with

$$K_{\odot B}^2 - G(M_\odot + M_\oplus + M_{\rotatebox{0}{)}})a_B(1 - e_B^2) = n_B^2 a_B^4 (1 - e_B^2); \tag{5.26}$$

n_B is the mean motion of barycentre B and a_B and e_B are elements of its orbit. We shall now put $n_B = n_\oplus$, $a_B = a_\oplus$, $e_B = e_\oplus$ ($n_\oplus$, $a_\oplus$, $e_\oplus$ are the mean motion and orbital elements of the Earth). Since the Moon's rotation is in resonance with its orbital motion, for practical purposes $\mathrm{d}\omega_{\rotatebox{0}{)}}/\mathrm{d}t = \mathrm{d}n_{\rotatebox{0}{)}}/\mathrm{d}t$; the tidal deceleration of the Sun's rotation, i.e. $\mathrm{d}\omega_\odot/\mathrm{d}t$, will now be neglected, and the variation in the Moon's and Sun's tensor of inertia will not be taken into account either, i.e. we shall put $\mathrm{d}\mathbf{I}_{\rotatebox{0}{)}}/\mathrm{d}t = 0$, $\mathrm{d}\mathbf{I}_\odot/\mathrm{d}t = 0$.

Hence,

$$\mathbf{I}_\oplus \, \mathrm{d}\boldsymbol{\omega}_\oplus/\mathrm{d}t + \boldsymbol{\omega}_\oplus \, \mathrm{d}\mathbf{I}_\oplus/\mathrm{d}t + \mathbf{I}_{\rotatebox{0}{)}} \, \mathrm{d}n_{\rotatebox{0}{)}}/\mathrm{d}t$$

$$= - M_\oplus M_{\rotatebox{0}{)}} (M_\oplus + M_{\rotatebox{0}{)}})^{-1} \, \mathrm{d}\mathbf{K}_{\oplus\rotatebox{0}{)}}/\mathrm{d}t$$

$$\qquad - M_\odot (M_\oplus + M_{\rotatebox{0}{)}})(M_\odot + M_\oplus + M_{\rotatebox{0}{)}})^{-1} \, \mathrm{d}\mathbf{K}_{\odot B}/\mathrm{d}t. \tag{5.27}$$

We shall now express the component of (5.27) acting in the direction of vector $\boldsymbol{\omega}_\oplus$. Since we are only interested in the secular tidal part of variations $\mathrm{d}\omega_\oplus/\mathrm{d}t$ and $\mathrm{d}n_{\rotatebox{0}{)}}/\mathrm{d}t$, and since the components of rotation vector $\boldsymbol{\omega}_\oplus$ along the equatorial axes do not apparently have a secular character, which also applies to time variations $\mathrm{d}I_{31}/\mathrm{d}t$, $\mathrm{d}I_{32}/\mathrm{d}t$ of the products of inertia, we may simplify this to

$$(I_{33})_\oplus \, \mathrm{d}\omega_\oplus/\mathrm{d}t + \omega_\oplus \, \mathrm{d}(I_{33})_\oplus/\mathrm{d}t = - M_\oplus M_{\rotatebox{0}{)}} (M_\oplus + M_{\rotatebox{0}{)}})^{-1}$$

$$\times [\cos i_{\rotatebox{0}{)}} \, \mathrm{d}K_{\oplus\rotatebox{0}{)}}/\mathrm{d}t - \sin i_{\rotatebox{0}{)}} K_{\oplus\rotatebox{0}{)}} \, \mathrm{d}i_{\rotatebox{0}{)}}/\mathrm{d}t]$$

$$- M_\odot (M_\oplus + M_{\rotatebox{0}{)}})(M_\odot + M_\oplus + M_{\rotatebox{0}{)}})^{-1} [\cos \tilde{\varepsilon}_0 \, \mathrm{d}K_{\odot B}/\mathrm{d}t$$

$$- \sin \tilde{\varepsilon}_0 K_{\odot B} \, \mathrm{d}\tilde{\varepsilon}_0/\mathrm{d}t] - (I_{33})_{\rotatebox{0}{)}} \cos i_{\oplus\rotatebox{0}{)}} \, \mathrm{d}n_{\rotatebox{0}{)}}/\mathrm{d}t; \tag{5.28}$$

$i_{\rotatebox{0}{)}}$ is the angle between the Moon's orbital plane L and the plane of the Earth's equator $A_\oplus$ (Fig. 5.2); $\tilde{\varepsilon}_0 = 23° 26'$ is the angle between the plane of the Earth's

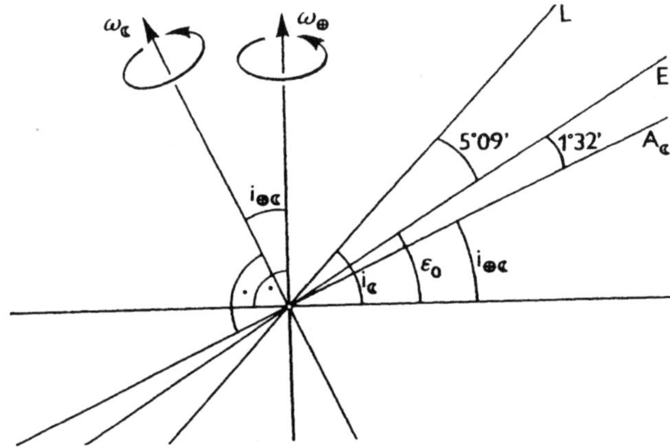

Fig. 5.2. The planes and directions of reference of the Earth–Moon system

equator and the plane of the ecliptic E; $i_{\oplus\mathrm{D}}$ is the angle between the Earth's and Moon's axes of rotation or between the planes of the Earth's, $A_\oplus$, and Moon's, A_D, equators; $(I_{33})_\oplus = C$ is the largest principal moment of inertia of the Earth, of which we shall continue to assume that it has no secular time component, i.e. we shall put $dC/dt = 0$:

$$
\begin{aligned}
dK_{\oplus\mathrm{D}}/dt &= -\tfrac{1}{3}a_\mathrm{D}^2(1 - e_\mathrm{D}^2)^{1/2}\,dn_\mathrm{D}/dt \\
&\quad - n_\mathrm{D}a_\mathrm{D}^2 e_\mathrm{D}(1 - e_\mathrm{D}^2)^{-1/2}\,de_\mathrm{D}/dt \\
&= [G(M_\oplus + M_\mathrm{D})]^{1/2}[\tfrac{1}{2}a_\mathrm{D}^{-1/2}(1 - e_\mathrm{D}^2)^{1/2}\,da_\mathrm{D}/dt \\
&\quad - e_\mathrm{D}a_\mathrm{D}^{1/2}(1 - e_\mathrm{D}^2)^{-1/2}\,de_\mathrm{D}/dt]\,,
\end{aligned}
\tag{5.29}
$$

where de_D/dt and di_D/dt are variations of the Moon's orbital elements e_D and i_D due to tidal deformations. They can be determined from Lagrange's planetary equations of motion by substituting the additional potential, δV, generated by the tidal deformations of the Earth, for the perturbation function. Analogously,

$$
\begin{aligned}
dK_{\odot B}/dt &= -\tfrac{1}{3}a_\oplus^2(1 - e_\oplus^2)^{1/2}\,dn_\oplus/dt \\
&\quad - n_\oplus a_\oplus^2 e_\oplus(1 - e_\oplus^2)^{-1/2}\,de_\oplus/dt \\
&= [G(M_\odot + M_\oplus + M_\mathrm{D})]^{1/2}[\tfrac{1}{2}a_\oplus^{-1/2}(1 - e_\oplus^2)^{1/2}\,da_\oplus/dt \\
&\quad - e_\oplus a_\oplus^{1/2}(1 - e_\oplus^2)^{-1/2}\,de_\oplus/dt]\,.
\end{aligned}
\tag{5.30}
$$

Hereinafter we shall neglect the effect of tidal deformations $de_\oplus/dt$ and $d\tilde\varepsilon_0/dt$, which have been proved to be quite negligible (Kaula 1964; MacDonald

1964), and we shall draw on the simplified relation

$$
\begin{aligned}
(I_{33})_\oplus \, d\omega_\oplus/dt = {} & [G(M_\oplus + M_\leftmoon)]^{1/2} M_\oplus M_\leftmoon (M_\oplus + M_\leftmoon)^{-1} \{\cos i_\leftmoon [-\tfrac{1}{2} a_\leftmoon^{-1/2} \\
& \times (1 - e_\leftmoon^2)^{1/2} \, da_\leftmoon/dt + a_\leftmoon^{1/2} e_\leftmoon (1 - e_\leftmoon^2)^{-1/2} \, de_\leftmoon/dt] \\
& + \sin i_\leftmoon a_\leftmoon^{1/2} (1 - e_\leftmoon^2)^{1/2} \, di_\leftmoon/dt\} \\
& - \tfrac{1}{2}[G(M_\odot + M_\oplus + M_\leftmoon)]^{1/2} \\
& \times M_\odot (M_\oplus + M_\leftmoon)(M_\odot + M_\oplus + M_\leftmoon)^{-1} \\
& \times \cos \tilde{\varepsilon}_0 a_\oplus^{-1/2} (1 - e_\oplus^2)^{1/2} \\
& \times da_\oplus/dt + \tfrac{3}{2}(I_{33})_\leftmoon [G(M_\oplus + M_\leftmoon)]^{1/2} \cos i_{\oplus\leftmoon} a_\leftmoon^{-5/2} \, da_\leftmoon/dt
\end{aligned}
$$

$$\tag{5.31}$$

For further balance considerations we shall modify it to read

$$
\begin{aligned}
d\omega_\oplus/dt = {} & a_1 \, da_\leftmoon/dt + a_2 \, de_\leftmoon/dt + a_3 \, di_\leftmoon/dt + a_4 \, da_\oplus/dt, \\
a_1 = {} & - \tfrac{1}{2}[G(M_\oplus + M_\leftmoon)]^{1/2} (I_{33})_\oplus^{-1} \\
& \times [M_\oplus M_\leftmoon (M_\oplus + M_\leftmoon)^{-1} a_\leftmoon^{-1/2} \cos i_\leftmoon (1 - e_\leftmoon^2)^{1/2} \\
& - 3(I_{33})_\leftmoon a_\leftmoon^{-5/2} \cos i_{\oplus\leftmoon}] = -4.2 \times 10^{-13} \, \mathrm{m}^{-1}\mathrm{s}^{-1}, \\
a_2 = {} & e_\leftmoon [G(M_\oplus + M_\leftmoon)]^{1/2} (I_{33})_\oplus^{-1} M_\oplus M_\leftmoon (M_\oplus + M_\leftmoon)^{-1} \\
& \times a_\leftmoon^{1/2} \cos i_\leftmoon (1 - e_\leftmoon^2)^{-1/2} = 1.7 \times 10^{-5} \, \mathrm{s}^{-1}, \\
a_3 = {} & [G(M_\oplus + M_\leftmoon)]^{1/2} (I_{33})_\oplus^{-1} M_\oplus M_\leftmoon (M_\oplus + M_\leftmoon)^{-1} \\
& \times a_\leftmoon^{1/2} \sin i_\leftmoon (1 - e_\leftmoon^2)^{1/2} = 1.7 \times 10^{-4} \, \mathrm{s}^{-1}, \\
a_4 = {} & - \tfrac{1}{2}[G(M_\odot + M_\oplus + M_\leftmoon)]^{1/2} (I_{33})_\oplus^{-1} M_\odot (M_\oplus + M_\leftmoon) \\
& \times (M_\odot + M_\oplus + M_\leftmoon)^{-1} a_\oplus^{-1/2} (1 - e_\oplus^2)^{1/2} \cos \tilde{\varepsilon}_0 \\
& = -1.0 \times 10^{-9} \, \mathrm{m}^{-1}\mathrm{s}^{-1}.
\end{aligned}
$$

$$\tag{5.32}$$

Tidal secular variations $da_\leftmoon/dt$ and $da_\oplus/dt$ can be expressed using Lagrange's planetary equation (1.108); however, the perturbation function, which is the cause of the Moon's deceleration and generates the variations in question, has to be known. This is function (5.9) but here we shall present it in more detail:

$$
\delta V = \delta V_\leftmoon + \delta V_\odot,
\tag{5.33}
$$

$$
\delta V_\leftmoon = k \frac{GM_\leftmoon}{\varDelta_{\oplus\leftmoon}^*} \left(\frac{R}{\varDelta_{\oplus\leftmoon}^*}\right)^2 \left(\frac{R}{\varDelta_{\oplus\leftmoon}}\right)^3 P_2^{(0)}(\cos \varepsilon_\leftmoon),
\tag{5.34}
$$

$$
\delta V_\odot = k \frac{GM_\odot}{\varDelta_{\oplus\odot}^*} \left(\frac{R}{\varDelta_{\oplus\odot}^*}\right)^2 \left(\frac{R}{\varDelta_{\oplus\odot}}\right)^3 P_2^{(0)}(\cos \varepsilon_\odot),
\tag{5.35}
$$

$$P_2^{(0)}(\cos \varepsilon_{\mathbb{D}}) = P_2^{(0)}(\sin \delta_{O'}^*)\,P_2^{(0)}(\sin \delta_{O'})$$

$$+ 2\sum_{m=1}^{2} \frac{(2-m)!}{(2+m)!}\,P_2^{(m)}(\sin \delta_{O'}^*)\,P_2^{(m)}(\sin \delta_{O'})\cos m(T_{O'}^* - T_{O'})\,, \tag{5.36}$$

$$P_2^{(0)}(\cos \varepsilon_{\odot}) = P_2^{(0)}(\sin \delta_{O''}^*)\,P_2^{(0)}(\sin \delta_{O''})$$

$$+ 2\sum_{m=1}^{2} \frac{(2-m)!}{(2+m)!}\,P_2^{(m)}(\sin \delta_{O''}^*)\,P_2^{(m)}(\sin \delta_{O''})\cos m(T_{O''}^* - T_{O''})\,; \tag{5.37}$$

$\varepsilon_{\mathbb{D}} = v_{\mathbb{D}}^* - v_{\mathbb{D}}$; $\varepsilon_{\odot} = v_{\odot}^* - v_{\odot}$; k is the Love number characterizing the overall reaction of the elastic Earth, angles $\varepsilon_{\mathbb{D}}$, $\varepsilon_{\odot}$ depend in general on the orbital elements; here we shall put $\varepsilon_{\mathbb{D}} = \varepsilon_{\odot} = \varepsilon$; $\delta_{O'}$ and $\delta_{O''}$ are the geocentric declinations of the Moon and Sun; $T_{O'}$ and $T_{O''}$ are their hour angles reckoned from the prime meridian; geocentric coordinates $\Delta_{\oplus\mathbb{D}}^*$, $\Delta_{\oplus\odot}^*$, $\delta_{O'}^*$, $\delta_{O''}^*$, $T_{O'}^*$, $T_{O''}^*$ define the fictitious position of the tide-generating body, corresponding to the maximum tide at the given point; $v_{\mathbb{D}}$ and $v_{\oplus}$ are the true anomalies of the tide-generating bodies at their actual positions and $v_{\mathbb{D}}^*$ and $v_{\odot}^*$ are at the said fictitious positions. Hereinafter we shall designate the Love number as k and not k_2. Note that this parameter and the other Love numbers depend on the frequency of the phenomenon observed (Zschau 1978).

Perturbing potential (5.33) generates the variations of the Moon's and Earth's orbital elements. The variations of the semimajor axes read

$$da_{\mathbb{D}}/dt = 2kGM_{\mathbb{D}}(M_{\oplus} + M_{\mathbb{D}})M_{\oplus}^{-1}[G(M_{\oplus} + M_{\mathbb{D}})]^{-1/2}$$

$$\times a_{\mathbb{D}}^{-1/2}\left(\frac{R}{a_{\mathbb{D}}}\right)^5 P_2^{(1)}(\sin \varepsilon_{\mathbb{D}}) + \text{the terms containing } e_{\mathbb{D}}\,, \tag{5.38}$$

$$da_{\oplus}/dt = 2kGM_{\odot}(M_{\odot} + M_{\oplus} + M_{\mathbb{D}})(M_{\oplus} + M_{\mathbb{D}})^{-1}$$

$$\times [G(M_{\odot} + M_{\oplus} + M_{\mathbb{D}})]^{-1/2} a_{\oplus}^{-1/2}\left(\frac{R}{a_{\oplus}}\right)^5$$

$$\times P_2^{(1)}(\sin \varepsilon_{\odot}) + \text{the terms containing } e_{\oplus}\,. \tag{5.39}$$

After substituting $P_2^{(1)}(\sin \varepsilon_{\mathbb{D}}) \sim 3\varepsilon_{\mathbb{D}}$, $P_2^{(1)}(\sin \varepsilon_{\odot}) \sim 3\varepsilon_{\odot}$, and neglecting the terms which depend on eccentricity, $\varepsilon_{\mathbb{D}}$, $\varepsilon_{\odot}$ being expressed in degrees, we arrive at

$$da_{\mathbb{D}}/dt = 1.7(k\varepsilon_{\mathbb{D}}^0)\times 10^{-9}\,\mathrm{m\,s}^{-1} = 5.3(k\varepsilon_{\mathbb{D}}^0)\,\mathrm{cm\,y}^{-1}\,, \tag{5.40}$$

$$da_{\oplus}/dt = 1.4(k\varepsilon_{\odot}^0)\times 10^{-13}\,\mathrm{m\,s}^{-1} = 4.6(k\varepsilon_{\odot}^0)\times 10^{-4}\,\mathrm{cm\,y}^{-1}\,. \tag{5.41}$$

Quantity (5.40) was measured directly by lunar laser ranging (5.17). By comparing (5.40) with (5.17) we find that

$$k\varepsilon_{\mathbb{D}}^0 = 0.70 \pm 0.04\,. \tag{5.42}$$

Assuming that $\varepsilon_\odot = \varepsilon_{\mathbb{D}}$, and by substituting (5.42) into (5.41) we obtain

$$da_\oplus/dt = (9.8 \pm 0.6) \times 10^{-14}\,\mathrm{m\,s}^{-1} = (3.2 \pm 0.2) \times 10^{-4}\,\mathrm{cm\,y}^{-1}. \tag{5.43}$$

Secular variations $de_{\mathbb{D}}/dt$ and $di_{\mathbb{D}}/dt$, caused by tidal friction, are small: according to MacDonald (1964) and Kaula (1964),

$$de_{\mathbb{D}}/dt = +1.2 \times 10^{-11}\,\mathrm{y}^{-1}, \tag{5.44}$$

$$di_{\mathbb{D}}/dt = -9.35 \times 10^{-12}\,\mathrm{rad\,y}^{-1}. \tag{5.45}$$

With a view to (5.32) their contributions to variations $d\omega_\oplus/dt$ are

$$a_2\,de_{\mathbb{D}}/dt = +6.5 \times 10^{-24}\,\mathrm{rad\,s}^{-2},$$
$$a_3\,di_{\mathbb{D}}/dt = -5.0 \times 10^{-23}\,\mathrm{rad\,s}^{-2}. \tag{5.46}$$

The rhs of the resultant balance equation thus displays the following terms:

$$a_1\,da_{\mathbb{D}}/dt = -5.02 \times 10^{-22}\,\mathrm{rad\,s}^{-2},$$
$$a_2\,de_{\mathbb{D}}/dt = +0.06 \times 10^{-22}\,\mathrm{rad\,s}^{-2},$$
$$a_3\,di_{\mathbb{D}}/dt = -0.50 \times 10^{-22}\,\mathrm{rad\,s}^{-2},$$
$$a_4\,da_\oplus/dt = -1.01 \times 10^{-22}\,\mathrm{rad\,s}^{-2},$$

$$\text{Sum:} \qquad -6.47 \times 10^{-22}\,\mathrm{rad\,s}^{-2}. \tag{5.47}$$

The absolute value of resultant quantity (5.47) is larger than the observed value by $1.1 \times 10^{-22}\,\mathrm{rad\,s}^{-2}$, and this is the residual which has to be explained as due to non-tidal effects. In other words, the Earth's rotation is decelerated less than the tidal mechanism would indicate. Moreover, we have not considered the effect of ocean tides which, as a result of water friction on ocean beds and on shallow sea beds, should, according to Pariiskii et al. (1972) and Pariiskii (1978), decrease the deceleration of the Earth's rotation significantly.

In other words, a non-tidal mechanism should exist which would accelerate the Earth's rotation so that the resultant deceleration is smaller than the tides would indicate. This is a problem being considered by an international cooperation of astronomers, geophysicists, meteorologists and oceanographers in the IUGG.

First, however, it is necessary to determine the part played by ocean and sea tides more accurately. The deceleration of the Earth's rotation due to ocean tides can be determined numerically by calculating the water friction on sea and ocean beds, or by calculating the moments of forces with which the Moon and Sun act on the water particles displaced by tides.

The latter method has a considerable advantage over the former in that it does not require the knowledge of the ocean bed topography or of the respective water friction coefficient; these are the limiting factors in the former method. The method's principle is simple. Every mass unit at any point of the tidally perturbed sea or ocean is acted upon by a component of the tidal force in the plane of the prime vertical (4.102) which has two non-zonal terms, (4.104) and (4.105), given $n = 2$. It is their moment which is responsible for the variations in the angular velocity of rotation of the body. To carry out a numerical calculation it is necessary to know the amplitude of the tidal sea wave, i.e. data from co-tidal maps, for every surface element of mean levels of oceans and seas, undisturbed by tides or otherwise.

The dynamics of the system is also affected by radial lunar tides caused by the eccentricity of the Moon's orbit (Fig. 5.3). If the Moon's orbit were strictly circular ($e_{\mathrm{D}} = 0$), the Earth's gravitational effect (under ideal synchronous rotation) would be constant at all its points. The Moon would thus be radially deformed 'once and for ever', and its tides would thus be constant. However, $e_{\mathrm{D}} \neq 0$; hence the tidal deformations when the Moon is in its perigee P ($\Delta_{\oplus\mathrm{D}} = \min. = 364\,400$ km) are larger than when it is in its apogee A ($\Delta_{\oplus\mathrm{D}} = \max. = 406\,730$ km). As a result of viscosity, the radial tidal deformations do not occur immediately and another additional moment is generated, because the mass centres of the two bodies are not aligned with the tide axis. The lunar radial tides do not affect the system's angular momentum; however, the system loses mechanical energy due to friction.

The Earth–Moon–Sun system may thus be considered a complicated three-body gyroscope whose bodies are connected by tidal forces which redistribute the total angular momentum of the system, $\mathbf{L}_{\oplus\mathrm{D}\odot}$, as a result of acting on them. The force moments affect the angular momentum of the system and cause the system to lose mechanical energy. The kinetic energy of the Earth's rotation is partly subject to dissipation and is partly transformed to thermal energy (the Earth is heated by tides) and partly to the potential energy of the Moon's orbital motion (the Moon recedes from the Earth). The Earth's angular momentum decreases,

$$C\frac{\mathrm{d}\omega_{\oplus}}{\mathrm{d}t} = -4.34 \times 10^{16}\ \mathrm{kg\,m^2\,rad\,s^{-2}}, \tag{5.48}$$

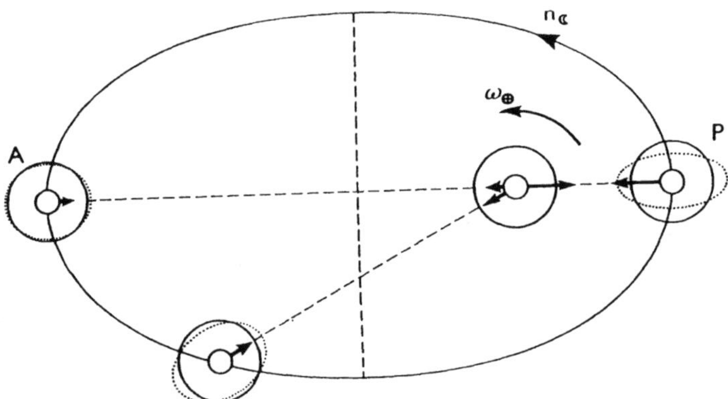

Fig. 5.3. Radial tides of the Moon

and the Moon's orbital angular momentum (i.e. its component in the direction of $\omega_\oplus$) increases,

$$\frac{\mathrm{d}L_{\oplus\mathbb{D}}}{\mathrm{d}t} = \frac{M_\oplus M_\mathbb{D}}{M_\oplus + M_\mathbb{D}} \frac{\mathrm{d}}{\mathrm{d}t}(K_{\oplus\mathbb{D}}\cos i_\mathbb{D})$$

$$= \frac{M_\oplus M_\mathbb{D}}{M_\oplus + M_\mathbb{D}} \frac{\mathrm{d}}{\mathrm{d}t}[n_\mathbb{D} a_\mathbb{D}^2 (1 - e_\mathbb{D}^2)^{1/2}\cos i_\mathbb{D}]$$

$$= -\frac{M_\oplus M_\mathbb{D}}{M_\oplus + M_\mathbb{D}} K_{\oplus\mathbb{D}}\left[\frac{1}{3} n_\mathbb{D}^{-1}\cos i_\mathbb{D}\frac{\mathrm{d}n_\mathbb{D}}{\mathrm{d}t} + e_\mathbb{D}(1 - e_\mathbb{D}^2)^{-1}\right.$$

$$\left.\times \cos i_\mathbb{D}\frac{\mathrm{d}e_\mathbb{D}}{\mathrm{d}t} + \sin i_\mathbb{D}\frac{\mathrm{d}i_\mathbb{D}}{\mathrm{d}t}\right] \sim 4.4 \times 10^{16}\,\mathrm{kg\,m^2\,rad\,s^{-2}}. \tag{5.49}$$

The orbital angular momentum (i.e. its component in the direction of $\omega_\oplus$) also increases, approximately at a rate of

$$\frac{\mathrm{d}L_{\odot B}}{\mathrm{d}t} = \frac{M_\odot (M_\oplus + M_\mathbb{D})}{M_\odot + M_\oplus + M_\mathbb{D}} \frac{\mathrm{d}}{\mathrm{d}t}(K_{\odot B}\cos \tilde{\varepsilon}_0)$$

$$= -\frac{1}{3}\frac{M_\odot (M_\oplus + M_\mathbb{D})}{M_\odot + M_\oplus + M_\mathbb{D}} K_{\odot B} n_\odot^{-1}\cos \tilde{\varepsilon}_0\frac{\mathrm{d}n_\odot}{\mathrm{d}t}$$

$$= 0.8 \times 10^{16}\,\mathrm{kg\,m^2\,rad\,s^{-2}}. \tag{5.50}$$

The sum $\mathrm{d}L_{\oplus\mathbb{D}}/\mathrm{d}t + \mathrm{d}L_{\odot B}/\mathrm{d}t$ does not, therefore, balance the decrement $C\,\mathrm{d}\omega_\oplus/\mathrm{d}t$ in full. The difference amounts to $0.9 \times 10^{16}\,\mathrm{kg\,m^2\,s^{-2}}$, and has to be explained in terms of non-tidal phenomena. It could easily be explained by the decrease of the Earth's principal moment of inertia (5.136), corresponding to the observed decrease of the second zonal geopotential coefficient (5.137) under the incompressibility condition.

The numerical values given are only approximate and can only be used in estimating the momentum balance. Assuming homogeneous rotation of the Earth, of the said tidal momentum $-5 \times 10^{15}\,\mathrm{kg\,m^2\,s^{-2}}$ is attributable to the core, $-3.8 \times 10^{16}\,\mathrm{kg\,m^2\,s^{-2}}$ to the mantle and $-4 \times 10^{14}\,\mathrm{kg\,m^2\,s^{-2}}$ to the crust. The Earth's mantle thus plays the most important part in this particular case, and its decelerating tidal momentum is relatively the largest of all components involved.

In Fig. 3.15 (bottom graph) the right-hand scale shows the angular acceleration, i.e. the overall time variation of the angular velocity of the Earth's rotation, but multiplied by the value, C_p, of the principal moment of inertia of the Earth's mantle relative to axis x_3:

$$C_p = 7.04 \times 10^{37}\,\mathrm{kg\,m^2}. \tag{5.51}$$

This is then the corresponding moment of the force acting on the mantle. For example, around epoch 1900 the amplitude of this moment increased to $10^{18}\,\mathrm{kg\,m^2\,s^{-2}}$ over a period of a mere 5 years. This is substantially higher than the secular term (5.48) which is mostly of tidal origin. Its magnitude is shown schematically in Fig. 3.15 as the arrow (tidal moment).

The Earth's total rotational kinetic energy,

$$E_\oplus = \tfrac{1}{2}(A\omega_1^2 + B\omega_2^2 + C\omega_3^2) \sim \tfrac{1}{2}C\omega_\oplus^2 = 2.137 \times 10^{29}\,\mathrm{kg\,m^2\,s^{-2}}, \tag{5.52}$$

also decreases with the angular velocity of the Earth's rotation:

$$\frac{\mathrm{d}E_\oplus}{\mathrm{d}t} = C\omega_\oplus \frac{\mathrm{d}\omega_\oplus}{\mathrm{d}t} = -3.2 \times 10^{12}\,\mathrm{kg\,m^2\,s^{-3}}, \tag{5.53}$$

$$\frac{1}{E_\oplus}\frac{\mathrm{d}E_\oplus}{\mathrm{d}t} = -1.5 \times 10^{-17}\,\mathrm{s^{-1}} = -4.7 \times 10^{-8}\,\mathrm{cy^{-1}}. \tag{5.54}$$

The Moon's total orbital kinetic energy, which is approximately (the energy integral in the two-body problem)

$$E_\mathrm{D} = \tfrac{1}{2}M_\mathrm{D} n_\mathrm{D}^2 a_\mathrm{D}^2 - G\frac{M_\oplus M_\mathrm{D}}{2a_\mathrm{D}} \sim -G\frac{M_\oplus M_\mathrm{D}}{2a_\mathrm{D}} = 3.8 \times 10^{28}\,\mathrm{kg\,m^2\,s^{-2}}, \tag{5.55}$$

in contrast increases

$$\frac{\mathrm{d}E_\mathrm{D}}{\mathrm{d}t} = \frac{1}{2}G\frac{M_\oplus M_\mathrm{D}}{a_\mathrm{D}^2}\frac{\mathrm{d}a_\mathrm{D}}{\mathrm{d}t} = 1.2 \times 10^{11}\,\mathrm{kg\,m^2\,s^{-3}}, \tag{5.56}$$

$$\frac{1}{E_\mathrm{D}}\frac{\mathrm{d}E_\mathrm{D}}{\mathrm{d}t} = -3.1 \times 10^{-18}\,\mathrm{s^{-1}} = 9.8 \times 10^{-9}\,\mathrm{cy^{-1}}. \tag{5.57}$$

The difference,

$$\left|\frac{\mathrm{d}E_\oplus}{\mathrm{d}t}\right| - \left|\frac{\mathrm{d}E_\mathrm{D}}{\mathrm{d}t}\right| = 3.1 \times 10^{12}\,\mathrm{kg\,m^2\,s^{-3}}, \tag{5.58}$$

should thus represent the dissipation rate of tidal energy spent within the Earth (heating), provided that both the quantities used in the estimate, $\mathrm{d}\omega_\oplus/\mathrm{d}t$ and $\mathrm{d}a_\mathrm{D}/\mathrm{d}t$, are exclusively of tidal origin.

The mechanism of tidal deformation thus affects the dynamics of the rotational and orbital motion of the Earth–Moon system, and has a considerable influence on the dynamic evolution of the system. Indeed, the tidal deformations of planets and satellites played an important part in moulding the Solar System into its present form. In general, tides are responsible for the synchronization of rotational and orbital periods, and may also be the source of the internal heat in the bodies of the Solar System. The dissipation of tidal energy in planet cores may play an important part in the generation of their magnetic fields.

As regards the tidal evolution of the Earth–Moon system, it is possible to calculate the time at which the ideal state with no tidal friction will occur, i.e. when the Earth's rotational motion and the Moon's orbital motion become synchronized,

$$\omega_\oplus = n_\mathrm{D} = \bar{\omega}.$$

If the Earth–Moon system is considered to be isolated, then the equation of angular momentum balance (5.19) in simplified form reads

$$(I_{33})_{\oplus}\,\omega_{\oplus} + (I_{33})_{\mathbb{D}}\,n_{\mathbb{D}}\cos i_{\oplus\mathbb{D}} + L_{\oplus\mathbb{D}}\cos i_{\mathbb{D}} = L_0 = 3.2085 \times 10^{34}\ \mathrm{kg\,m^2\,s^{-1}},$$

where L_0 is a constant for the whole duration of the system provided that it is isolated. Assume that synchronization occurs at time

$$\bar{t} = t_0 + \Delta t,$$

t_0 being the present epoch. At epoch $\bar{t}$,

$$\bar{\omega}\,[(I_{33}) + (I_{33})_{\mathbb{D}}\cos i_{\oplus\mathbb{D}}]$$

$$+\ \bar{\omega}^{-1/3}\,\frac{M_{\oplus}M_{\mathbb{D}}}{M_{\oplus} + M_{\mathbb{D}}}\,[G(M_{\oplus} + M_{\mathbb{D}})]^{2/3}(1 - e^2)^{1/2}\cos i_{\mathbb{D}} = L_0.$$

After inserting numerical values and substituting $x = 10^2\,\omega^{1/3}$, we arrive at the equation

$$x^4 - 399x + 452 - 0,$$

the solution of which is

$$\bar{\omega} = 1.468 \times 10^{-6}\ \mathrm{rad\,s^{-1}}\,;$$

the Earth's rotation period and the Moon's orbital period under synchronization will then be $\bar{T} = 49.5$ days. If the process of deceleration of the Earth's rotation were linear, i.e. if value (5.15) were constant over the whole interval of tidal evolution,

$$\Delta t = [\bar{\omega} - \omega_{\oplus}(t_0)][d\omega_{\oplus}/dt]^{-1} = 4.1 \times 10^9\ \mathrm{y}\,.$$

The semimajor axis of the Moon's orbit at time $t = \bar{t}$ should be $a_{\mathbb{D}}(\bar{t}) = 572\,091$ km, which follows from Kepler's third law:

$$a_{\mathbb{D}}(\bar{t}) = [G(M_{\oplus} + M_{\mathbb{D}})]^{1/3}\,\bar{\omega}^{-2/3}\,.$$

5.3 Deformations of the Earth due to the Variations in the Earth's Rotation

5.3.1 Variations in the Potential of Centrifugal Forces; Perturbing Forces

Apart from the tidal forces we discussed in Chapter 4, also centrifugal forces, i.e. variations in the vector of the Earth's instantaneous rotation, cause deformations of the Earth. From a theoretical point of view this phenomenon is very similar to the viscoelastic deformation phenomena of the tides.

The motion of poles is responsible for the change in the mutual positions of the instantaneous axis of rotation (vector of instantaneous rotation $\omega_{\oplus}$) and of the Earth's ellipsoid of inertia with axes x_j. This causes variations in the potential of centrifugal force Q at the point of observation $M(\varrho, \phi, \Lambda)$.

We shall consider a general geocentric coordinate system $x_j (j = 1, 2, 3)$ and assume that it is fixed with the Earth. In the ideal case in which the vector of instantaneous rotation $\boldsymbol{\omega}$ falls in with axis x_3, the potential of centrifugal force Q at point $M(\varrho, \phi, \Lambda)$ of the Earth would be

$$Q = \frac{1}{2} \omega^2 \varrho^2 \cos^2 \phi = \frac{1}{3} \omega^2 \varrho^2 [1 - P_2^{(0)}(\sin \phi)]$$

$$= \frac{1}{3} \frac{GM_\oplus}{\varrho} \left(\frac{a_0}{\varrho} \right)^{-3} q [1 - P_2^{(0)}(\sin \phi)] ; \quad q = \frac{\omega_\oplus^2 a_0^3}{GM_\oplus} , \tag{5.59}$$

where ϱ is the geocentric radius-vector of point M, ϕ is its geocentric latitude and Λ is its geocentric longitude.

If vector $\boldsymbol{\omega}$ deviates from $\mathbf{e}_3$ (unit vector along axis x_3), the potential, Q, at point M will change. If Θ (note that the same symbol is used for sidereal time) is the angle between vector $\boldsymbol{\omega}$ and the radius-vector of point M, the potential of the centrifugal force at point M

$$Q = \frac{1}{2} \omega^2 \varrho^2 \sin^2 \Theta = \frac{1}{3} \omega^2 \varrho^2 [1 - P_2^{(0)}(\cos \Theta)]$$

$$= \frac{1}{3} \frac{GM_\oplus}{\varrho} \left(\frac{a_0}{\varrho} \right)^{-3} q [1 - P_2^{(0)}(\cos \Theta)] , \tag{5.60}$$

and (Fig. 5.4)

$$\cos \Theta = \sum_j \frac{x_j \omega_j}{\varrho \, \omega} = \frac{1}{\omega} (\omega_1 \cos \phi \cos \Lambda + \omega_2 \cos \varphi \sin \Lambda + \omega_3 \sin \phi)$$

$$= \sin \phi \sin \phi_\omega + \cos \phi \cos \phi_\omega \cos(\Lambda - \Lambda_\omega) ; \tag{5.61}$$

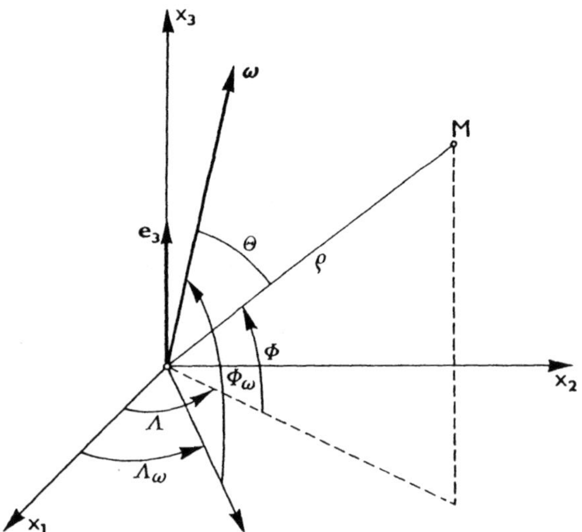

Fig. 5.4. Diagram used to derive the variations of the potential of centrifugal forces

ω_j are the components of vector $\boldsymbol{\omega}$ along axes x_j, ϕ_ω and Λ_ω are the geocentric latitude and longitude of the end-point of vector $\boldsymbol{\omega}$, i.e. of the quantity defining the direction of vector $\boldsymbol{\omega}$:

$$\sin \phi_\omega = \frac{\omega_3}{\omega}, \quad \cos \phi_\omega \cos \Lambda_\omega = \frac{\omega_1}{\omega}, \quad \cos \phi_\omega \sin \Lambda_\omega = \frac{\omega_2}{\omega}. \tag{5.62}$$

Using the formula for developing spherical harmonics we obtain

$$\begin{aligned}
\mathrm{P}_2^{(0)}(\cos \Theta) = \ & \mathrm{P}_2^{(0)}(\sin \phi_\omega)\,\mathrm{P}_2^{(0)}(\sin \phi) \\
& + \tfrac{1}{3}\mathrm{P}_2^{(1)}(\sin \phi_\omega)\,\mathrm{P}_2^{(1)}(\sin \phi)\cos(\Lambda - \Lambda_\omega) \\
& + \tfrac{1}{12}\mathrm{P}_2^{(2)}(\sin \phi_\omega)\,\mathrm{P}_2^{(2)}(\sin \phi)\cos 2(\Lambda - \Lambda_\omega),
\end{aligned} \tag{5.63}$$

which enables us to express the variable part, δQ, of the potential of centrifugal forces as the sum of the variations of zonal, $\delta Q_2^{(0)}$, tesseral, $\delta Q_2^{(1)}$, and sectorial, $\delta Q_2^{(2)}$, components:

$$Q = \frac{1}{3}\frac{GM}{\varrho}\left(\frac{a_0}{\varrho}\right)^{-3} q[1 - \mathrm{P}_2^{(0)}(\sin \phi)] + \delta Q, \tag{5.64}$$

$$\delta Q = \delta Q_2^{(0)} + \delta Q_2^{(1)} + \delta Q_2^{(2)}, \tag{5.65}$$

$$\begin{aligned}
\delta Q_2^{(0)} &= \frac{1}{3}\frac{GM}{\varrho}\left(\frac{a_0}{\varrho}\right)^{-3} q\,\mathrm{P}_2^{(0)}(\sin \phi)[1 - \mathrm{P}_2^{(0)}(\sin \phi_\omega)] \\
&= \tfrac{1}{2}\varrho^2(\omega_1^2 + \omega_2^2)\,\mathrm{P}_2^{(0)}(\sin \phi),
\end{aligned} \tag{5.66}$$

$$\begin{aligned}
\delta Q_2^{(1)} &= -\frac{1}{9}\frac{GM}{\varrho}\left(\frac{a_0}{\varrho}\right)^{-3} q\,\mathrm{P}_2^{(1)}(\sin \phi)\,\mathrm{P}_2^{(1)}(\sin \phi_\omega)\cos(\Lambda - \Lambda_\omega) \\
&= -\tfrac{1}{3}\varrho^2\omega_3(\omega_1\cos \Lambda + \omega_2\sin \Lambda)\,\mathrm{P}_2^{(1)}(\sin \phi),
\end{aligned} \tag{5.67}$$

$$\begin{aligned}
\delta Q_2^{(2)} &= -\frac{1}{36}\frac{GM}{\varrho}\left(\frac{a_0}{\varrho}\right)^{-3} q\,\mathrm{P}_2^{(2)}(\sin \phi)\,\mathrm{P}_2^{(2)}(\sin \phi_\omega)\cos 2(\Lambda - \Lambda_\omega) \\
&= -\tfrac{1}{12}\varrho^2(\omega_1^2 + \omega_2^2)\,\mathrm{P}_2^{(2)}(\sin \phi)\cos 2(\Lambda - \Lambda_\omega) \\
&= -\tfrac{1}{12}\varrho^2[(\omega_1^2 - \omega_2^2)\cos 2\Lambda + 2\omega_1\omega_2\sin 2\Lambda]\,\mathrm{P}_2^{(2)}(\sin \phi).
\end{aligned} \tag{5.68}$$

Zonal perturbation $\delta Q_2^{(0)}$ depends on the squares of the components of the rotation vector, ω_1^2 and ω_2^2. Their relative magnitude in units of the fundamental term of the zero-degree potential, GM/ϱ, is approximately

$$\begin{aligned}
\delta Q_2^{(0)}(GM)^{-1}\varrho &= \tfrac{1}{3}q\,\mathrm{P}_2^{(0)}(\sin \phi)[1 - \mathrm{P}_2^{(0)}(\sin \phi_\omega)] \\
&= \tfrac{1}{2}q\,\mathrm{P}_2^{(0)}(\sin \phi)\sin^2(\boldsymbol{\omega}, \mathbf{e}_3).
\end{aligned} \tag{5.69}$$

Some values of angle $(\boldsymbol{\omega}, \mathbf{e}_3)$, given $\phi = \pm 90°$, are shown in Table 5.1. Values $0.1''$ and $0.3''$ correspond to the present motion of the ellipsoid of inertia about vector $\boldsymbol{\omega}$. Values of $1°$ and larger, even tens of degrees, are expected to occur in palaeogeography and palaeogeology. They are given to illustrate the

Table 5.1. Zonal, tesseral and sectorial perturbations of the geopotential due to polar motion

(ω, e_3)	$\delta Q_2^{(0)}$ $[GM/\varrho]$	$\delta Q_2^{(1)}$ $[GM/\varrho]$	$-\delta Q_2^{(2)}$ $[GM/\varrho]$
$0.1''$	4×10^{-16}	8×10^{-10}	2×10^{-16}
$0.3''$	4×10^{-15}	3×10^{-9}	2×10^{-15}
$1°$	5×10^{-7}	3×10^{-5}	3×10^{-7}
$10°$	5×10^{-5}	3×10^{-4}	3×10^{-5}
$30°$	4×10^{-4}	7×10^{-4}	2×10^{-4}

phenomenon, without saying whether in fact this phenomenon could occur. In harmony with the theory of lithospheric plates, it is more probable that only the outer layers of the body, i.e. continents and oceanic crust, but not the body as a whole, underwent larger displacements. If they had moved along surfaces close to equipotential surfaces, no variations δQ could have occurred.

The pattern of the signs of the zonal deformations of equipotential surfaces of the geopotential is shown schematically in Fig. 5.5a. The deformations are zero only for $\phi = \pm 35° 16'$ (if $\phi_\omega \neq 0$); they are maximum at the poles, $\phi = \pm 90°$, $P_2^{(0)}(\sin \phi) = 1$, and minimum at the equator, $P_2^{(0)}(\sin \phi) = -\frac{1}{2}$.

In units of GM/ϱ the tesseral perturbations are approximately

$$\delta Q_2^{(1)}(GM)^{-1} \varrho = -\tfrac{1}{4} q \sin 2\phi \sin 2(\omega, e_3) \cos(\Lambda - \Lambda_\omega). \tag{5.70}$$

The values for selected angles $(\Lambda = \Lambda_\omega, \phi = -45°)$ are given in Table 5.1. Tesseral deformations are zero for $\phi = 0°$, $\pm 90°$, or $\Lambda = \Lambda_\omega + 90°$, $\Lambda = \Lambda_\omega + 270°$. They take maximum values along parallels $\phi = \pm 45°$. The respective sign pattern is shown in Fig. 5.5b.

Sectorial perturbations, $\delta Q_2^{(2)}$, again expressed in units of GM/ϱ, are approximately

$$\delta Q_2^{(2)}(GM)^{-1} \varrho = -\tfrac{1}{4} q \cos^2 \phi \sin^2(\omega, e_3) \cos 2(\Lambda - \Lambda_\omega). \tag{5.71}$$

They are zero for $\phi = \pm 90°$, or $\Lambda_\omega = \Lambda \pm 45°$ and $\Lambda = \Lambda_\omega \pm 135°$, and display maximum absolute values at the equator. Their values for selected values of the angle between rotation vector ω and axis x_3 (and for $\phi = 0$, $\Lambda_\omega = \Lambda$) are given in Table 5.1. The respective sign pattern is shown in Fig. 5.5c.

The periodicity of deformations $\delta Q_2^{(0)}$, $\delta Q_2^{(1)}$ and $\delta Q_2^{(2)}$ depends on the periodicity of components ω_1 and ω_2. Long-period or secular variations in the direction of vector ω, provided that they existed in geological epochs, caused systematic changes in the geopotential at a given point, and theoretically one must admit their effect on geotectonic phenomena; although the variations are small, they could have acted for a long time.

Perturbation forces determine the perturbations of the geopotential, δQ:

$$\delta \mathbf{F} = -\operatorname{grad} \delta Q. \tag{5.72}$$

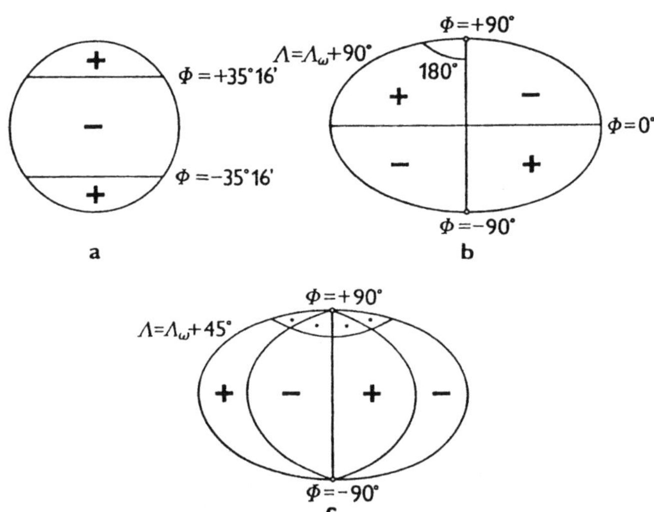

Fig. 5.5. The **a** zonal, **b** tesseral and **c** sectorial perturbations of the geopotential due to variations in the Earth's rotation

Here, and hereinafter (Tables 5.2–5.4), the perturbation forces are reckoned per unit mass, i.e. in units of acceleration.

We shall calculate the component along the external normal **n** to the equipotential surface of the geopotential, i.e. the vertical component,

$$\delta F_n = - \partial\delta Q/\partial n , \tag{5.73}$$

and the components acting in the two fundamental horizontal directions, i.e. the component in the plane of the meridian (positive to the south),

$$\delta F_\phi = - \partial\delta Q/(\varrho\,\partial\phi) , \tag{5.74}$$

and the component in the plane of the prime vertical (positive to the west),

$$\delta F_\Lambda = - \partial\delta Q/(\varrho\cos\phi\,\partial\Lambda) . \tag{5.75}$$

In view of the size of the angle between normal **n** to the equipotential surface of the geopotential and the geocentric radius-vector of the given point, M, vertical component (5.73) is practically equal to the radial component:

$$\delta F_\varrho = - \partial\delta Q/\partial\varrho . \tag{5.76}$$

In view of (5.66) its zonal part

$$(\delta F_\varrho)_2^{(0)} = - \partial\delta Q_2^{(0)}/\partial\varrho = - \frac{2}{3}\frac{GM}{\varrho^2}\left(\frac{a_0}{\varrho}\right)^{-3} q\, \mathrm{P}_2^{(0)}(\sin\phi)[1 - \mathrm{P}_2^{(0)}(\sin\phi_\omega)]$$

$$= - \frac{GM}{\varrho^2}\left(\frac{a_0}{\varrho}\right)^{-3} q\, \mathrm{P}_2^{(0)}(\sin\phi)\sin^2(\boldsymbol{\omega}, \mathbf{e}_3)$$

$$= - \varrho\omega^2 \sin^2(\boldsymbol{\omega}, \mathbf{e}_3)\, \mathrm{P}_2^{(0)}(\sin\phi) = - \varrho(\omega_1^2 + \omega_2^2)\, \mathrm{P}_2^{(0)}(\sin\phi) . \tag{5.77}$$

Table 5.2. Radial components of the perturbing force due to polar motion

$(\mathbf{\omega}, \mathbf{e}_3)$	$-(\delta F_\varrho)_2^{(0)}$ $(10^{-5}\,\mathrm{m\,s^{-2}})$	$(\delta F_\varrho)_2^{(1)}$ $(10^{-5}\,\mathrm{m\,s^{-2}})$	$(\delta F_\varrho)_2^{(2)}$ $(10^{-5}\,\mathrm{m\,s^{-2}})$
$0.1''$	8×10^{-10}	2×10^{-3}	4×10^{-10}
$0.3''$	7×10^{-9}	5×10^{-3}	4×10^{-9}
$1°$	1	60	0.5
$10°$	100	580	50
$30°$	850	1470	425

The zero values of the zonal radial force component are the same as in Fig. 5.5a. The magnitude of component (5.77) for particular angles between vectors $\mathbf{\omega}$ and $\mathbf{e}_3$ (given $\phi = \pm 90°$) is shown in Table 5.2.

The tesseral perturbation of the radial component of the force, $(\delta F_\varrho)_2^{(1)}$, follows from Eq. (5.67):

$$(\delta F_\varrho)_2^{(1)} = -\partial \delta Q_2^{(1)}/\partial \varrho = \frac{2}{9}\frac{GM}{\varrho^2}\left(\frac{a_0}{\varrho}\right)^{-3} q P_2^{(1)}(\sin \phi)$$

$$\times P_2^{(1)}(\sin \phi_\omega)\cos(\Lambda - \Lambda_\omega)$$

$$= \frac{1}{2}\frac{GM}{\varrho^2}\left(\frac{a_0}{\varrho}\right)^{-3} q \sin 2(\mathbf{\omega}, \mathbf{e}_3)\cos(\Lambda - \Lambda_\omega)\sin 2\phi$$

$$= \varrho\omega_3(\omega_1 \cos \Lambda + \omega_2 \sin \Lambda)\sin 2\phi .\tag{5.78}$$

The sign pattern is the same as in Fig. 5.5b. Table 5.2 shows the appropriate value for selected angles $(\mathbf{\omega}, \mathbf{e}_3)$ and $\phi = +45°$, $\Lambda_\omega = \Lambda$. It indicates that, given the present values of ω_1 and ω_2, these perturbations should be taken into account.

In view of (5.68), the sectorial perturbation, $(\delta F_\varrho)_2^{(2)}$, of the radial component of the force due to the variations of the potential of centrifugal forces is

$$(\delta F_\varrho)_2^{(2)} = -\partial \delta Q_2^{(2)}/\partial \varrho = \frac{1}{18}\frac{GM}{\varrho^2}\left(\frac{a_0}{\varrho}\right)^{-3} q P_2^{(2)}(\sin \phi)$$

$$\times P_2^{(2)}(\sin \phi_\omega)\cos 2(\Lambda - \Lambda_\omega)$$

$$= \frac{1}{2}\frac{GM}{\varrho^2}\left(\frac{a_0}{\varrho}\right)^{-3} q \cos^2 \phi \sin^2(\mathbf{\omega}, \mathbf{e}_3)\cos 2(\Lambda - \Lambda_\omega)$$

$$= \frac{1}{2}\frac{GM}{\varrho^2}\left(\frac{a_0}{\varrho}\right)^{-3} q(\omega_1^2 + \omega_2^2)\cos^2 \phi \cos 2(\Lambda - \Lambda_\omega) .\tag{5.79}$$

The sign pattern is the same as in Fig. 5.5c. The values for the angles $(\mathbf{\omega}, \mathbf{e}_3)$ being considered and for $\phi = 0°$, $\Lambda = \Lambda_\omega$ are given in Table 5.2, and the interpretation of the phenomenon is the same as for the zonal perturbations.

In view of (5.66)–(5.68) the horizontal component in the plane of the meridian (5.74) of point M of the Earth's crust is

$$\delta F_\phi = -\frac{1}{2}\frac{GM}{\varrho^2}\left(\frac{a_0}{\varrho}\right)^{-3} q\, \mathrm{P}_2^{(1)}(\sin\phi)\sin^2(\boldsymbol{\omega}, \mathbf{e}_3)$$

$$+\frac{1}{6}\frac{GM}{\varrho^2}\left(\frac{a_0}{\varrho}\right)^{-3} q[1 - 4\mathrm{P}_2^{(0)}(\sin\phi)]\sin 2(\boldsymbol{\omega}, \mathbf{e}_3)\cos(\Lambda - \Lambda_\omega)$$

$$-\frac{1}{6}\frac{GM}{\varrho^2}\left(\frac{a_0}{\varrho}\right)^{-3} q\,\mathrm{P}_2^{(1)}(\sin\phi)\sin^2(\boldsymbol{\omega}, \mathbf{e}_3)\cos 2(\Lambda - \Lambda_\omega). \qquad (5.80)$$

It has a zonal part, independent of Λ,

$$(\delta F_\phi)_2^{(0)} = -\frac{3}{4}\frac{GM}{\varrho^2}\left(\frac{a_0}{\varrho}\right)^{-3} q\,\sin^2(\boldsymbol{\omega}, \mathbf{e}_3)\sin 2\phi\,, \qquad (5.81)$$

and two tesseral terms, which depend on ϕ as well as on Λ, the argument in one case being Λ,

$$(\delta F_\phi)_2^{(\Lambda)} = \frac{1}{2}\frac{GM}{\varrho^2}\left(\frac{a_0}{\varrho}\right)^{-3} q(1 - 2\sin^2\phi)\sin 2(\boldsymbol{\omega}, \mathbf{e}_3)\cos(\Lambda - \Lambda_\omega) \qquad (5.82)$$

and in the other case 2Λ,

$$(\delta F_\phi)_2^{(2\Lambda)} = -\frac{1}{4}\frac{GM}{\varrho^2}\left(\frac{a_0}{\varrho}\right)^{-3} q\,\sin 2\phi\,\sin^2(\boldsymbol{\omega}, \mathbf{e}_3)\cos 2(\Lambda - \Lambda_\omega). \qquad (5.83)$$

In this case there is no sectorial part.

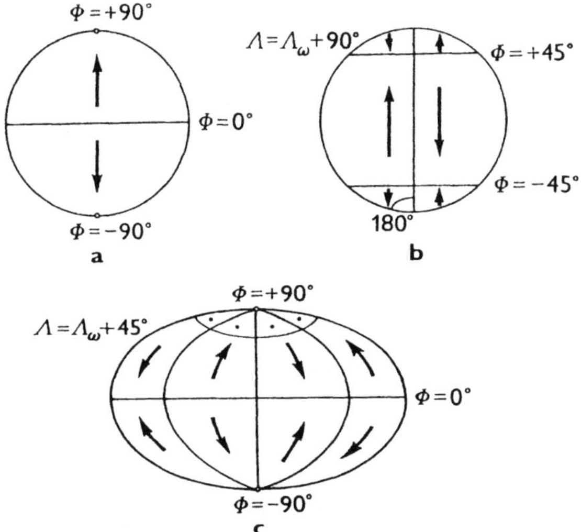

Fig. 5.6. The directions in which the horizontal components of the perturbing force, generated by polar motion, act along the meridian. **a** Zonal component; **b** and **c** non-zonal components

The zero values and signs of the zonal horizontal perturbations along the meridian (5.81) are shown in Fig. 5.6a. This deformation is maximum for $\phi = \pm 45°$. Some of the values for this latitude are given in Table 5.3. Hence, these horizontal perturbations for $\phi = 45°$ are of the same order as the radial perturbations.

Tesseral perturbation (5.82) is zero for $\phi = \pm 45°$. The respective sign pattern is shown in Fig. 5.6b and some values for $\phi = 0$ and $\Lambda = \Lambda_\omega$ are given in Table 5.3. Tesseral perturbation (5.83) is zero at the equator; its signs are shown schematically in Fig. 5.6c and values for $\phi = 45°$ and $\Lambda = \Lambda_\omega$ are given in Table 5.3.

The horizontal perturbations along the meridian may thus be of the same order as the radial perturbations. Any long-period or secular drift in the variations of angle $(\omega, \mathbf{e}_3)$ which may have occurred in geological epochs would have been small, so that the perturbations at point M would have increased gradually at relatively very long intervals of time.

The horizontal component in the plane of the prime vertical (5.75) of the point being considered is, in view of (5.66)–(5.68),

$$\delta F_\Lambda = (\delta F_\Lambda)_1^{(1)} + (\delta F_\Lambda)_2^{(2)}, \tag{5.84}$$

$$(\delta F_\Lambda)_1^{(\Lambda)} = -\frac{1}{9}\frac{GM}{\varrho^2}\left(\frac{a_0}{\varrho}\right)^{-3} q P_2^{(1)}(\sin\phi) P_2^{(1)}(\sin\phi_\omega)\cos^{-1}\phi \sin(\Lambda - \Lambda_\omega),$$

$$\tag{5.85}$$

$$(\delta F_\Lambda)_2^{(2\Lambda)} = -\frac{1}{18}\frac{GM}{\varrho^2}\left(\frac{a_0}{\varrho}\right)^{-3} q P_2^{(2)}(\sin\phi) P_2^{(2)}(\sin\phi_\omega)\cos^{-1}\phi \sin 2(\Lambda - \Lambda_\omega).$$

$$\tag{5.86}$$

There is thus no zonal part. The sectorial part (5.85), which can be expressed as

$$(\delta F_\Lambda)_1^{(1)} = -\frac{1}{2}\frac{GM}{\varrho^2}\left(\frac{a_0}{\varrho}\right)^{-3} q \sin\phi \sin 2(\omega, \mathbf{e}_3)\sin(\Lambda - \Lambda_\omega), \tag{5.87}$$

Table 5.3. Horizontal component of the perturbing force along the meridian due to polar motion

$(\omega, \mathbf{e}_3)$	$-(\delta F_\phi)_2^{(0)}$ $(\mathrm{m\,s}^{-2})$	$(\delta F_\phi)_2^{(\Lambda)}$ $(\mathrm{m\,s}^{-2})$	$-(\delta F_\phi)_2^{(2\Lambda)}$ $(\mathrm{m\,s}^{-2})$
$0.1''$	6×10^{-15}	10^{-8}	3×10^{-15}
$0.3''$	5×10^{-14}	3×10^{-8}	3×10^{-14}
$1°$	8×10^{-6}	4×10^{-4}	4×10^{-6}
$10°$	8×10^{-4}	4×10^{-3}	4×10^{-4}
$30°$	6×10^{-3}	10^{-2}	3×10^{-3}

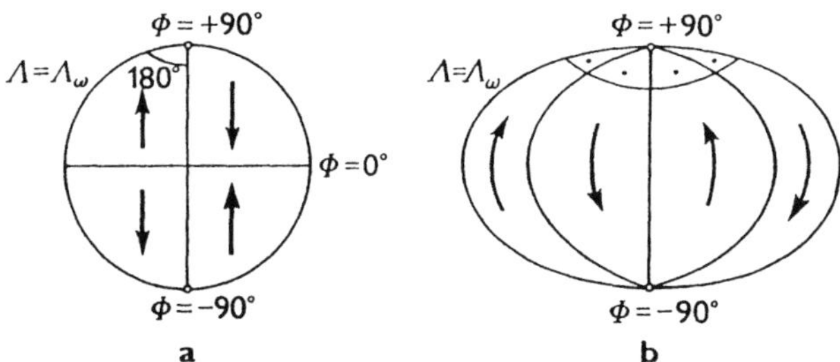

Fig. 5.7. The directions in which the horizontal components of the perturbing force, generated by polar motion, act along the prime vertical. **a** illustrates Eq. (5.85), **b** illustrates Eq. (5.86)

Table 5.4. Horizontal component of the perturbing force along the prime vertical due to polar motion

$(\omega, \mathbf{e}_3)$	$(\delta F_\Lambda)^{(1)}_1$ $(\mathrm{m\,s}^{-2})$	$(\delta F_\Lambda)^{(2)}_2$ $(\mathrm{m\,s}^{-2})$
$0.1''$	2×10^{-8}	4×10^{-15}
$0.3''$	5×10^{-8}	4×10^{-14}
$1°$	6×10^{-4}	5×10^{-6}
$10°$	6×10^{-3}	5×10^{-4}
$30°$	10^{-2}	4×10^{-3}

is a spherical harmonic of the first degree and first order. The sign pattern can be seen in Fig. 5.7a. The values for particular angles $(\omega, \mathbf{e}_3)$ and $\phi = 90°$, $\Lambda = \Lambda_\omega + 90°$ are given in Table 5.4.

Part (5.86), which can be expressed as

$$(\delta F_\Lambda)^{(2)}_2 = -\frac{1}{2}\frac{GM}{\varrho^2}\left(\frac{a_0}{\varrho}\right)^{-3} q\cos\phi\sin^2(\omega, \mathbf{e}_3)\sin 2(\Lambda - \Lambda_\omega), \qquad (5.88)$$

is a sectorial function in character, although spherical sectorial harmonics of the second degree and second order contain $\cos^2\phi$ and not $\cos\phi$. The sign pattern is shown in Fig. 5.7b and some of the values for $\phi = 0°$, $\Lambda = \Lambda_\omega + 45°$ are given in Table 5.4.

5.3.2 Deformations of Equipotential Surfaces due to Polar Motion for a Perfectly Rigid Earth

Force $\delta\mathbf{F}$ (5.72) causes radial displacements of equipotential surfaces $W = \mathrm{const}$ of the geopotential. If the angle between the normal to surface $W = \mathrm{const}$, passing through the given point, M, and its geocentric radius-vector, is neglected, then the radial displacement according to Bruns' theorem is

$$\delta\varrho = -\,\delta Q/(\partial W/\partial\varrho) \sim \delta Q/g\,, \tag{5.89}$$

g being the gravity at M.

Displacement $\delta\varrho$ can be expressed as the sum of the zonal, tesseral and sectorial parts:

$$\delta\varrho = (\delta\varrho)_2^{(0)} + (\delta\varrho)_2^{(1)} + (\delta\varrho)_2^{(2)}\,. \tag{5.90}$$

In view of (5.66) the zonal part is approximately

$$(\delta\varrho)_2^{(0)} = \tfrac{1}{2}\varrho q\, P_2^{(0)}(\sin\phi)\sin^2(\omega, e_3)\,. \tag{5.91}$$

The zero-value and sign pattern are shown in Fig. 5.5a and the values for selected angles between rotation vector $\boldsymbol{\omega}$ and the x_3-axis for $\phi = \pm\,90^\circ$ (maxima) are given in Table 5.5. The data related to angles $(\boldsymbol{\omega}, \mathbf{e}_3) = 1^\circ, 10^\circ, 30^\circ$ only illustrate the tremendous rise in the local equipotential surfaces and, consequently, also the change in potential energy which would occur at the given point under the deviation mentioned. However, if a larger displacement of the ellipsoid of inertia relative to the axis of rotation did actually occur in geological epochs, it was gradual and the ellipsoid of inertia simultaneously adjusted to the change. Also, the radius-vector of the point of observation, which is assumed to be constant in these estimates, changed in accordance with the theory of hydrostatic equilibrium. Nevertheless, the data in Table 5.5 can serve to balance mechanically some of the phenomena assumed in palaeogeology, and to support the opinion that only the upper layers of the Earth (lithospheric plates) wandered and that no larger deflections of the Earth's axis of rotation, relative to the body as a whole, occurred.

The same deliberations apply to tesseral perturbations which, in view of (5.67), are

$$(\delta\varrho)_2^{(1)} = -\tfrac{1}{4}\varrho q\sin 2\phi\,\sin 2(\boldsymbol{\omega}, \mathbf{e}_3)\cos(\Lambda - \Lambda_\omega): \tag{5.92}$$

their zero-value and sign patterns are given in Fig. 5.5b and values for particular angles and $\phi = -45^\circ$, $\Lambda = \Lambda_\omega$ are given in Table 5.5.

Table 5.5. Radial displacements of equipotential surfaces due to polar motion

$(\boldsymbol{\omega}, \mathbf{e}_3)$	$(\delta\varrho)_2^{(0)}$ (m)	$(\delta\varrho)_2^{(1)}$ (m)	$-(\delta\varrho)_2^{(2)}$ (m)
$0.1''$	3×10^{-9}	5.3×10^{-3}	10^{-9}
$0.3''$	2×10^{-8}	1.6×10^{-2}	10^{-8}
1°	3	190	2
10°	330	1880	170
30°	2750	4770	1380

With regard to (5.68) the sectorial perturbations are

$$(\delta\varrho)_2^{(2)} = -\tfrac{1}{4}\varrho q \cos^2\phi \sin^2(\boldsymbol{\omega}, \mathbf{e}_3)\cos 2(\Lambda - \Lambda_\omega); \tag{5.93}$$

their zero-value and sign patterns are identical to those in Fig. 5.5c and the values for $\phi = 0°$, $\Lambda = \Lambda_\omega$ are given in Table 5.5.

The perturbations of equipotential surfaces of the geopotential, due to forces $\delta\mathbf{F}$, are not only generally of a radial nature, QED, but they also change the shape of the equipotential surface, which can be expressed in terms of variations of the deflections of the vertical. These can be obtained from the horizontal components of the perturbing force (5.74) and (5.75). The variation of component $\delta\xi$ in the plane of the meridian

$$\delta\xi = \delta F_\phi\left(\frac{\partial W}{\partial n}\right)^{-1} = \delta F_\phi\left(\frac{\partial W}{\partial\varrho}\right)^{-1} \tag{5.94}$$

and the component in the plane of the prime vertical

$$\delta\eta = \delta F_\Lambda\left(\frac{\partial W}{\partial\varrho}\right)^{-1}. \tag{5.95}$$

The meridional component has a zonal term,

$$(\delta\xi)_2^{(0)} = -\tfrac{3}{4}q\,\sin^2(\boldsymbol{\omega}, \mathbf{e}_3)\sin 2\phi \tag{5.96}$$

and two non-zonal terms,

$$(\delta\xi)_2^{(\Lambda)} = \tfrac{1}{2}q(1 - 2\sin^2\phi)\sin 2(\boldsymbol{\omega}, \mathbf{e}_3)\cos(\Lambda - \Lambda_\omega) \tag{5.97}$$

and

$$(\delta\xi)_2^{(2\Lambda)} = -\tfrac{1}{4}q\sin 2\phi\,\sin^2(\boldsymbol{\omega}, \mathbf{e}_3)\cos 2(\Lambda - \Lambda_\omega). \tag{5.98}$$

As regards their character, distribution of zero values and signs, the same applies as to the force components (5.82) and (5.83). Table 5.6 gives the values for particular angles $(\boldsymbol{\omega}, \mathbf{e}_3)$, i.e. the extreme values for the same arguments as in Table 5.5.

Table 5.6. Deflections of the vertical along the meridian due to polar motion

$(\boldsymbol{\omega}, \mathbf{e}_3)$	$-(\delta\xi)_2^{(0)}$ ($''$)	$(\delta\xi)_2^{(\Lambda)}$ ($''$)	$-(\delta\xi)_2^{(2\Lambda)}$ ($''$)
$0.1''$	10^{-10}	2×10^{-4}	6×10^{-11}
$0.3''$	10^{-9}	7×10^{-4}	6×10^{-10}
$1°$	0.16	8	0.08
$10°$	16	81	8
$30°$	134	206	67

Table 5.7. Deflections of the vertical along the prime
vertical due to polar motion

$(\omega, \mathbf{e}_3)$	$(\delta\eta)_1^{(1)}$ $('')$	$(\delta\eta)_2^{(2)}$ $('')$
$0.1''$	3×10^{-4}	8×10^{-11}
$0.3''$	0.001	8×10^{-10}
$1°$	12	0.11
$10°$	122	11
$30°$	309	89

As in (5.87) and (5.88) the component in the prime vertical has two non-zonal
terms:

$$(\delta\eta)_1^{(1)} = -\tfrac{1}{2} q \sin\phi \sin 2(\omega, \mathbf{e}_3) \sin(\Lambda - \Lambda_\omega) , \tag{5.99}$$

$$(\delta\eta)_2^{(2)} = -\tfrac{1}{2} q \cos\phi \sin^2(\omega, \mathbf{e}_3) \sin 2(\Lambda - \Lambda_\omega) . \tag{5.100}$$

It has no zonal term. The character of functions (5.99) and (5.100) is identical
with that of (5.87) and (5.88), which also applies to the distribution of zero values
and signs. The values for the same arguments as in Table 5.6 are given in
Table 5.7.

5.3.3 Deformations of Equipotential Surfaces due to Polar Motion for a Perfectly Elastic Earth

Assuming the Earth to be perfectly elastic, the radial displacements of the
equipotential surface of the geopotential, $\delta\varrho'$, are larger than (5.89):

$$\delta\varrho' = \left(1 + \frac{\delta Q_{\mathrm{def}}}{\delta Q_\omega}\right)\delta\varrho = (1 + k_2)\delta\varrho \sim 1.3\,\delta\varrho , \tag{5.101}$$

where $k_2 \sim 0.3$ is the Love number (see Sects. 4.6 and 4.7).

In this case, however, not only the equipotential surfaces of the geopotential
are perturbed but also the Earth, namely, its crust. Its radial displacement is

$$\delta\bar\varrho = h_2\,\delta\varrho , \tag{5.102}$$

where $h_2 \sim 0.6$ is the so-called second Love number which expresses the
proportionality of the deformation of the Earth's crust to that of the
equipotential surface if the Earth were perfectly rigid (see Sects. 4.6 and 4.7).

The radial components of the forces acting in this case are larger than (5.76),

$$\delta\bar F_\varrho = (1 - \tfrac{3}{2}k_2 + h_2)\delta F_\varrho = 1.17\,\delta F_\varrho . \tag{5.103}$$

In contrast the horizontal components are smaller than (5.74) and (5.75):

$$\delta \bar{F}_\phi = (1 + k_2 - h_2)\delta F_\phi = 0.7\delta F_\phi \,,$$

$$\delta \bar{F}_A = (1 + k_2 - h_2)\delta F_A = 0.7\delta F_A \,. \tag{5.104}$$

The components of the deflections of the vertical, relative to the surface of the deforming Earth's crust, are

$$\delta \bar{\xi} = (1 + k_2 - l_2)\delta \xi = 1.2\,\delta \xi \,,$$

$$\delta \bar{\eta} = (1 + k_2 - l_2)\delta \eta = 1.2\delta \eta \,, \tag{5.105}$$

where $l_2 \sim 0.08$ is Shida's number (see Sects. 4.6 and 4.7).

Only the sectorial components $(1 + k - h)\delta \xi_2^{(1)}$, $(1 + k - h)\delta \eta_2^{(1)}$, $(1 + k - l)\delta \xi_2^{(1)}$ and $(1 + k - l)\delta \eta_2^{(1)}$, which may amount to thousandths of a second of arc ($\sim 0.002''$), are of any practical significance. The other components are several orders of magnitude smaller and have no practical significance.

All the data in Tables 5.1–5.7 have to be adjusted by ratios (5.101)–(5.105) if they are to reflect a perfectly elastic Earth.

On the whole, one can summarize that the perturbations of the Earth's equipotential surfaces, namely the deformations of the Earth's crust due to the wandering of the Earth's poles, are similar to tidal perturbations. However, the periods of both phenomena differ. The tidal perturbations are semi-annual, semi-monthly, diurnal and semi-diurnal, where the perturbations due to the variations of the centrifugal force, e.g. due to Chandler's pole wandering, have a period of about 430 days.

Both phenomena affect the deflections of the vertical and, consequently, all astronomical and geodetic observations in which instruments with levels or mercury horizons are used.

5.3.4 Deformations due to Variations in the Earth's Angular Velocity

In the foregoing part of this chapter we have discussed the perturbations of the geopotential caused exclusively by changes in the direction of the rotation vector within the Earth. In this case the system of reference consists of the axes of the Earth's central ellipsoid of inertia. If $\omega_1 = 0$ and $\omega_2 = 0$, i.e. if the vector of instantaneous rotation falls in with the smallest axis of the ellipsoid of inertia, all perturbations are zero.

Adopting this frame of reference has no effect on the perturbations caused by the changes in the modulus of the velocity vector, i.e. $d\omega/dt$. If the rotation perturbations are to be reflected in full, we shall have to adopt Munk's and MacDonald's (1960) concept in which the perturbing potential δQ is defined as follows:

$$\delta Q = \tfrac{1}{2}\omega^2[\varrho \sin(\boldsymbol{\omega}, \boldsymbol{\varrho})]^2 - \tfrac{1}{3}\omega^2\varrho^2 \,. \tag{5.106}$$

The radial component $\omega^2 \varrho^2 / 3$ has been excluded completely from this definition and, consequently, the surface of reference is the sphere. Tesseral $\delta Q_2^{(1)}$ and sectorial $\delta Q_2^{(2)}$ components remain unchanged as in our case, but the zonal term is different:

$$\delta Q_2^{(0)} = -\tfrac{1}{3}\omega^2 \varrho^2 \, \mathrm{P}_2^{(0)}(\sin \phi) \, \mathrm{P}_2^{(0)}(\sin \phi_\omega)$$
$$= -\tfrac{1}{3}\varrho^2 \left[\omega_3^2 - \tfrac{1}{2}(\omega_1^2 + \omega_2^2)\right] \mathrm{P}_2^{(0)}(\sin \phi). \tag{5.107}$$

Since $\omega_3 \gg \omega_1, \omega_2$, i.e. $\omega_3 \sim \omega$, it is then useful to introduce $\tilde{\Omega} = \mathrm{const} \sim \omega_3$, and to express the components of vector $\boldsymbol{\omega}$ as small dimensionless quantities:

$$m_1 = \omega_1/\tilde{\Omega}, \quad m_2 = \omega_2/\tilde{\Omega}, \quad m_3 = \frac{\omega_3 - \tilde{\Omega}}{\tilde{\Omega}} = \frac{\omega_3}{\tilde{\Omega}} - 1. \tag{5.108}$$

Hence,

$$\delta Q_2^{(0)} = -\tfrac{1}{3}\varrho^2 \tilde{\Omega}^2 \left\{ \left(\frac{\omega_3}{\tilde{\Omega}}\right)^2 - \frac{1}{2}\left[\left(\frac{\omega_1}{\tilde{\Omega}}\right)^2 + \left(\frac{\omega_2}{\tilde{\Omega}}\right)^2\right]\right\} \mathrm{P}_2^{(0)}(\sin \phi)$$
$$= \tfrac{1}{6}\varrho^2 \tilde{\Omega}^2 \left[m_1^2 + m_2^2 - 2(m_3 + 1)^2\right] \mathrm{P}_2^{(0)}(\sin \phi). \tag{5.109}$$

In this definition component m_3 reflects the variations in the Earth's rotational velocity; their effect on potential Q, which we shall denote $(\delta Q_2^{(0)})_\omega$, can be expressed as

$$(\delta Q_2^{(0)})_\omega = -\tfrac{1}{3}\varrho^2 \tilde{\Omega}^2 (2m_3 + m_3^2) \, \mathrm{P}_2^{(0)}(\sin \phi). \tag{5.110}$$

Neglecting the quadratic term, we get

$$(\delta Q_2^{(0)})_\omega = -\tfrac{2}{3}\varrho^2 \tilde{\Omega}(\omega_3 - \tilde{\Omega}) \, \mathrm{P}_2^{(0)}(\sin \phi) \tag{5.111}$$

and, consequently,

$$\mathrm{d}(\delta Q_2^{(0)})_\omega/\mathrm{d}t = -\tfrac{2}{3}\varrho^2 \tilde{\Omega} \, \mathrm{P}_2^{(0)}(\sin \phi)\,\mathrm{d}\omega_3/\mathrm{d}t$$
$$= -\frac{2}{3}\frac{GM}{\varrho} q \left(\frac{a_0}{\varrho}\right)^{-3} \tilde{\Omega}^{-1} \, \mathrm{P}_2^{(0)}(\sin \phi)\,\mathrm{d}\omega_3/\mathrm{d}t \,;$$
$$q = \tilde{\Omega}^2 a_0^3/(GM) = 3461.39 \times 10^{-6},$$
$$\tilde{\Omega} = 7.292\,115 \times 10^{-5} \,\mathrm{rad\,s}^{-1}. \tag{5.112}$$

We discussed variations $\mathrm{d}\omega_3/\mathrm{d}t$ in Chapter 3. To be able to assess the relative magnitude of the changes in the geopotential they cause, we shall express (5.112) in units of GM/ϱ, and for points of the Earth's surface ($a_0/\varrho = 1$),

$$(GM/\varrho)^{-1} \, \mathrm{d}(\delta Q_2^{(0)})_\omega/\mathrm{d}t = -31.6 \, \mathrm{P}_2^{(0)}(\sin \phi)\,\mathrm{d}\omega_3/\mathrm{d}t \,. \tag{5.113}$$

Here $d\omega_3/dt$ has to be substituted in rad s^{-2} and the relative changes will come out in s^{-1}. For example, secular change $d\omega_3/dt = -5.4 \times 10^{-22}$ rad s^{-2} generates a relative change in the geopotential on the Earth's surface of about $+5 \times 10^{-11}$ $P_2^{(0)}$ $(\sin\phi)$ per century. However, the irregular changes $\delta\omega/\omega$ are much larger (by as much as two to three orders of magnitude) and amount to as much as 5×10^{-8}, even over an interval of a few years. A relative change of 10^{-8} per year yields an annual change in the geopotential of 3×10^{-11} $P_2^{(0)}$ $(\sin\phi)$.

The changes in geopotential (5.111) can be used to derive the displacements of equipotential surfaces and of the Earth's crust in very much the same way as in Sections 5.3.1 and 5.3.2.

5.3.5 Comparison with Tidal Deformations

Let us compare the amplitudes of the perturbations generated by recent variations in the vector of the Earth's rotation with the amplitudes of tidal perturbations generated by the Moon. For this purpose it is sufficient to compare amplitude A_{tide} of the tidal potential (Sect. 4.2) with the amplitudes of potential (5.65). Since tesseral term $\delta Q_2^{(1)}$ is the largest, we shall adopt its amplitude and denote it $A_2^{(1)}$. Hence,

$$\frac{A_2^{(1)}}{A_{\text{tide}}} = \left[\frac{GM_\oplus}{\varrho}\left(\frac{a_0}{\varrho}\right)^{-3} q\,\frac{\omega_1\omega_3}{\omega^2}\right] : \left[\frac{GM_{\mathbb{D}}}{\varDelta_{\oplus\mathbb{D}}}\left(\frac{\varrho}{\varDelta_{\oplus\mathbb{D}}}\right)^2\right] \sim \frac{1}{10}, \qquad (5.114)$$

where $GM_{\mathbb{D}} = 4902.8 \times 10^9$ m^3 s^{-2} is the selenocentric gravitational constant and $\varDelta_{\oplus\mathbb{D}} = 384\,400$ km is the mean distance between the Earth's and Moon's centres of mass. This means that the perturbations being studied may amount to as much as 10% of the tidal perturbations. However, in the estimate we considered the present amplitudes of functions $\omega_1(t)$ and $\omega_2(t)$. In geological epochs these amplitudes could have been many times larger, and the effects being studied could have exceeded the tidal deformations. These were also larger in the past because the Moon was substantially closer to the Earth. The terms of the second order, containing ω_1^2 and ω_2^2, and short-term variations in modulus ω seem to be negligible with regard to the Earth's deformations.

To conclude, it can be said that recent variations due to the variations in the Earth's rotation are not negligible with regard to the present accuracy in monitoring geodynamic phenomena. In the geological past these perturbations could have been much larger, but their cause should be sought in internal forces. Vector $\boldsymbol{\omega}$ was apparently always close in direction to the vector of angular momentum $\mathbf{L}$, and time variations $d\mathbf{L}/dt$ in the system of a fixed ecliptic must have balanced the angular momentum of the external forces which were responsible for precession and nutation of the body as a whole. That is why preference cannot be given to time variations $d\omega/dt$ in geological deliberations, but they should rather be considered as the consequence of the internal dynamics of terrestrial masses.

5.4 Dynamics of the Earth's Ellipsoid of Inertia

The Earth is not perfectly rigid. Perturbing forces cause transfer of mass and, consequently, also time variations in elements I_{ik} of the tensor of inertia, $\mathbf{I}$:

$$I_{ik} = \int\limits_M \left[\sum_{j=1}^{3} (\delta_{ik} x_j^2 - x_i x_k) \right] dm;$$

$$i, k = 1, 2, 3; \quad \delta_{ik} = \begin{matrix} 1 \\ 0 \end{matrix} \ \text{for} \ \begin{matrix} i = k \\ i \neq k \end{matrix} \quad \text{(Kronecker symbol)}. \tag{5.115}$$

This is responsible for the Earth's central ellipsoid of inertia changing its shape in time and for its axes changing their direction in space.

The time variations of all elements of the geocentric ellipsoid of inertia can best be derived from the variations of geopotential coefficients $\delta J_n^{(k)}$ and $\delta S_n^{(k)}$, which can be used to express the external 'additional' perturbing potential,

$$\delta W = \frac{GM_\oplus}{\varrho} \sum_{n=2}^{\infty} \left(\frac{a_0}{\varrho} \right)^n \sum_{k=0}^{n} (\delta J_n^{(k)} \cos k\Lambda + \delta S_n^{(k)} \sin k\Lambda) \, P_n^{(k)}(\sin \phi), \tag{5.116}$$

where ϱ, ϕ, Λ are the geocentric spherical coordinates of the external potential point and $a_0 = 6\,378\,140$ m.

The required variations of the geopotential coefficients, $\delta J_n^{(k)}, \delta S_n^{(k)}$, can be derived from the condition that $\delta W = \delta \bar{W}$, where $\delta \bar{W}$ is the perturbing potential generated by tidal and rotational deformations, on the boundary surface which will be considered a sphere with radius a_0. The tidal part was derived in Section 4.4 [Eq. (4.67)], the rotational part (Sect. 5.3) is equal to $k_2 \delta Q$. On aggregate, for point $a_0, \bar{\phi}, \bar{\Lambda}$ on the surface of the sphere,

$$
\begin{aligned}
\delta \bar{W} = {} & k_2 \frac{GM_{\mathcouldnt}}{\varDelta_{\oplus\mathcouldnt}} \sum_{n=2}^{\infty} \left(\frac{a_0}{\varDelta_{\oplus\mathcouldnt}} \right)^n \sum_{k=0}^{n} (2 - \delta_{k,0}) \frac{(n-k)!}{(n+k)!} \\
& \times P_n^{(k)}(\sin \bar{\phi}) \, P_n^{(k)}(\sin \delta_{\mathcouldnt}) \cos k(\bar{\Lambda} + T_{\mathcouldnt}) \\
& + k_2 \frac{GM_\odot}{\varDelta_{\oplus\odot}} \sum_{n=2}^{\infty} \left(\frac{a_0}{\varDelta_{\oplus\odot}} \right)^n \sum_{k=0}^{n} (2 - \delta_{k,0}) \frac{(n-k)!}{(n+k)!} \\
& \times P_n^{(k)}(\sin \bar{\phi}) \, P_n^{(k)}(\sin \delta_\odot) \cos k(\bar{\Lambda} + T_\odot) \\
& - \tfrac{1}{3} k_2 a_0^2 [\omega_3^2 - \tfrac{1}{2}(\omega_1^2 + \omega_2^2)] \, P_2^{(0)}(\sin \bar{\phi}) \\
& - \tfrac{1}{3} k_2 a_0^2 \omega_3 (\omega_1 \cos \bar{\Lambda} + \omega_2 \sin \bar{\Lambda}) \, P_2^{(1)}(\sin \bar{\phi}) \\
& - \tfrac{1}{12} k_2 a_0^2 [(\omega_1^2 - \omega_2^2) \cos 2\bar{\Lambda} + 2\omega_1 \omega_2 \sin 2\bar{\Lambda}] \, P_2^{(2)}(\sin \bar{\phi}).
\end{aligned} \tag{5.117}
$$

For $\varrho = a_0, \phi = \bar{\phi}, \Lambda = \bar{\Lambda}$, expansion (5.116) must be identically equal to (5.117), which yields the conditions for the variations of the geopotential coefficients finally used in expressing (4.68):

$$\delta J_2^{(0)} = k_2 \left\{ \frac{GM_{\mathrm{D}}}{GM_{\oplus}} \left(\frac{a_0}{\Delta_{\oplus\mathrm{D}}} \right)^3 \mathrm{P}_2^{(0)}(\sin \delta_{\mathrm{D}}) + \frac{GM_{\odot}}{GM_{\oplus}} \left(\frac{a_0}{\Delta_{\oplus\odot}} \right)^3 \mathrm{P}_2^{(0)}(\sin \delta_{\odot}) \right.$$

$$\left. - \tfrac{1}{6}q[2(1 + m_3)^2 - (m_1^2 + m_2^2)] \right\}, \tag{5.118}$$

$$\delta J_2^{(1)} = \frac{1}{3} k_2 \left[\frac{GM_{\mathrm{D}}}{GM_{\oplus}} \left(\frac{a_0}{\Delta_{\oplus\mathrm{D}}} \right)^3 \mathrm{P}_2^{(1)}(\sin \delta_{\mathrm{D}}) \cos T_{\mathrm{D}} \right.$$

$$\left. + \frac{GM_{\odot}}{GM_{\oplus}} \left(\frac{a_0}{\Delta_{\oplus\odot}} \right)^3 \mathrm{P}_2^{(1)}(\sin \delta_{\odot}) \cos T_{\odot} - qm_1(1 + m_3) \right], \tag{5.119}$$

$$\delta S_2^{(1)} = -\frac{1}{3} k_2 \left[\frac{GM_{\mathrm{D}}}{GM_{\oplus}} \left(\frac{a_0}{\Delta_{\oplus\mathrm{D}}} \right)^3 \mathrm{P}_2^{(1)}(\sin \delta_{\mathrm{D}}) \sin T_{\mathrm{D}} \right.$$

$$\left. + \frac{GM_{\odot}}{GM_{\oplus}} \left(\frac{a_0}{\Delta_{\oplus\odot}} \right)^3 \mathrm{P}_2^{(1)}(\sin \delta_{\odot}) \sin T_{\odot} + qm_2(1 + m_3) \right], \tag{5.120}$$

$$\delta J_2^{(2)} = \frac{1}{12} k_2 \left[\frac{GM_{\mathrm{D}}}{GM_{\oplus}} \left(\frac{a_0}{\Delta_{\oplus\mathrm{D}}} \right)^3 \mathrm{P}_2^{(2)}(\sin \delta_{\mathrm{D}}) \cos 2T_{\mathrm{D}} \right.$$

$$\left. + \frac{GM_{\odot}}{GM_{\oplus}} \left(\frac{a_0}{\Delta_{\oplus\odot}} \right)^3 \mathrm{P}_2^{(2)}(\sin \delta_{\mathrm{D}}) \cos 2T_{\odot} - q(m_1^2 - m_2^2) \right], \tag{5.121}$$

$$\delta S_2^{(2)} = -\frac{1}{12} k_2 \left[\frac{GM_{\mathrm{D}}}{GM_{\oplus}} \left(\frac{a_0}{\Delta_{\oplus\mathrm{D}}} \right)^3 \mathrm{P}_2^{(2)}(\sin \delta_{\mathrm{D}}) \sin 2T_{\mathrm{D}} \right.$$

$$\left. + \frac{GM_{\odot}}{GM_{\oplus}} \left(\frac{a_0}{\Delta_{\oplus\odot}} \right)^3 \mathrm{P}_2^{(2)}(\sin \delta_{\odot}) \sin 2T_{\odot} + 2qm_1 m_2 \right], \tag{5.122}$$

$$\frac{\delta J_n^{(k)}}{\delta S_n^{(k)}} = (2 - \delta_{k,0}) \frac{(n-k)!}{(n+k)!} k_n \left[\frac{GM_{\mathrm{D}}}{GM_{\oplus}} \left(\frac{a_0}{\Delta_{\oplus\mathrm{D}}} \right)^{n+1} \mathrm{P}_n^{(k)}(\sin \delta_{\mathrm{D}}) \right.$$

$$\left. \begin{matrix} \cos kT_{\mathrm{D}} \\ -\sin kT_{\mathrm{D}} \end{matrix} + \frac{GM_{\odot}}{GM_{\oplus}} \left(\frac{a_0}{\Delta_{\oplus\odot}} \right)^{n+1} \mathrm{P}_n^{(k)}(\sin \delta_{\odot}) \begin{matrix} \cos kT_{\odot} \\ -\sin kT_{\odot} \end{matrix} \right], \quad n \geq 3, \tag{5.123}$$

$$m_1 = \omega_1/\tilde{\Omega}, \quad m_2 = \omega_2/\tilde{\Omega}, \quad m_3 = (\omega_3 - \tilde{\Omega})/\tilde{\Omega}, \quad q = \tilde{\Omega}^2 a_0^3/(GM_{\oplus}), \tag{5.124}$$

where $\tilde{\Omega} = \mathrm{const}$, k_n is an n-th order Love number (Sect. 4.7) and δ_{D}, T_{D}, $\delta_{\odot}$ and $T_{\odot}$ are geocentric coordinates of the Moon's and Sun's centre of mass, respectively (denoted $\delta_{0'}$, $T_{0'}$ $\delta_{0''}$, $T_{0''}$ in Chaps. 3 and 4). Strictly speaking, the Love numbers depend on the periodicity of the phenomenon, which is not indicated in this schematic expression.

The variations of the tensor of inertia, δI_{ik}, are functions of variations $\delta J_2^{(k)}$ and $\delta S_2^{(k)}$, i.e. of the variations of only second-degree geopotential coefficients:

$$\delta I_{12} = -2\delta S_2^{(2)} M_{\oplus} a_0^2 ,$$

$$\delta I_{13} = -\delta J_2^{(1)} M_{\oplus} a_0^2 .$$

$$\delta I_{23} = - \delta S_2^{(1)} M_\oplus a_0^2 \,,$$

$$\delta I_{33} - \delta I_{11} = - (\delta J_2^{(0)} - 2\delta J_2^{(2)}) \, M_\oplus a_0^2 \,,$$

$$\delta I_{33} - \delta I_{22} = - (\delta J_2^{(0)} + 2\delta J_2^{(2)}) \, M_\oplus a_0^2 \,,$$

$$\delta I_{22} - \delta I_{11} = 4\delta J_2^{(2)} M_\oplus a_0^2 \,. \tag{5.125}$$

Given the condition that the total variation of the principal moments of inertia should be zero (Munk and MacDonald 1960),

$$\delta I_{11} + \delta I_{22} + \delta I_{33} = 0 \,; \tag{5.126}$$

in view of (4.70)

$$\delta I_{11} = (\tfrac{1}{3}\delta J_2^{(0)} - 2\delta J_2^{(2)}) \, M_\oplus a_0^2 \,,$$

$$\delta I_{22} = (\tfrac{1}{3}\delta J_2^{(0)} + 2\delta J_2^{(2)}) \, M_\oplus a_0^2 \,, \tag{5.127}$$

$$\delta I_{33} = - \tfrac{2}{3}\delta J_2^{(0)} M_\oplus a_0^2 \,.$$

Table 5.8 shows the approximate magnitude of the maximum values of the variations of the second-degree geopotential coefficients.

The variations of geopotential coefficients cause variations in the mean polar ($\bar{\alpha}$) and equatorial (α_1) flattening of the Earth's triaxial ellipsoid of inertia. The radial displacement of the external equipotential surface of the geopotential, close to the Earth's surface,

$$\delta \varrho = (1 + k_2) \frac{\delta W}{g}, \tag{5.128}$$

can be used directly to express

$$\delta \alpha = [\delta \varrho_{(\phi = 0^\circ)} - \delta \varrho_{(\phi = 90^\circ)}]/\varrho$$

$$= - \tfrac{3}{2}(1 + k_2) \left\{ \frac{GM_{\mathbb{D}}}{GM_\oplus} \left(\frac{a_0}{\Delta_{\oplus \mathbb{D}}} \right)^3 \mathrm{P}_2^{(0)}(\sin \delta_{\mathbb{D}}) \right.$$

$$\left. + \frac{GM_\odot}{GM_\oplus} \left(\frac{a_0}{\Delta_{\oplus \odot}} \right)^3 \mathrm{P}_2^{(0)}(\sin \delta_\odot) - \tfrac{1}{6}q[2(1 + m_3)^2 - (m_1^2 + m_2^2)] \right\}, \tag{5.129}$$

$$\delta \alpha_1 = 6(\delta J_2^{(2)} \cos 2\mu + \delta S_2^{(2)} \sin 2\mu) \, (1 + k_2)/k_2 \,, \tag{5.130}$$

Table 5.8. Order-of-magnitude variations of second-degree geopotential coefficients

Variation	Origin		
	Moon (10^{-9})	Sun (10^{-9})	Rotation (10^{-9})
$\delta J_2^{(0)}$	8.2	3.7	2×10^{-6}
$\delta J_2^{(1)}, \delta S_2^{(1)}$	7.0	3.2	0.5
$\delta J_2^{(2)}, \delta S_2^{(2)}$	4.1	1.9	10^{-7}

where $\mu = -14.9°(W)$ is the geographical longitude of the major axis of the equatorial ellipse (see Fig. 3.2).

The directions of axes x_j ($j = 1, 2, 3$) of the geocentric ellipsoid of inertia can be referred to the directions of axes x_j' of the geocentric system of reference used in the satellite determination of geopotential coefficients $J_2^{(k)}$ and $S_2^{(k)}$. Axes x_3 and x_3' do not deviate by more than tenths of a second of arc, but axes x_1 and x_1' (x_2 and x_2') display considerable deviations (14.9° on average). The mutual transformation of systems x_j and x_j', which we shall start with, will therefore be carried out in terms of Cardan (and not Euler) angles, denoted v, ε, μ; they correspond to the following rotation matrices:

$$\mathbf{v} = \begin{pmatrix} 1 & 0 & -v \\ 0 & 1 & 0 \\ v & 0 & 1 \end{pmatrix}, \quad \boldsymbol{\varepsilon} = \begin{pmatrix} 1 & 0 & 0 \\ 0 & 1 & \varepsilon \\ 0 & -\varepsilon & 1 \end{pmatrix},$$

$$\boldsymbol{\mu} = \begin{pmatrix} \cos\mu & \sin\mu & 0 \\ -\sin\mu & \cos\mu & 0 \\ 0 & 0 & 1 \end{pmatrix}. \tag{5.131}$$

All rotations are positive counterclockwise when viewed from the positive ends of the axes of rotation. Squares v^2, ε^2 and product $v\varepsilon$ have been neglected.

The variations of angles v, ε, μ, due to tidal (4.71) and rotational deformations of the Earth (in seconds of arc) (Burša 1983), read

$$\delta v = -1.03\, P_2^{(1)}(\sin\delta_{\mathbb{D}})\cos T_{\mathbb{D}} - 0.47\, P_2^{(1)}(\sin\delta_{\odot})\cos T_{\odot} + 0.31\,\delta x,$$

$$\delta\varepsilon = 1.03\, P_2^{(1)}(\sin\delta_{\mathbb{D}})\sin T_{\mathbb{D}} + 0.47\, P_2^{(1)}(\sin\delta_{\odot})\sin T_{\odot} + 0.31\,\delta y,$$

$$\delta v = 67.1\, P_2^{(2)}(\sin\delta_{\mathbb{D}})\sin 2T_{\mathbb{D}} - 38.7\, P_2^{(2)}(\sin\delta_{\mathbb{D}})\cos 2T_{\mathbb{D}}$$

$$+ 30.7\, P_2^{(2)}(\sin\delta_{\odot})\sin 2T_{\odot} - 17.7\, P_2^{(2)}(\sin\delta_{\odot})\cos 2T_{\odot}$$

$$+ \{1.9\,\delta x\,\delta y + 0.6[(\delta x)^2 - (\delta y)^2]\}\,10^{-4}, \tag{5.132}$$

where δx and δy are variations of the pole coordinates expressed in seconds of arc. Vondrák (1984) presented a detailed analysis of the variations in the directions of the axes of the Earth's ellipsoid of inertia.

The total values of angles ε, v, μ can be calculated from the second-degree geopotential coefficients for a given time:

$$\varepsilon = \frac{S_2^{(1)}}{J_2^{(0)} + 2J_2^{(2)}}, \quad v = \frac{J_2^{(1)}}{J_2^{(0)} - 2J_2^{(2)}}, \quad \tan 2\mu = -S_2^{(2)}/J_2^{(2)}. \tag{5.133}$$

5.5 On the Hypothesis of an Expanding Earth

We shall now refer back to Section 5.2, in particular to the angular momentum balance of the tidal forces decelerating the Earth's rotation and accelerating the Moon in its orbit. We shall use the angular momentum balance equation to test the probability of the expanding Earth hypothesis. If this phenomenon does exist, as is being constantly assumed in various applications, it must be accompanied by an increase of the Earth's principal moment of inertia $C = (I_{33})_\oplus$ and this should be reflected in the angular momentum balance, at least as regards the trend.

Hence, we shall admit to $dI_{33}/dt \neq 0$, and add this term to Eq. (5.32). The change will be in putting $d(C\omega_\oplus)/dt = C\,d\omega/dt + \omega_\oplus\,dC/dt$ instead of the previous $d(C\omega_\oplus)/dt = C\,d\omega_\oplus/dt$. The resultant balance equation (5.32) will now read

$$d\omega_\oplus/dt = a_1\,da_\mathbb{D}/dt + a_2\,de_\mathbb{D}/dt + a_3\,di_\mathbb{D}/dt$$
$$+ a_4\,da_\oplus/dt + a_5\,dC/dt\,. \tag{5.134}$$

The coefficient in the new term with dC/dt

$$a_5 = -\omega_\oplus/C = -9.1 \times 10^{-43}\ \mathrm{kg^{-1}\,m^{-2}\,s^{-1}};$$

Eq. (5.134) then yields

$$a_5\,dC/dt = (d\omega_\oplus/dt)_{\text{observed}} - (a_1\,da_\mathbb{D}/dt + a_2\,de_\mathbb{D}/dt$$
$$+ a_3\,di_\mathbb{D}/dt + a_4\,da_\oplus/dt)\,. \tag{5.135}$$

The observed value $d\omega_\oplus/dt = -(5.4 \pm 0.5) \times 10^{-22}\ \mathrm{rad\,s^{-2}}$, and the sum of the terms with coefficients $a_1, \ldots, a_4$ (5.47) amounts to $-6.47 \times 10^{-22}\ \mathrm{rad\,s^{-2}}$. Consequently,

$$a_5\,dC/dt = +1.1 \times 10^{-22}\ \mathrm{rad\,s^{-2}}$$

and thus

$$dC/dt = -(1.2 \pm 0.4) \times 10^{20}\ \mathrm{kg\,m^2\,s^{-1}}$$
$$= -(3.8 \pm 1.2) \times 10^{29}\ \mathrm{kg\,m^2\,cy^{-1}}\,. \tag{5.136}$$

This means that, in the light of contemporary astronomical and satellite data, the largest Earth's moment of inertia should be decreasing and not increasing, which is contrary to the expanding Earth hypothesis.

In addition, the effect of ocean tides should produce an even larger negative value of the residual being studied. This lends even more support to the conclusions that the assumption of an expanding Earth has no justification in relation to astronomical data (over an interval of at least the last 2500 years). The same conclusion is also indicated by the analyses of fossil formations, which enable the number of days in a tropical year to be determined and which yield approximately value (5.15). If these palaeontological studies are sufficiently accurate, as claimed by their authors, then the Earth should not have in fact expanded over approximately the last 450 million years.

5.6 Decrease in the Maximum Principal Moment of the Earth's Inertia and Its Effect on Polar Motion

The analyses of the orbit of the geodynamic satellite LAGEOS, ranged since 1976 by a network of satellite laser range-finders with a high accuracy (of a few centimetres), have disclosed the secular decrease of the absolute value of the second zonal geopotential coefficient $J_2 = -J_2^{(0)}$ (Yoder et al. 1983):

$$\frac{\mathrm{d}J_2}{\mathrm{d}t} = -(2.6 \pm 0.6) \times 10^{-9}\,\mathrm{cy}^{-1}. \tag{5.137}$$

This phenomenon has dynamic consequences, particularly with regard to the Earth's rotation. Indeed, the principal moments of the Earth's inertia change and this in turn affects, for example, free nutation (pole wandering). We shall study this phenomenon in more detail using a simplified model which assumes that the principal moments of inertia in the plane of the equator are equal $(B = A)$. We shall reproduce the solution of Burša and Šidlichovský (1985).

Assuming incompressibility, i.e. $\delta A + \delta B + \delta C = 0$ (Munk and MacDonald 1960), (5.137) will yield

$$\frac{\mathrm{d}C}{\mathrm{d}t} = \frac{2}{3}\frac{\mathrm{d}J_2}{\mathrm{d}t}\,Ma_0^2 = -(4.2 \pm 1.0) \times 10^{29}\,\mathrm{kg\,m^2\,cy^{-1}}, \tag{5.138}$$

$$\frac{\mathrm{d}A}{\mathrm{d}t} = -\frac{1}{3}\frac{\mathrm{d}J_2}{\mathrm{d}t}\,Ma_0^2 = (2.1 \pm 0.5) \times 10^{29}\,\mathrm{kg\,m^2\,cy^{-1}}. \tag{5.139}$$

Value (5.138) corresponds relatively well to (5.136) which we calculated quite independently in Section 5.5.

In solving the problem we shall only consider the linear terms (5.138) and (5.139), i.e. we shall put

$$A(t) = A_0 + \frac{\mathrm{d}A}{\mathrm{d}t}\,t, \quad C(t) = C_0 + \frac{\mathrm{d}C}{\mathrm{d}t}\,t. \tag{5.140}$$

In the absence of external forces, the time variation of the resultant kinetic angular momentum of the Earth, $\mathbf{L}$ (3.4), must be equal to zero:

$$\left(\frac{\mathrm{d}\mathbf{L}}{\mathrm{d}t}\right)_r + [\boldsymbol{\omega} \times \mathbf{L}] = 0. \tag{5.141}$$

We shall solve the equation in the geocentric coordinate system of the axes of the ellipsoid of inertia x_j $(j = 1, 2, 3)$ in which the products of inertia D, E and F are zero. In this simplified model the components of the angular momentum (dots are used to indicate derivatives with respect to time) are

$$
L_1 = A\omega_1 + \int_M (x_2\dot{x}_3 - x_3\dot{x}_2)\,\mathrm{d}m\,,
$$

$$
L_2 = A\omega_2 + \int_M (x_3\dot{x}_1 - x_1\dot{x}_3)\,\mathrm{d}m\,,
$$

$$
L_3 = C\omega_3 + \int_M (x_1\dot{x}_2 - x_2\dot{x}_1)\,\mathrm{d}m\,, \tag{5.142}
$$

and the integrals on the rhs are components of the additional angular momentum $\delta\mathbf{L}$, generated by mass transfer:

$$
\delta\mathbf{L} = \int_M [\boldsymbol{\varrho} \times \dot{\boldsymbol{\varrho}}]\,\mathrm{d}m\,. \tag{5.143}
$$

However, since we have no data to calculate $\delta\mathbf{L}$ we shall put $\delta\mathbf{L} = 0$ in this model, assuming thereby that mass is displaced only in the radial direction. The initial equation describing the free nutation in this model reads

$$
\frac{\mathrm{d}L_1}{\mathrm{d}t} + [C(t) - A(t)]\frac{L_2}{A(t)}\frac{L_3}{C(t)} = 0\,,
$$

$$
\frac{\mathrm{d}L_2}{\mathrm{d}t} - [C(t) - A(t)]\frac{L_1}{A(t)}\frac{L_3}{C(t)} = 0\,,
$$

$$
\frac{\mathrm{d}L_3}{\mathrm{d}t} = 0\,. \tag{5.144}
$$

The third equation of (5.144) immediately yields

$$
L_3 = \omega_3 C(t) = \mathrm{const}\,, \tag{5.145}
$$

i.e. in view of (5.140)

$$
\dot{\omega}_3 + \frac{k_1}{1 + k_1 t}\,\omega_3 = 0\,, \tag{5.146}
$$

$$
k_1 = \dot{C}/C_0 = -6.1 \times 10^{-9}\,\mathrm{cy}^{-1}\,;
$$

$$
C_0 = 8.036 \times 10^{37}\,\mathrm{kg\,m^2}\,; \tag{5.147}
$$

and

$$
\omega_3 = \frac{L_3}{C(t)} = \frac{L_3}{C_0(1 + k_1 t)} = \frac{\omega_0}{1 + k_1 t}\,, \tag{5.148}
$$

where $\omega_0 = \omega_3\,(t = 0) = 7.292\,115 \times 10^{-5}\,\mathrm{rad\,s}^{-1}$ is an integration constant.

The first integral of system (5.144) is known:

$$
L_1^2 + L_2^2 + L_3^2 = L^2 = \mathrm{const}\,; \tag{5.149}
$$

it and Eq. (5.145) yield

$$
L_1^2 + L_2^2 = L^2 - L_3^2 = K^2 = \mathrm{const}\,, \tag{5.150}
$$

and then, regardless of the sign, which is not detrimental to generality,

$$L_1 = K\left[1 - \left(\frac{L_2}{K}\right)^2\right]^{1/2},$$ (5.151)

$$L_2 = K\left[1 - \left(\frac{L_1}{K}\right)^2\right]^{1/2}.$$ (5.152)

We now substitute (5.151) into the first equation of (5.144)

$$\frac{1}{K}\frac{dL_1}{dt} + L_3\left[\frac{1}{A(t)} - \frac{1}{C(t)}\right]\left[1 - \left(\frac{L_1}{K}\right)^2\right]^{1/2} = 0,$$ (5.153)

and, analogously, (5.152) into the second equation of (5.144)

$$\frac{1}{K}\frac{dL_2}{dt} - L_3\left[\frac{1}{A(t)} - \frac{1}{C(t)}\right]\left[1 - \left(\frac{L_2}{K}\right)^2\right]^{1/2} = 0.$$ (5.154)

We now arrive at the exact solution,

$$L_1 = K\sin\left\{L_3\int_0^t\left[\frac{1}{C(t)} - \frac{1}{A(t)}\right]dt\right\},$$

$$L_2 = K\cos\left\{L_3\int_0^t\left[\frac{1}{C(t)} - \frac{1}{A(t)}\right]dt\right\},$$ (5.155)

the time being reckoned from the instant that $L_1 = 0$.

Equations (5.155) hold true for any time functions $A(t)$ and $C(t)$, i.e. they are much more general than required by assumption (5.140); consequently,

$$\omega_1 = \frac{K}{A(t)}\sin\left\{L_3\int_0^t\left[\frac{1}{C(t)} - \frac{1}{A(t)}\right]dt\right\},$$

$$\omega_2 = \frac{K}{A(t)}\cos\left\{L_3\int_0^t\left[\frac{1}{C(t)} - \frac{1}{A(t)}\right]dt\right\}.$$ (5.156)

In the special case of (5.140), in which the variations of the moments are linear, we get

$$\frac{C(t)}{C_0} = 1 + \frac{\dot{C}}{C_0}t = 1 + k_1 t,$$ (5.157)

$$\frac{A(t)}{A_0} = 1 + \frac{\dot{A}}{A_0}t = 1 + k_2 t;$$ (5.158)

$$k_2 = \dot{A}/A_0 = -\frac{1}{2}k_1\frac{C_0}{A_0} = 3.0 \times 10^{-9}\text{ cy}^{-1};$$

$$A_0 = 8.010 \times 10^{37}\text{ kg m}^2;$$ (5.159)

k_1 and k_2 may be considered to be small parameters. The exact solution for the special case of linear variations of the momenta reads

$$L_1 = K \sin\left[\frac{L_3}{k_1 C_0} \ln(1 + k_1 t) - \frac{L_3}{k_2 A_0} \ln(1 + k_2 t)\right],$$

$$L_2 = K \cos\left[\frac{L_3}{k_1 C_0} \ln(1 + k_1 t) - \frac{L_3}{k_2 A_0} \ln(1 + k_2 t)\right],$$

$$L_3 = C_0 \omega_0 ; \tag{5.160}$$

$$\omega_1 = \frac{K}{A_0}(1 + k_2 t)^{-1} \sin\left\{\omega_0\left[\frac{\ln(1 + k_1 t)}{k_1 t} - \frac{C_0}{A_0}\frac{\ln(1 + k_2 t)}{k_2 t}\right]t\right\},$$

$$\omega_2 = \frac{K}{A_0}(1 + k_2 t)^{-1} \cos\left\{\omega_0\left[\frac{\ln(1 + k_1 t)}{k_1 t} - \frac{C_0}{A_0}\frac{\ln(1 + k_2 t)}{k_2 t}\right]t\right\}. \tag{5.161}$$

By putting $k_1 = 0$ and $k_2 = 0$ we arrive at Euler's free nutation:

$$\omega_1 = \frac{K}{A_0} \sin\left[\omega_0(1 - C_0 A_0^{-1})t\right] = -\frac{K}{A_0} \sin \sigma_E t ,$$

$$\omega_2 = \frac{K}{A_0} \cos\left[\omega_0(1 - C_0 A_0^{-1})t\right] = \frac{K}{A_0} \cos \sigma_E t ; \tag{5.162}$$

$$\sigma_E = \omega_0 \frac{C_0 - A_0}{A_0} = k_0 \ (3.50) \text{ is Euler's circular frequency.}$$

It is convenient to modify solution (5.161) by introducing the 'generalized frequency'

$$\sigma_E' = \sigma_E + \Delta\sigma_E , \tag{5.163}$$

where $\Delta\sigma_E$ is the correction generated by $k_1 \neq 0$ and $k_2 \neq 0$. For this purpose we shall put

$$\frac{\ln(1 + k_1 t)}{k_1 t} - \frac{C_0}{A_0}\frac{\ln(1 + k_2 t)}{k_2 t} = \frac{1}{k_1 t}\int_0^{k_1 t}\frac{1}{1 + k_1 t}\,d(k_1 t)$$

$$-\frac{C_0}{A_0}\frac{1}{k_2 t}\int_0^{k_2 t}\frac{1}{1 + k_2 t}\,d(k_2 t) = \frac{1}{k_1 t}\int_0^{k_1 t}\sum_{n=0}^{\infty}(-k_1 t)^n d(k_1 t)$$

$$-\frac{C_0}{A_0}\frac{1}{k_2 t}\int_0^{k_2 t}\sum_{n=0}^{\infty}(-k_2 t)^n d(k_2 t)$$

$$= \sum_{n=0}^{\infty}(-1)^n \frac{1}{n+1}\left[(k_1 t)^n - \frac{C_0}{A_0}(k_2 t)^n\right]$$

$$= \sum_{n=0}^{\infty}(-1)^n \frac{1}{n+1}\left[(-2)^n\left(\frac{A_0}{C_0}\right)^n - \frac{C_0}{A_0}\right](k_2 t)^n$$

$$= \frac{A_0 - C_0}{A_0} + \sum_{n=1}^{\infty}\frac{1}{n+1}\left[2^n\left(\frac{A_0}{C_0}\right)^{n+1} - (-1)^n\right]\frac{C_0}{A_0}k_2^n t^n ; \tag{5.164}$$

after substituting into (5.161) we get

$$\omega_1 = -\frac{K}{A_0}(1 + k_2 t)^{-1} \sin(\sigma_E + \Delta\sigma_E)T ,$$

$$\omega_2 = \frac{K}{A_0}(1 + k_2 t)^{-1} \cos(\sigma_E + \Delta\sigma_E)t , \tag{5.165}$$

$$\Delta\sigma_E = -\omega_0 \frac{C_0}{A_0} \sum_{n=1}^{\infty} \frac{1}{n+1}\left[2^n\left(\frac{A_0}{C_0}\right)^{n+1} - (-1)^n \right] k_2^n t^n . \tag{5.166}$$

If k_2 is sufficiently small to be able to neglect $(k_2 t)^2$, we arrive at the linear variation of Euler's frequency,

$$\Delta\sigma_E = -\omega_0\left(\frac{A_0}{C_0} - \frac{C_0}{2A_0}\right)k_2 t = \frac{1}{3}\omega_0 \frac{Ma_0^2}{A_0}\left(\frac{A_0}{C_0} + \frac{C_0}{2A_0}\right)\frac{dJ_2}{dt} t , \tag{5.167}$$

caused by the linear variation of the second zonal geopotential coefficient (5.137):

$$\omega_1 = -\frac{K}{A_0}\left(1 + \frac{1}{3}\frac{Ma_0^2}{A_0}\frac{dJ_2}{dt} t\right)^{-1} \sin\left[\sigma_E + \frac{1}{3}\omega_0 \frac{Ma_0^2}{A_0}\left(\frac{A_0}{C_0} + \frac{C_0}{2A_0}\right)\frac{dJ_2}{dt} t\right]t ,$$

$$\omega_2 = \frac{K}{A_0}\left(1 + \frac{1}{3}\frac{Ma_0^2}{A_0}\frac{dJ_2}{dt} t\right) \cos\left[\sigma_E + \frac{1}{3}\omega_0 \frac{Ma_0^2}{A_0}\left(\frac{A_0}{C_0} + \frac{C_0}{2A_0}\right)\frac{dJ_2}{dt} t\right]t .$$

$$\tag{5.168}$$

The solution yields the following conclusions: The secular decrease of the absolute value of the second zonal geopotential coefficient (5.137) is responsible for the secular diminishing of Euler's frequency $\sigma_E = k_0/(2\pi)$ (3.50). The observed Chandler frequency,

$$\sigma_{CH} = 2.3956 \times 10^{-7} \, \mathrm{rad\,s}^{-1} = 433° \, \mathrm{y}^{-1} = 1.186° \, \mathrm{d}^{-1} , \tag{5.169}$$

diminishes in the linear approximation and, according to (5.167), by about $3.1 \times 10^{-9} \, \mathrm{rad\,s}^{-1}$ every 10^6 years. This means that Euler's period (3.80) increases in the model being considered by 8.4 days every 10^6 years.

The amplitude of the motion of the axis of rotation in this model diminishes. In the linear approximation this decrease amounts to about $3 \times 10^{-9} \, \mathrm{cy}^{-1}$.

This example alone illustrates the new options that contemporary satellite data afford geodynamics only a few years after the first geodynamic satellite was launched (1976). The time variations of the gravitational field, generated by mass transfer within the Earth and in the atmosphere and detected by satellites, represent one of the most topical problems now being treated within the scope of international cooperation. In the near future these results should yield more accurate descriptions of the Earth's recent dynamics as a 'live' planet whose structure is being changed by external and internal forces. This requires satellite tracking with the maximum accuracy possible. Evidence of how high this accuracy is is clearly provided by Table 1.4, adopted from Marsh et al. (1990).

5.7 Secular Decrease in the Earth's Angular Momentum and Kinetic Energy

This phenomenon has already been mentioned in Sections 5.2, 5.5, and 5.6, but only in relation to detailed approaches to the description of specific problems. We shall now attempt a synthesis.

The angular momentum of the Earth $\mathbf{L}_\omega$ due to its rotation is

$$\mathbf{L}_\omega = \mathbf{I} \cdot \boldsymbol{\omega}_\oplus ,\tag{5.170}$$

where $\mathbf{I}$ is the tensor of the Earth's inertia (3.3) and $\boldsymbol{\omega}_\oplus$ is the vector of rotation. Its magnitude is

$$L_\omega = [(A\omega_1)^2 + (B\omega_2)^2 + (C\omega_3)^2]^{1/2} \sim C\omega_\oplus$$
$$= 5.86 \times 10^{33} \,\mathrm{kg\,m^2\,s^{-1}} = 1.85 \times 10^{43} \,\mathrm{kg\,m^2\,cy^{-1}} .\tag{5.171}$$

In view of (5.31) the time variations generated by the tidal effects of the Moon and Sun [the last term in (5.31) is negligibly small] are

$$\frac{\mathrm{d}L_\omega}{\mathrm{d}t} = -\frac{1}{2}\frac{M_\oplus M_{\mathbb{D}}}{M_\oplus + M_{\mathbb{D}}} a_{\mathbb{D}}^{-1/2} [G(M_\oplus + M_{\mathbb{D}})(1 - e_{\mathbb{D}}^2)]^{1/2} \cos i_{\mathbb{D}} \frac{\mathrm{d}a_{\mathbb{D}}}{\mathrm{d}t}$$

$$+ \frac{M_\oplus M_{\mathbb{D}}}{M_\oplus + M_{\mathbb{D}}} e_{\mathbb{D}}(1 - e_{\mathbb{D}}^2)^{-1/2} [G(M_\oplus + M_{\mathbb{D}})a_{\mathbb{D}}]^{1/2} \cos i_{\mathbb{D}} \frac{\mathrm{d}e_{\mathbb{D}}}{\mathrm{d}t}$$

$$+ \frac{M_\oplus M_{\mathbb{D}}}{M_\oplus + M_{\mathbb{D}}} [G(M_\oplus + M_{\mathbb{D}})a_{\mathbb{D}}(1 - e_{\mathbb{D}}^2)]^{1/2} \sin i_{\mathbb{D}} \frac{\mathrm{d}i_{\mathbb{D}}}{\mathrm{d}t}$$

$$- \frac{1}{2}\frac{M_\odot(M_\oplus + M_{\mathbb{D}})}{M_\odot + M_\oplus + M_{\mathbb{D}}} a_\oplus^{-1/2} [G(M_\odot + M_\oplus + M_{\mathbb{D}})(1 - e_\oplus^2)]^{1/2}$$

$$\times \cos \tilde{\varepsilon}_0 \frac{\mathrm{d}a_\oplus}{\mathrm{d}t} .\tag{5.172}$$

By substituting values (5.17), (5.44), (5.45) and (5.43) into (5.172) we get

$$\frac{\mathrm{d}L_\omega}{\mathrm{d}t} = -5.2 \times 10^{16} \,\mathrm{kg\,m^2\,s^{-2}} = -5.2 \times 10^{35} \,\mathrm{kg\,m^2\,cy^{-2}} .\tag{5.173}$$

The observed secular decrease in the angular velocity of the Earth's rotation (5.15),

$$C\frac{\mathrm{d}\omega_\oplus}{\mathrm{d}t} = -4.3 \times 10^{16} \,\mathrm{kg\,m^2\,s^{-2}} = -4.3 \times 10^{35} \,\mathrm{kg\,m^2\,cy^{-2}} ,\tag{5.174}$$

and the secular diminishing of the largest principal moment of the Earth's inertia (5.138),

$$\omega_\oplus \frac{\mathrm{d}C}{\mathrm{d}t} = -1.0 \times 10^{16} \,\mathrm{kg\,m^2\,s^{-2}} = -1.0 \times 10^{35} \,\mathrm{kg\,m^2\,cy^{-2}} ,\tag{5.175}$$

contribute to value (5.173). The difference between (5.173) and the sum of (5.174) plus (5.175) is within the limits of the standard errors of observed quantities (5.15) and (5.17).

The decrease in the angular momentum of the Earth (5.173) is compensated by the increase in the orbital angular momentum of the Moon, $L_{\oplus\mathbb{D}}$, within the limits of the standard errors of the observed quantities involved

$$\frac{dL_{\oplus\mathbb{D}}}{dt} = 4.3 \times 10^{35}\ \mathrm{kg\,m^2\,cy^{-2}}, \tag{5.176}$$

and by the increase in the orbital momentum of the barycentre of the Earth–Moon system, $L_{\odot B}$,

$$\frac{dL_{\odot B}}{dt} = 0.8 \times 10^{35}\ \mathrm{kg\,m^2\,cy^{-2}}. \tag{5.177}$$

The magnitude of tidal force $\mathbf{f}$ acting along the tangent to the Moon's orbit (Sect. 5.2, Fig. 5.1), neglecting the eccentricity of the Moon's orbit, is roughly

$$f - \frac{1}{a_{\mathbb{D}}} \frac{dL_{\oplus\mathbb{D}}}{dt} - 1.1 \times 10^8\ \mathrm{kg\,m\,s^{-2}}, \tag{5.178}$$

and acceleration $w_{\mathbb{D}}$, which is imparted to the Moon thereby,

$$w_{\mathbb{D}} = 1.5 \times 10^{-15}\ \mathrm{m\,s^{-2}}. \tag{5.179}$$

The kinetic energy of the Earth, $E_{\oplus}$, due to rotation is given by (5.52). Its secular change due, on the one hand, to the decrease (5.54) of the Earth's rotational velocity (5.15) and, on the other hand, to the diminishing of the moment of inertia (5.138),

$$\frac{dE_{\oplus}}{dt} = C\omega_{\oplus} \frac{d\omega_{\oplus}}{dt} + \frac{1}{2}\omega_{\oplus}^2 \frac{dC}{dt} = -3.6 \times 10^{12}\ \mathrm{kg\,m^2\,s^{-3}}$$

$$= -1.1 \times 10^{41}\ \mathrm{kg\,m^2\,cy^{-3}}. \tag{5.180}$$

The systematic decrease in the Earth's angular momentum (5.173) and in its kinetic energy (5.180) are principal geodynamic phenomena which have a substantial effect on the dynamic evolution of the Earth and generate a number of other geodynamic phenomena of a secular nature. They are also of cosmogonic significance. Their substantial part, generated by the tidal effects of the Moon, will gradually decrease and last until the Earth's rotational period and the Moon's orbital period are in a $1:1$ resonance, which will occur in about 4×10^9 years (Sect. 5.2). The smaller part ($\sim 1/5$), generated by the tidal effect of the Sun, will survive and become the principal factor with a tendency to achieve resonance of the Earth's rotational period with the Earth's orbital period about the Sun.

Satellite methods have made a significant contribution to solving this problem by discovering the systematic decrease (5.137) of the second zonal geopotential coefficient, $J_2 = -J_2^{(0)}$, and the related decrease in the largest principal moment of the Earth's inertia (5.138). The origin of this phenomenon has still not been found. It is necessary to seek it in the processes occurring in the Earth's mantle, the duration of which applies to the same interval to which the secular term in the variations of the Earth's rotation (5.15), and secular term in the variations of the Moon's orbit semi-axis (5.138), apply. This is one of the fundamental geodynamic problems which has yet to be solved by the cooperation of numerous sciences of the Earth and of the Universe, especially of astronomy, geophysics, physics of the atmosphere, oceanology and geology (Brosche and Sündermann 1982).

5.8 Long-Term Variations in the Earth's Gravity Field due to Variations in the Earth's Rotation Vector and in the Second Zonal Geopotential Coefficient

Geodynamic phenomena, which have a long-term effect on the Earth's dynamics and on the distribution of its masses, generated changes in the Earth's gravity field. Some of the phenomena of this nature have already been discussed in Chapters 3 and 4. In this section we shall summarily discuss changes of a secular nature caused by the secular diminishing of the angular velocity of the Earth's rotation (5.15), by the variation of the second zonal geopotential coefficient $J_2 = -J_2^{(0)}$ in (5.137), and by the secular motion of the pole (see Table 3.3) described by Vondrák (1985):

$$\frac{\mathrm{d}\gamma}{\mathrm{d}t} = 0.0033''\mathrm{y}^{-1}, \quad \Lambda_0 = 281.8°\mathrm{E}. \tag{5.181}$$

The secular change in the geopotential $(\mathrm{d}W/\mathrm{d}t)_s$, caused by the phenomena mentioned and neglecting terms of the second order (containing, for example, $\gamma \mathrm{d}\gamma/\mathrm{d}t$), is

$$\begin{aligned}
\left(\frac{\mathrm{d}W}{\mathrm{d}t}\right)_s = &-\frac{GM}{\varrho}\left\{\left(\frac{a_0}{\varrho}\right)^2 \frac{\mathrm{d}J_2}{\mathrm{d}t} \mathrm{P}_2^{(0)}(\sin\phi) \right.\\
&+ \frac{2}{3}(1 + k_s - h_s)q\left(\frac{a_0}{\varrho}\right)^{-3}\left[\frac{1}{\omega}\frac{\mathrm{d}\omega}{\mathrm{d}t}\mathrm{P}_2^{(0)}(\sin\phi)\right.\\
&\left.\left. + \frac{1}{2}\frac{\mathrm{d}\gamma}{\mathrm{d}t}\mathrm{P}_2^{(1)}(\sin\phi)\cos(\Lambda - \Lambda_0)\right]\right\};
\end{aligned} \tag{5.182}$$

$GM = 398\,600.44 \times 10^9\,\mathrm{m}^3\,\mathrm{s}^{-2}$, $a_0 = 6\,378\,140$ m, ρ, ϕ, Λ are geocentric spherical coordinates of the potential point, k_s, h_s are secular Love numbers corresponding to long-term deformations and $q = \omega^2 a_0^3/(GM)$.

The long-term variations of the radius-vector of the equipotential surface of geopotential, implied by (5.182), are

$$\begin{aligned}
\frac{1}{\varrho}\frac{\mathrm{d}\varrho}{\mathrm{d}t} = &-\left(\frac{a_0}{\varrho}\right)^2 \frac{\mathrm{d}J_2}{\mathrm{d}t}\mathrm{P}_2^{(0)}(\sin\phi) - \frac{2}{3}(1 + k_s)q\left(\frac{a_0}{\varrho}\right)^{-3}\\
&\times\left[\frac{1}{\omega}\frac{\mathrm{d}\omega}{\mathrm{d}t}\mathrm{P}_2^{(0)}(\sin\phi) + \frac{1}{2}\frac{\mathrm{d}\gamma}{\mathrm{d}t}\mathrm{P}_2^{(1)}(\sin\phi)\cos(\Lambda - \Lambda_0)\right].
\end{aligned} \tag{5.183}$$

Hence, assuming $a_0/\varrho = 1$, the long-term variation of the polar flattening

$$\frac{d\alpha}{dt} = \frac{3}{2}\frac{dJ_2}{dt} + (1 + k_s)\frac{q}{\omega}\frac{d\omega}{dt}, \tag{5.184}$$

i.e.

$$\frac{d\alpha}{dt} = -4.1 \times 10^{-9}\,\mathrm{cy}^{-1}, \quad \frac{d\left(\dfrac{1}{\alpha}\right)}{dt} = 3.7 \times 10^{-4}\,\mathrm{cy}^{-1}.$$

The long-term variation of gravity, derived from (5.182), is analogously

$$\frac{1}{g}\frac{dg}{dt} = -3\left(\frac{a_0}{\varrho}\right)^2\frac{dJ_2}{dt}P_2^{(0)}(\sin\phi) - \frac{4}{3}\left(1 - \frac{3}{2}k_s + h_s\right)q\left(\frac{a_0}{\varrho}\right)^{-3}$$

$$\times\left[\frac{1}{\omega}\frac{d\omega}{dt}P_2^{(0)}(\sin\phi) + \frac{1}{2}\frac{d\gamma}{dt}P_2^{(1)}(\sin\phi)\cos(\Lambda - \Lambda_0)\right], \tag{5.185}$$

i.e. in units of $10^{-9}\,\mathrm{cy}^{-1}$,

$$\frac{1}{g}\frac{dg}{dt} = 7.9 P_2^{(0)}(\sin\phi) - 3.7(1 - \tfrac{3}{2}k_s + h_s)P_2^{(1)}(\sin\phi)\cos(\Lambda - \Lambda_0). \tag{5.186}$$

Equation (5.186) indicates that the maximum secular variation of gravity, generated by the three geodynamic phenomena mentioned, is $\sim 10^{-7}\,\mathrm{m\,s}^{-2}$ $(10^{-2}\,\mathrm{mgal})\mathrm{cy}^{-1}$. Its principal part is caused by the variation in the second zonal geopotential coefficient (5.137) and by the secular motion of the pole (5.181). The effect of decelerating the Earth's rotation (5.15) is one order of magnitude smaller.

6 The Earth in the Solar System

6.1 Introduction

In this concluding chapter we shall briefly discuss the dynamic system of which
the Earth is a part. This will not be an interpretation of the dynamics of the Solar
System and of cosmogony. These fundamental problems would require a thorough account of the foundations of the orbital rotational and tidal dynamics of
the system of general bodies by way of introduction, and this has been dealt with
elsewhere.

We shall only mention some of the basic facts, dynamic parameters and
relations required with regard to the broader connection between geodynamic
phenomena discussed in this book.

6.2 Structure of the Solar System

Its central body is the Sun whose heliocentric gravitational constant

$$GM_\odot = 13\,271\,244.0 \times 10^{13}\,\mathrm{m}^3\,\mathrm{s}^{-2}$$

and mass

$$M_\odot = 1.9885 \times 10^{30}\,\mathrm{kg}\,,$$

thus representing about 99.86% of the total mass of the system.

Nine planets are known at this time; their basic parameters are given in
Tables 6.1–6.3. Hereinafter we shall omit the planet Pluto, since it is the system
of a planet and relatively massive and close satellites whose parameters are still
known only approximately.

The Solar System also includes asteroids, amounting to several tens of
thousands in number, millions of comets and a tremendous number of small
bodies, particles and cosmic dust (diameter 1–10 µm).

Several of the planets have their own satellites: Jupiter 16, Saturn 18, Uranus
15 and Neptune 8. Rings of small particles orbit four planets (Saturn, Uranus,
Jupiter and Neptune). Also, the Sun has a ring in a sense: this is formed by the
asteroids scattered in the belt between Mars and Jupiter.

Table 6.1. Centric gravitational constants of planets GM, masses M (given $G = 6673 \times 10^{-14}\,\mathrm{kg^{-1}\,m^3\,s^{-2}}$) and second-degree geopotential coefficients $J_2^{(0)}, J_2^{(2)}\,S_2^{(2)}$

Body	GM $(10^9\,\mathrm{m^3\,s^{-2}})$	$GM/GM_\oplus$	$GM_\odot/GM$	M $(10^{24}\,\mathrm{kg})$	$-J_2^{(0)}$ (10^{-6})	$J_2^{(2)}$ (10^{-6})	$S_2^{(2)}$ (10^{-6})
Sun	$13\,271\,244.0 \times 10^4$	332 946.0	1	$1.988\,8 \times 10^6$	10		
Mercury	22 032.09	0.055 274	6 023 597	0.330 17	60		
Venus	324 858.60	0.814 998	408 523.7	4.868 3	5.972	0.91	0.24
Earth	398 600.440	1	332 946.04	5.973 3	1 082.626	1.574	− 0.904
Moon	4 902.799	1/81.300 59	27 068 709	0.073 472	202.151	22.1	0.0
Mars	42 828.3	0.107 447	3 098 709	0.641 81	1 959.2	− 54.9	31.3
Jupiter	126 686 537	317.828	1 047.565	1898.5	14 697	− 0.2	− 0.3
Saturn	37 931 272	95.161	3 498.761	568.43	16 298		
Uranus	5 793 939	14.536	22 905.4	86.827	3 343		
Neptune	6 835 096	17.148	19 416.3	102.43	3 710		

Table 6.2. Parameters of the best-fitting triaxial $(a, \alpha, \alpha_1, \Lambda_a)$ and rotational $(\bar{a}, \bar{\alpha})$ ellipsoids of planets, mean values of planetoid radius-vectors $(\bar{\rho})$, and values of the gravity potential on them (W_0)

Body	a (km)	$1/\alpha$	$1/\alpha_1$	Λ_a	$\bar{a}$ (km)	$1/\bar{\alpha}$	$\bar{\rho}$ (km)	$R_0 = GM/W_0$ (km)	W_0 $(\mathrm{m^2\,s^{-2}})$
Sun					696×10^3	10^5	696×10^3	696×10^3	1.91×10^{11}
Mercury					2 439.5	8.3×10^3	2 439	2 439	9.033×10^6
Venus	6 051 476	113 800	253 200	− 6.2°	6 051 464	146 800	6 051.45	6 051.45	53 682 700
Earth	6 378.172	297.776	91 650	− 14.95°	6 378.137	298.253	6 370.995	6 363.673	62 636 856
Moon	1 735 554	2 670	7 490	0.03°	1 735 438	3 250	1 737.53	1 737.53	2 821 710
Mars	3 396 510	183.9	2 630	75.0°	3 395 860	190.5	3 389.9	3 384.76	12 653 300
Jupiter			4.6×10^5		71 492	15.42	69 911	67 797	1.8686×10^9
Saturn					60 268	10.21	58 232	55 227	6.8683×10^8
Uranus					22 559	43.61	25 362	25 112	2.3072×10^8
Neptune					25 269	60	25 112	24 923	2.7424×10^8

Table 6.3. Heliocentric distances and orbital velocities of planets

Planet	Heliocentric distance		Orbital velocity	
	In perihelion (10^6 km)	In aphelion (10^6 km)	In perihelion (km s^{-1})	In aphelion (km s^{-1})
Mercury	46.0	69.8	59.0	38.9
Venus	107.5	108.9	35.3	34.8
Earth + Moon	147.1	152.1	30.3	29.3
Mars	206.6	249.2	26.5	22.0
Jupiter	740.9	815.8	13.8	12.5
Saturn	1357	1503	10.2	9.2
Uranus	2752	3023	7.1	6.5
Neptune	4501	4565	5.4	5.4

6.3 Orbital Elements of the Planets

The dynamics of the Solar System has to be described in a coordinate system which is independent of the motion of the bodies forming the System, i.e. inertial in Newton's sense. However, in comparison with the planets' masses, that of the Sun is large:

$$\sum_p GM_p = 178\,130.7 \times 10^{12}\,\mathrm{m^3 s^{-2}}, \sum_p M_p = 2.669 \times 10^{27}\,\mathrm{kg},$$

$$\sum_p M_p/M_\odot = 1.34 \times 10^{-3}.$$

It is therefore convenient to describe the motion of planets in heliocentric system $\bar{x}_j$, the fundamental plane $(\bar{x}_1, \bar{x}_2)$ being the plane of the ecliptic at a particular epoch, and axis $\bar{x}_1$ pointing to the vernal equinox.

We shall simplify the problem from the very beginning by considering all the bodies of the Solar System to be perfectly rigid and their gravitational fields to be spherically symmetrical. Let us denote the Sun and its mass $M_\odot$, and the p-th planet and its mass M_p, $p = 1, 2, \ldots, 8$ (Pluto's system being omitted). Using a procedure similar to that used in Chapter 1 we shall derive the equations of motion for the p-th planet (see Sect. 1.2):

$$\frac{d^2\bar{x}_{j,p}}{dt^2} + G(M_\odot + M_p)\frac{\bar{x}_{j,p}}{\Delta_{\odot p}^3} = \sum_i GM_i\left(\frac{\bar{x}_{j,i} - \bar{x}_{j,p}}{\Delta_{pi}^3} - \frac{\bar{x}_{j,i}}{\Delta_{\odot i}^3}\right);$$

index i takes values from 1 to 8, $i = p$ being omitted; Δ stands for the distance between the mass centres of the bodies with the respective indices: for example, $\Delta_{\odot p}$ is the heliocentric distance of the mass centre of the p-th planet, and Δ_{pi} is the distance between the mass centres of the p-th and i-th planets.

The components of perturbing forces on the rhs are of planetary origin. They can be expressed in terms of the perturbing function (Sect. 1.3.1),

$$R_p = \sum_i GM_i \left(\frac{1}{\Delta_{pi}} - \frac{\sum_{j=1}^{3} \bar{x}_{j,p}\,\bar{x}_{j,i}}{\Delta_{\odot i}^3} \right),$$

i.e.

$$\frac{\mathrm{d}^2 \bar{x}_{j,p}}{\mathrm{d}t^2} + G(M_\odot + M_p)\frac{\bar{x}_{j,p}}{\Delta_{\odot p}^3} = \frac{\partial R_p}{\partial \bar{x}_{j,p}}.$$

The perturbing function R_p has a specific form for each planet, to which the respective equation of motion applies. It is not the same for the system as a whole. It is, therefore, more convenient to express the equation of motion of the p-th planet in terms of Jacobi's coordinates (Subbotin 1937, 1941, 1949),

$$\mu_p \frac{\mathrm{d}^2 \bar{\bar{x}}_{j,p}}{\mathrm{d}t^2} = \frac{\partial U}{\partial \bar{\bar{x}}_p},$$

where·

$$U = GM_\odot \sum_{p=1}^{8} \frac{M_p}{\Delta_{\odot p}} + GM_1 \sum_{i=2}^{8} \frac{M_i}{\Delta_{2i}} + GM_2 \sum_{i=3}^{8} \frac{M_i}{\Delta_{3i}} + \ldots + G\frac{M_7 M_8}{\Delta_{78}}.$$

Coefficient μ_p represents the reduced mass and is defined for each (p-th) planet as follows:

$$\mu_1 = \frac{M_1 M_\odot}{M_\odot + M_1}, \quad \mu_2 = \frac{M_2(M_\odot + M_1)}{M_\odot + M_1 + M_2}, \quad \mu_3 = \frac{M_3(M_\odot + M_1 + M_2)}{M_\odot + M_1 + M_2 + M_3},$$

$$\vdots$$

$$\mu_8 = \frac{M_8(M_\odot + M_1 + M_2 + \ldots + M_7)}{M_\odot + M_1 + M_2 + \ldots + M_8}.$$

Since $M_\odot \gg M_p$, the separate reduced masses are practically equal to the mass of the separate planets, i.e. $\mu_p \sim M_p$ (the difference is largest for Jupiter, but only amounts to about 0.1%).

The advantage of introducing the relative coordinate system described is that the first integrals have a form formally identical to that in the inertial system, with only the planet mass M_p being replaced by its reduced mass μ_p. For example, the components of the resultant angular momentum of the System are

$$L_1 = \sum_p \mu_p(\bar{\bar{x}}_{2,p}\dot{\bar{\bar{x}}}_{3,p} - \bar{\bar{x}}_{3,p}\dot{\bar{\bar{x}}}_{2,p}),$$

$$L_2 = \sum_p \mu_p(\bar{\bar{x}}_{3,p}\dot{\bar{\bar{x}}}_{1,p} - \bar{\bar{x}}_{1,p}\dot{\bar{\bar{x}}}_{3,p}),$$

$$L_3 = \sum_p \mu_p(\bar{\bar{x}}_{1,p}\dot{\bar{\bar{x}}}_{2,p} - \bar{\bar{x}}_{2,p}\dot{\bar{\bar{x}}}_{1,p}).$$

The uniform expression of the perturbing function is of principal importance in solving the system. Of course, the perturbations due to each i-th planet affecting the p-th planet, whose motion is being dealt with, are small compared with the effect of the central force. Consequently, the problem can also be solved by successive approximations, i.e. the positions and velocity components of the perturbing i-th planets can be considered known, only the motion of the p-th planet can be solved, and the accuracy of the orbital elements can thus be gradually and repeatedly improved.

The 'mean' positions of planets can be calculated using 'mean' orbital elements which define a kind of mean orbit that would exist without periodic perturbations over a relatively long time span of several centuries. These mean orbits then form a basis for the general analytical solution of the perturbed motion.

The mean orbital elements should not be confused with the osculating elements (Sect. 1.4) which define the position and components of the planet's velocity at a particular instant. We shall adopt the mean elements from Seidelmann et al. (1974). These are referred to epoch 1950.0, i.e. to the position of the ecliptic and vernal equinox on that date: L is the mean longitude, $\tilde{\omega}$ the longitude of the perihelion, Ω the longitude of the ascending node, i the inclination of the orbital plane to the plane of the ecliptic, e the eccentricity, a the semimajor axis, T the time in tropical years (as of 1950.0), n the mean motion per tropical century, 'ob' is the abbreviation used for 'orbit', $A_{\odot} = \mathrm{AU} = 1.495\,9787 \times 10^{11}$ m.

Mercury

$$L = 34°53'58.19'' + (415^{\mathrm{ob}} + 250\,133.74'')\,T - 0.033''\,T^2$$

$$\tilde{\omega} = 76°40'42.56'' + 575.17''\,T - 0.050'\,T^2$$

$$\Omega = 47°44'19.32'' - 452.13''\,T - 0.325''\,T^2$$

$$i = 7°00'13.60'' - 21.68''\,T + 0.003''\,T^2$$

$$e = 0.205\,624\,41 + 0.000\,020\,42\,T - 0.000\,000\,03\,T^2$$

$$n = 538\,090\,133.74'' - 0.066''\,T.$$

Venus

$$L = 82°14'59.34'' + (162^{\mathrm{ob}} + 707\,636.81'')\,T + 0.005''\,T^2$$

$$\tilde{\omega} = 130°51'55.69'' + 29.45''\,T - 4.655''\,T^2$$

$$\Omega = 76°13'45.85'' - 1001.59''\,T - 0.369''\,T^2$$

$$i = 3°23'38.57'' - 3.71''\,T - 0.117''\,T^2$$

$$e = 0.006\,796\,76 - 0.000\,047\,73\,T + 0.000\,000\,09\,T^2$$

$$n = 210\,659\,636.81'' + 0.010''\,T.$$

Earth

$$L = 100°00'19.15'' + (99^{\text{ob}} + 1\,290\,974.35'')\,T - 0.021''\,T^2$$

$$\tilde{\omega} = 102°04'35.59'' + 1149.75''\,T + 0.57''\,T^2$$

$$\Omega = 174°24'58.95'' - 868.84''\,T + 0.043''\,T^2$$

$$i = 0° + 46.85''\,T - 0.054''\,T^2$$

$$e = 0.016\,730\,12 - 0.000\,041\,92\,T - 0.000\,000\,13\,T^2$$

$$n = 129\,594\,974.35'' - 0.042''\,T.$$

Mars

$$L = 144°33'07.94'' + (53^{\text{ob}} + 215\,635.84'')\,T$$

$$\tilde{\omega} = 335°08'16.56'' + 1594.75''\,T - 0.636''\,T^2$$

$$\Omega = 49°10'16.59'' - 1062.10''\,T - 2.284''\,T^2$$

$$i = 1°50'59.89'' - 29.99''\,T - 0.082''\,T^2$$

$$e = 0.093\,354\,26 + 0.000\,090\,56\,T - 0.000\,000\,07\,T^2$$

$$n = 68\,903\,635.84''.$$

Jupiter

$$L = 316°12'18.76'' + (8^{\text{ob}} + 557\,497.68'')\,T + 26.45''\,T^2$$

$$\tilde{\omega} = 13°17'43.83'' - 28.95''\,T + 5.83''\,T^2$$

$$\Omega = 99°46'51.76'' + 6.61''\,T + 1.940''\,T^2$$

$$i = 1°18'29.17'' + 0.161''\,T + 0.0763''\,T^2$$

$$e = 0.048\,270\,62 + 0.000\,047\,756\,T + 0.000\,022\,676\,T^2$$

$$a = 5.202\,833\,481\,A_{\odot}.$$

Saturn

$$L = 158°17'46.96'' + (3^{\text{ob}} + 511\,352.55'')\,T - 69.49''\,T^2$$

$$\tilde{\omega} = 91°31'54.33'' + 94.29''\,T + 43.09''\,T^2$$

$$\Omega = 113°29'17.38'' + 6.20''\,T - 11.67''\,T^2$$

$$i = 2°29'16.60'' + 1.794''\,T + 0.736''\,T^2$$

$$e = 0.056\,045\,08 - 0.000\,025\,595\,T - 0.000\,016\,172\,T^2$$

$$a = 9.538\,762\,055\,A_{\odot}.$$

Uranus

$$L = 99°05'12.28'' + (1^{\text{ob}} + 246\,428.77'')\,T + 3.54''\,T^2$$

$$\tilde{\omega} = 172°03'33.46'' - 357.23''\,T - 167.13''\,T^2$$

$$\Omega = 73°42'22.91'' + 132.62''\,T + 0.820''\,T^2$$

$$i = 0°46'24.92'' - 3.567''\,T - 0.1803''\,T^2$$

$$e = 0.046\,137\,34 - 0.000\,048\,118\,T + 0.000\,015\,396\,T^2$$

$$a = 19.191\,391\,28\,A_\odot\,.$$

Neptune

$$L = 194°25'32.09'' + 786\,544.04''\,T - 3.06''\,T^2$$

$$\tilde{\omega} = 38°18'31.13'' - 37\,373.57''\,T - 9977.14''\,T^2$$

$$\Omega = 131°14'21.79'' + 9.214''\,T - 3.804''\,T^2$$

$$i = 1°46'27.00'' - 0.619''\,T + 0.0747''\,T^2$$

$$e = 0.009\,714\,49 + 0.001\,095\,407\,T + 0.000\,362\,034\,T^2$$

$$a = 30.061\,069\,06\,A_\odot\,.$$

These data indicate that the motions of the planets have the following common features:

1. Their orbits display very small eccentricity.
2. Their osculating planes are close to one another.
3. Their osculating planes are close to the plane of the ecliptic; as we shall see later they are also close to Laplace's plane.

The direction of the planets' orbital motion and rotational motion is the same. Viewed from the north pole of the ecliptic the motion is counterclockwise (from west to east). The Sun rotates in the same direction. An exception is the slow rotation of Venus with a period of 243.018 days (Table 6.4), which is longer than the orbital period (224.701 days); the rotation also has the opposite sense (retrograde). Another exception is the rotation of Uranus: its axis of rotation is nearly in its orbital plane, forming an angle of about 8° with it. Since its deflection from the positive normal to the orbit amounts to 98°, i.e. it exceeds 90°, this rotation is also formally retrograde. It is interesting that most of the planets' satellites also orbit about them in the same direction.

The above indicates that the Solar System consists of fundamental bodies of apparently the same origin, i.e. these bodies are not alien to the System and were not captured randomly.

The planets form two distinctly different groups: (a) the terrestrial or inner type (Mercury, Venus, Earth and Mars) with relatively small dimensions (Table 6.2) but relatively large mean mass density (Table 6.5) and (b) the Jovian or outer type (Jupiter, Saturn, Uranus and Neptune) of substantially larger dimensions and masses but with densities similar to that of water (Saturn's mean density is in fact smaller than the density of water). However, with regard to the density of the inner cores, that of the outer planets is substantially larger than that of the terrestrial planets. In addition, the outer planets rotate considerably faster than the terrestrial ones (Table 6.4) and have a larger number of satellites.

Table 6.4. Direction of the north pole of rotation (J2000), rotational and orbital periods (Davies et al. 1989)

Body	α_0	δ_0	Rotational period	Orbital period	Inclination of equatorial plane to orbital plane
Sun	286.13°	63.87°	25.38 d		
Mercury	281.01° − 0.003°T	61.45° − 0.005°T	58.6462 d	87.969 d	2°
Venus	272.69°	67.17°	243.018 d*	224.701 d	3°
Earth	0.00° − 0.641°T	90.00° − 0.557°T	23 h 56 m 04.099 s + 0.002 sT	365.256 d	23°26′
Moon			27.32 d	27.32 d	6°41′
Mars	317.681° − 0.108°T	52.886° − 0.061°T	24 h 37 m 22.66 s	686.980 d	23°59′
Jupiter	268.05° − 0.009°T	64.49° + 0.003°T	9h 55 m 29.7 s	11.86 y	3°04′
Saturn	40.58° − 0.036°T	83.54° − 0.004°T	10 h 39 m 22.4 s	29.46 y	26°44′
Uranus	257.43°	− 15.10°	17 h 14 m 24 s**	84.02 y	98°
Neptune	298.00°	40.53°	17 h 52 m	164.79 y	29°

T, time interval in Julian ephemeris centuries (of 36 525 days) from epoch J2000;
* retrograde and nearly in synodic resonance with the Earth (243.16 days); ** retrograde.

Table 6.5. Mean density ($\bar{\sigma}$), angular velocity of rotation (ω), parameter $q = \omega^2 a^{-3}/(GM)$ and secular Love number $k_s = -3J_2^{(0)}/q$

Body	$\bar{\sigma}$ (kg m^{-3})	ω (rad s^{-1})	q (10^{-6})	k_s
Sun	1409	2.9×10^{-6}	21.5	1.4
Mercury	5433	1.2400×10^{-6}	1.01319	178
Venus	5245	2.9925×10^{-7}	0.06105	293
Earth	5514	7.292115×10^{-5}	3461.390	0.9383
Moon	3344	2.6619×10^{-6}	7.57424	80.1
Mars	3933	7.0882×10^{-5}	4599.2	1.278
Jupiter	1326	1.7585×10^{-4}	90716	0.486
Saturn	687	1.6378×10^{-4}	154815	0.316
Uranus	1271	1.013×10^{-4}	29570	0.340
Neptune	1544	9.769×10^{-5}	22530	0.494

The values of their secular Love numbers (k_s) indicate that their contemporary rotation and shape of their external equipotential surfaces correspond to hydrostatic equilibrium much more than is generally the case with the terrestrial planets.

Slower rotation may in general be assumed for planets close to the Sun. This is due to its tidal effects (Chap. 4), an effect which diminishes with the third power of distance from the central body.

These facts seem to support the hypothesis that the planets were created from a flattened (disk-shaped) cloud of interstellar matter, symmetrical with respect to Laplace's plane (see below), by equilibrium condensation, accumulation and non-uniform concentration (accretion).

6.4 Laplace's Invariable Plane of the Solar System

If the Solar System is considered to be isolated, i.e. if external effects, especially due to its motion about the Galaxy's mass centre and to the angular momentum of external forces linked with it are disregarded, the resultant vector of angular momentum, **L**, of the System is constant in time:

$$\frac{d\mathbf{L}}{dt} = 0 .$$

The plane perpendicular to vector **L**, i.e. Laplace's plane, has a constant position in space in this model.

The direction of vector **L** can be determined from the sum of the vectors of angular momentum of the planets, $\sum_n \mathbf{L}_{\odot p}$. We shall consider an approximate solution and we shall simplify the problem from the very beginning by determining vectors $\mathbf{L}_{\odot p}$ under the assumption that each planet or barycentre of its system orbits about the barycentre of the Sun-planet system along Kepler's ellipse. We shall therefore neglect the effects of perturbations due to the other planets, the effect of neither of the bodies being spherical and their tidal and rotational dynamic effects, and we shall round off the appropriate orbital elements of the planets at the same time. We shall also neglect the planets' angular momenta due to their rotation, which are always several orders of magnitude smaller.

Under this simplification the vector of angular momentum of the p-th planet (strictly speaking of the p-th Sun-planet system) $\mathbf{L}_{\odot p}$ is also invariant. It represents the first integral of Keplerian motion about the barycentre of the Sun-planet system and is defined by the sum of vector products

$$\mathbf{L}_{\odot p} = [\mathbf{r}_{\odot} \times M_{\odot} \dot{\mathbf{r}}_{\odot}] + [\mathbf{r}_p \times M_p \dot{\mathbf{r}}_p] ,$$

where $\mathbf{r}_{\odot}, \mathbf{r}_p$ are the radius-vectors of the bodies' mass centres and $M_{\odot}$ and M_p are their masses.

By definition of the barycentre

$$\frac{r_{\odot}}{r_p} = \frac{M_p}{M_{\odot}}$$

and if the radius-vector of the planet relative to the Sun's mass centre is denoted $\mathbf{r}_{\odot p}\{x_j\}$,

$$\mathbf{r}_p = \frac{M_{\odot}}{M_{\odot} + M_p} \mathbf{r}_{\odot p} ,$$

$$\mathbf{r}_{\odot} = \frac{M_p}{M_{\odot} + M_p} \mathbf{r}_{\odot p} .$$

Hence,

$$\mathbf{L}_{\odot p} = \frac{M_{\odot} M_p}{M_{\odot} + M_p} [\mathbf{r}_{\odot p} \times \dot{\mathbf{r}}_{\odot p}] = \frac{M_{\odot} M_p}{M_{\odot} + M_p} \mathbf{K}_{\odot p} ,$$

where $\mathbf{K}_{\odot p}$ is the vectorial integral of areas whose modulus

$$K_{\odot p} = [(x_2 \dot{x}_3 - x_3 \dot{x}_2)^2 + (x_3 \dot{x}_1 - x_1 \dot{x}_3)^2 + (x_1 \dot{x}_2 - x_2 \dot{x}_1)^2]^{1/2}$$
$$= [G(M_{\odot} + M_p) a_p (1 - e_p^2)]^{1/2} ;$$

G is the gravitational constant, a_p the semimajor axis and e_p the eccentricity of the orbit ellipse of the p-th planet. Numerical values are given in Tables 6.6 and 6.7. The distribution of the vectorial integrals of areas (Table 6.6) distinctly divides the planets into inner (terrestrial) and outer groups; moduli $K_{\odot p}$ differ by one order of magnitude. The angular momenta of the planets, due to their rotation, are in each case several orders of magnitude smaller (Table 6.8).

Table 6.6. Orders of magnitude of the orbital angular momenta of planets

$$L_{\odot p} = \frac{M_\odot M_p}{M_\odot + M_p}\left[G(M_\odot + M_p)a_p(1 - e_p^2)\right]^{1/2} = \frac{M_\odot M_p}{M_\odot + M_p}K_{\odot p}$$

Planet (p)	GM_p ($10^9\,\mathrm{m}^3\,\mathrm{s}^{-2}$)	$G(M_\odot + M_p)$ ($10^{13}\,\mathrm{m}^3\,\mathrm{s}^{-2}$)	$\dfrac{M_\odot M_p}{M_\odot + M_p}$ (kg)	a_p ($A_\odot$)	e_p	$K_{\odot p}$ ($\mathrm{m}^2\,\mathrm{s}^{-1}$)	$L_{\odot p}$ ($\mathrm{kg}\,\mathrm{m}^2\,\mathrm{s}^{-1}$)
Mercury	22 032.09	13 271 246.2	3.302×10^{23}	0.3871	0.2056	2.713×10^{15}	8.956×10^{38}
Venus	324 858.60	276.5	4.868×10^{24}	0.7233	0.0068	3.789×10^{15}	1.845×10^{40}
Earth–Moon system	403 503.22	284.4	6.046×10^{24}	1.0000	0.0167	4.455×10^{15}	2.694×10^{40}
Mars	42 828.3	248.3	6.418×10^{23}	1.524	0.0934	5.477×10^{15}	3.514×10^{39}
Jupiter	126 686 537	283 912.7	1.897×10^{27}	5.203	0.0485	1.02×10^{16}	1.93×10^{43}
Saturn	37 931 272	275 037.1	5.683×10^{26}	9.55	0.056	1.37×10^{16}	7.81×10^{42}
Uranus	5 793 939	271 823.4	8.682×10^{25}	19.2	0.046	1.95×10^{16}	1.69×10^{42}
Neptune	6 835 096	13 271 927.5	1.024×10^{26}	30.1	0.009	2.44×10^{16}	2.50×10^{42}
$\sum$		1 78 040.1 × 10³				8.42×10^{16}	3.13×10^{43}

$GM_\odot = 13\,271\,244.0 \times 10^{13}\,\mathrm{m}^3\,\mathrm{s}^{-2}$; $G = 6673 \times 10^{-14}\,\mathrm{m}^3\,\mathrm{s}^{-2}\,\mathrm{kg}^{-1}$; $A_\odot = 1.495\,9787 \times 10^{11}\,\mathrm{m}$

Table 6.7. Components of the planets' angular momentum vectors along axes x_j of the inertial ecliptical coordinate system (approximate mean values)

Planet (p)	Ω (°)	i (°)	$L_{\odot p}\cos(L_{\odot p},x_1)$ (kg m²s⁻¹)	$L_{\odot p}\cos(L_{\odot p},x_2)$ (kg m²s⁻¹)	$L_{\odot p}\cos(L_{\odot p},x_3)$ (kg m²s⁻¹)	$K_{\odot p}\cos(K_{\odot p},x_1)$ (m²s⁻¹)	$K_{\odot p}\cos(K_{\odot p},x_2)$ (m²s⁻¹)	$K_{\odot p}\cos(K_{\odot p},x_3)$ (m²s⁻¹)
Mercury	48.33	7.00	8.150×10^{37}	-7.254×10^{37}	8.889×10^{38}	2.469×10^{14}	-2.198×10^{14}	2.693×10^{15}
Venus	76.55	3.39	1.061×10^{39}	-2.538×10^{38}	1.842×10^{40}	2.180×10^{14}	-5.213×10^{13}	3.783×10^{15}
Earth + Moon	–	0.00	0.00	0.00	2.694×10^{40}	0.00	0.00	4.455×10^{15}
Mars	49.44	1.85	1.134×10^{38}	-7.377×10^{37}	3.512×10^{39}	1.768×10^{14}	-1.150×10^{14}	5.473×10^{15}
Jupiter	100.34	1.31	4.341×10^{41}	7.920×10^{40}	1.929×10^{43}	2.289×10^{14}	4.177×10^{13}	1.018×10^{16}
Saturn	113.55	2.49	3.079×10^{41}	1.342×10^{41}	7.723×10^{42}	5.482×10^{14}	2.389×10^{14}	1.375×10^{16}
Uranus	73.99	0.77	2.209×10^{40}	-6.338×10^{39}	1.710×10^{42}	2.520×10^{14}	-7.231×10^{13}	1.951×10^{16}
Neptune	131.64	1.77	5.817×10^{40}	5.172×10^{40}	2.519×10^{42}	5.659×10^{14}	5.031×10^{14}	2.450×10^{16}
Σ			8.234×10^{41}	2.585×10^{41}	3.129×10^{43}	1.990×10^{15}	5.443×10^{14}	8.165×10^{16}

Table 6.8. Planets' angular momenta due to their rotation, and their mean motion $n = [G(M_\odot + M_p)]^{1/2}a^{-3/2}$

Planet	$\dfrac{C}{M\bar{a}^2}$	$M\bar{a}^2$ (kg m²)	C (kg m²)	$C\omega$ (kg m²s⁻¹)	$\dfrac{C\omega}{L_{\odot p}}$	n (rad s⁻¹)
Mercury	0.34	1.965×10^{36}	6.6×10^{35}	8.2×10^{29}	9.2×10^{-10}	8.27×10^{-7}
Venus	0.336	1.782×10^{38}	5.99×10^{37}	1.8×10^{31}	9.9×10^{-10}	3.24×10^{-7}
Earth	0.330678	2.430×10^{38}	8.0358×10^{37}	5.860×10^{33}	2.18×10^{-7}	1.99×10^{-7}
Mars	0.377	7.408×10^{36}	2.79×10^{36}	1.9×10^{32}	5.5×10^{-8}	1.06×10^{-7}
Jupiter	0.257	9.678×10^{42}	2.49×10^{42}	4.38×10^{38}	2.27×10^{-5}	1.68×10^{-8}
Saturn	0.22	2.065×10^{42}	4.54×10^{41}	7.44×10^{37}	9.53×10^{-6}	6.75×10^{-9}
Uranus	0.26	5.672×10^{40}	1.28×10^{40}	1.30×10^{36}	7.68×10^{-7}	2.37×10^{-9}
Neptune	0.26	6.489×10^{40}	1.69×10^{40}	1.65×10^{36}	6.59×10^{-7}	1.21×10^{-9}

The direction cosines of vector $\mathbf{K}$, i.e. of the normal to the orbital plane, relative to axes $x_1 \equiv \Upsilon$, x_2 defining the plane of the ecliptic, and x_3 along the northern normal to it, are

$$\cos(\mathbf{K}_{\odot p}, x_1) = \sin\Omega_p \sin i_p ,$$

$$\cos(\mathbf{K}_{\odot p}, x_2) = -\cos\Omega_p \sin i_p ,$$

$$\cos(\mathbf{K}_{\odot p}, x_3) = \cos i_p ,$$

where Ω_p is the right ascension of the ascending node and i_p is the inclination of the planet's orbital plane.

The resultant vector, $\mathbf{L}$, of the Solar System can be determined from its components:

$$L_1 = \sum_p \frac{M_\odot M_p}{M_\odot + M_p} [G(M_\odot + M_p)a_p(1 - e_p^2)]^{1/2} \sin\Omega_p \sin i_p ,$$

$$L_2 = -\sum_p \frac{M_\odot M_p}{M_\odot + M_p} [G(M_\odot + M_p)a_p(1 - e_p^2)]^{1/2} \cos\Omega_p \sin i_p ,$$

$$L_3 = \sum_p \frac{M_\odot M_p}{M_\odot + M_p} [G(M_\odot + M_p)a_p(1 - e_p^2)]^{1/2} \cos i_p .$$

The data in Table 6.6 (GM_p values: Thomas 1991) yield:

$$L_1 = 8.23 \times 10^{41} \ \mathrm{kg\,m^2\,s^{-1}} ,$$

$$L_2 = 2.58 \times 10^{41} \ \mathrm{kg\,m^2\,s^{-1}} ,$$

$$L_3 = 3.13 \times 10^{43} \ \mathrm{kg\,m^2\,s^{-1}} ,$$

$$L \sim L_3 .$$

The position of Laplace's plane is defined by direction cosines L_j/L, or by inclination i_L to the plane of the ecliptic (x_1, x_2):

$$\tan i_L = \frac{(L_1^2 + L_2^2)^{1/2}}{L_3} ,$$

and the right ascension of its nodal line:

$$\tan\Omega_L = -L_1/L_2 .$$

In terms of numbers (if $\cos\Omega_L < 0$, $\sin\Omega_L > 0$)

$$i_L = 1.6° , \quad \Omega_L = 107.4° ;$$

these are values close to the orbital elements of Jupiter (Sect. 6.3) that, understandably, have a substantial effect on the position of Laplace's plane.

The ecliptical coordinates, β_L, λ_L, of the direction of vector $\mathbf{L}$ of the resultant angular momentum of the Solar System are

$$\beta_L = 90° - i_L = 88.4° , \quad \lambda_L = \Omega_L - 90° = 17.5° .$$

The equatorial coordinates, α_L, δ_L, can be calculated from the familiar formulae of spherical astronomy:

$$\cos \delta_L \cos \alpha_L = \cos \beta_L \cos \lambda_L \, ,$$

$$\cos \delta_L \sin \alpha_L = - \sin \beta_L \sin \varepsilon_0 + \cos \beta_L \operatorname{sn} \lambda_L \cos \varepsilon_0 \, ,$$

$$\sin \delta_L = \sin \beta_L \cos \varepsilon_0 + \cos \beta_L \sin \lambda_L \sin \varepsilon_0 \, ,$$

i.e. (given $\varepsilon_0 = 23.439\,291°$ for epoch J2000)

$$\alpha_L = 273.91°, \quad \delta_L = 66.99°.$$

The origin of the particles from which the separate bodies originated is of principal importance in answering the basic question of the origin of the System. The distribution of the angular momenta of the bodies is the key issue in this respect.

Since the modulus of the total angular momentum of the planets is substantially higher than the modulus of the Sun's angular momentum, one may conclude that the angular momentum was imparted to the planets essentially during the process of their origination, i.e. via the mass particles of which they are formed and which already orbited about the central part of the cloud (proto-sun) with the corresponding velocity.

The values are given in Table 6.6. The Sun's angular momentum $L_\odot$ is given by its rotation and by its orbital motion about the barycentre of the Solar System. We do not know the exact period of rotation of the Sun's heavy core and of its other inner layers. Given a rotation period of 25.38 days ($\omega_\odot = \sim 2.9 \times 10^{-6}$ rad s^{-1}), we get $L_\odot = 1.65 \times 10^{41}$ kg m^2 s^{-1}, and therefore $L_\odot / \Sigma L_{\odot p} = 0.005$, the principal moment of the Sun's inertia being $C_\odot = 5.7 \times 10^{46}$ kg m^2.

We are only able to estimate the orbital momentum of the Sun. Considering that it is mostly affected by Jupiter, i.e. that the Sun is moving about the

Table 6.9. Inclination of the planets' orbital planes to the invariable Laplace plane

Planet	I_L (°)
Mercury	6.33
Venus	2.19
Earth + Moon	1.57
Mars	1.68
Jupiter	0.33
Saturn	0.93
Uranus	1.03
Neptune	0.72

barycentre of the Sun–Jupiter system, the approximate value comes out as $2 \times 10^{40}\ \mathrm{kg\,m^2\,s^{-1}}$, i.e. a value which is even smaller than modulus $L_\odot$.

Table 6.9 gives the inclinations, I_L, of the orbital planes of the planets to Laplace's plane. These data are convincing evidence that the central plane of the rotating cloud of interstellar matter, from which the planets originated, must have been close to Laplace's plane and that the axis of rotation of the cloud pointed in the same direction as vector **L**.

6.5 Gravitational Forces Acting on the Earth

The Solar System as a whole is moving in the marginal part of the Galaxy, the distance from the centre of its core being

$$\Delta_{G\odot} = 3 \times 10^{20}\ \mathrm{m}\,,$$

and its velocity being

$$v_{G\odot} = 2.5 \times 10^5\ \mathrm{m\,s^{-1}}\,.$$

Its orbital period is thus approximately

$$T_{G\odot} = 2\pi\,\frac{\Delta_{G\odot}}{v_G} = 2.4 \times 10^8\ \mathrm{year}\,.$$

The galactocentric gravitational constant, i.e. the product of Newton's gravitational constant G and of the Galaxy's mass M_G, can be estimated using Kepler's third law,

$$GM_G = \left(\frac{2\pi}{T_{G\odot}}\right)^2 \Delta_{G\odot}^3 = 1.9 \times 10^{31}\ \mathrm{m^3\,s^{-2}}\,;$$

hence, given $G = 6673 \times 10^{-14}\ \mathrm{m^3\,s^{-2}\,kg^{-1}}$,

$$M_G = 2.8 \times 10^{41}\ \mathrm{kg}\,.$$

All these data are tentative and valid only for the order-of-magnitude estimates of the quantities on which they depend. They can be used to derive, for example, the gravitational acceleration, g_G, which the inner part of the Galaxy imparts to the Sun:

$$g_{G\odot} = \frac{GM_G}{\Delta_{G\odot}^2} = 2.1 \times 10^{-10}\ \mathrm{m\,s^{-2}}\,(= 2.1 \times 10^{-5}\ \mathrm{mgal}).$$

Compared with the gravity acceleration, $g \sim 9.81\ \mathrm{m\,s^{-2}}$, imparted by the Earth's masses to any point of the Earth, its value is very small:

$$\frac{g_{G\odot}}{g} = 2.1 \times 10^{-11}\,.$$

The acceleration, $g_{\odot\oplus}$, imparted to the Earth by the Sun ($GM_\odot = 13\,271\,244.0 \times 10^{13}\,\text{m}^3\,\text{s}^{-2}$; $A_\odot = \Delta_{\oplus\odot} = 1.495\,978\,7 \times 10^{11}$ m) is substantially larger,

$$g_{\odot\oplus} = \frac{GM_\odot}{A_\odot^2} = 0.005\,93\,\text{m}\,\text{s}^{-2}(=593\,\text{mgal})\,.$$

Its ratio to the acceleration due to the inner part of the Galaxy

$$\frac{g_{\odot\oplus}}{g_{G\odot}} = 3 \times 10^7\,,$$

and its ratio to the acceleration of gravity on Earth

$$\frac{g_{\odot\oplus}}{g} = 6 \times 10^{-4}\,.$$

The acceleration which the Sun imparts to the Earth is substantially larger than the acceleration due to the other bodies of the Solar System. The acceleration due to the Moon, for example, ($GM_{\mathbb{)}} = 4\,902.8 \times 10^9\,\text{m}^3\,\text{s}^{-2}$; $\Delta_{\oplus\mathbb{)}} = 3.844 \times 10^8$ m) is

$$g_{\mathbb{)}\oplus} = \frac{GM_{\mathbb{)}}}{\Delta_\oplus^2} = 0.000\,033\,\text{m}\,\text{s}^{-2}(=3.3\,\text{mgal});$$

hence

$$\frac{g_{\mathbb{)}\oplus}}{g_{\odot\oplus}} = 5.6 \times 10^{-3}\,.$$

The acceleration due to Jupiter $g_{4\oplus}$ ($GM_4 = 126\,686.9 \times 10^{12}\,\text{m}^3\,\text{s}^{-2}$; $\Delta_{\oplus 4} = 4.203\,A_\odot$) is

$$g_{4\oplus} = \frac{GM_4}{\Delta_{\oplus,4}^2} = 3.2 \times 10^{-7}\,\text{m}\,\text{s}^{-2}(=0.02\,\text{mgal})\,,$$

$$\frac{g_{4\oplus}}{g_{\odot\oplus}} = 5.4 \times 10^{-5}\,;$$

the acceleration due to Venus $g_{\venus\oplus}$ ($GM_\venus = 324\,858.8 \times 10^9\,\text{m}^3\,\text{s}^{-2}$; $\Delta_{\oplus\venus} = 0.277\,A_\odot$) is

$$g_{\venus\oplus} = \frac{GM_\venus}{\Delta_{\oplus\venus}^2} = 1.9 \times 10^{-7}\,\text{m}\,\text{s}^{-2}(=0.003\,\text{mgal})\,,$$

$$\frac{g_{\venus\oplus}}{g_{\odot\oplus}} = 3.2 \times 10^{-5}\,,$$

etc. The acceleration imparted to the Earth by other planets is even smaller.

Acceleration g_G, imparted to the Earth and to all the other planets of the Solar System, is therefore small compared with the acceleration imparted to the

Earth by the Sun and some of the other bodies of the Solar System:

$$\frac{g_{G\odot}}{g_{\odot\oplus}} = 3.5 \times 10^{-8}, \quad \frac{g_{G\odot}}{g_{\text{☽}\oplus}} = 6.4 \times 10^{-6},$$

$$\frac{g_{G\odot}}{g_{4\oplus}} = 6.6 \times 10^{-4}, \quad \frac{g_{G\odot}}{g_{\text{♀}\oplus}} = 1.1 \times 10^{-3}.$$

Of the orbital velocity with which the Earth moves in the region of our Galaxy 90% is due to the internal mass of the Galaxy ($\sim 250\ \text{km s}^{-1}$) and only a little less than 10% is due to the Sun ($\sim 30\ \text{km s}^{-1}$).

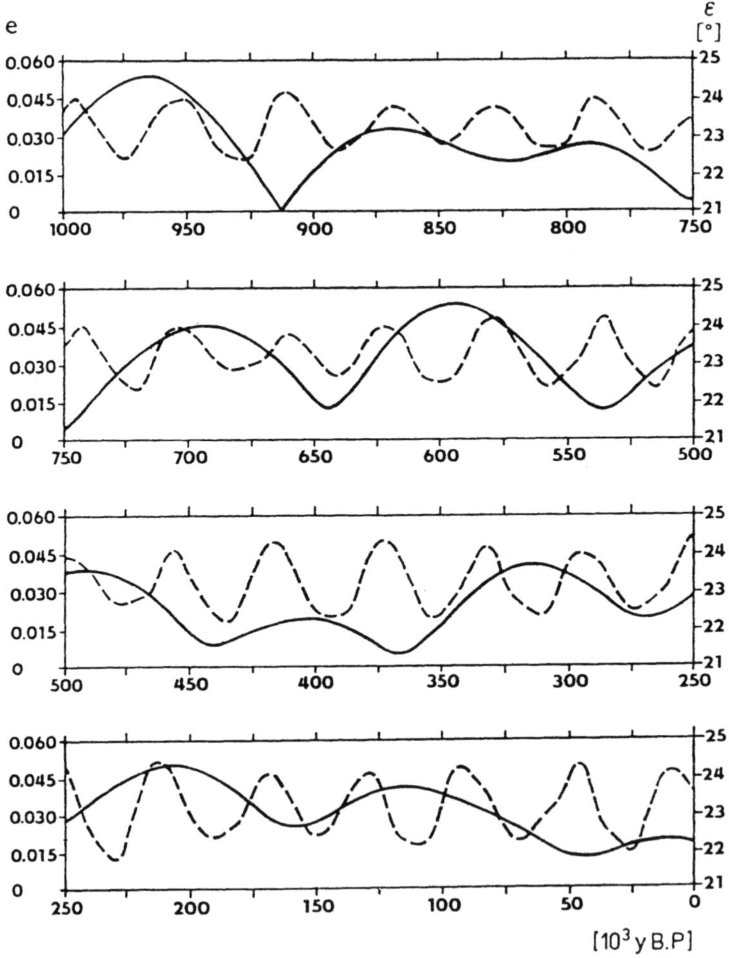

Fig. 6.1. Long-term variations of the Earth's orbital elements over the past 10^6 years and up to epoch 1950.0 (after Berger 1977). *Solid curve* Eccentricity (e); *dashed curve* inclination of the Earth's equatorial plane to the plane of the ecliptic (ε)

6.6 Orbital Elements of the Earth and Their Variations in Time

If the Earth were the only planet of the Sun, if it had no satellite, if the gravitational fields of both bodies were spherically symmetrical and if the Earth were perfectly rigid, then its motion would obey Kepler's laws and its rotation would be quite uniform. In other words, its orbital elements, vector integral of areas, energy integral and the other first integrals would be constant, i.e. they would be quantities invariant in time; this would also apply to the angular velocity of its rotation.

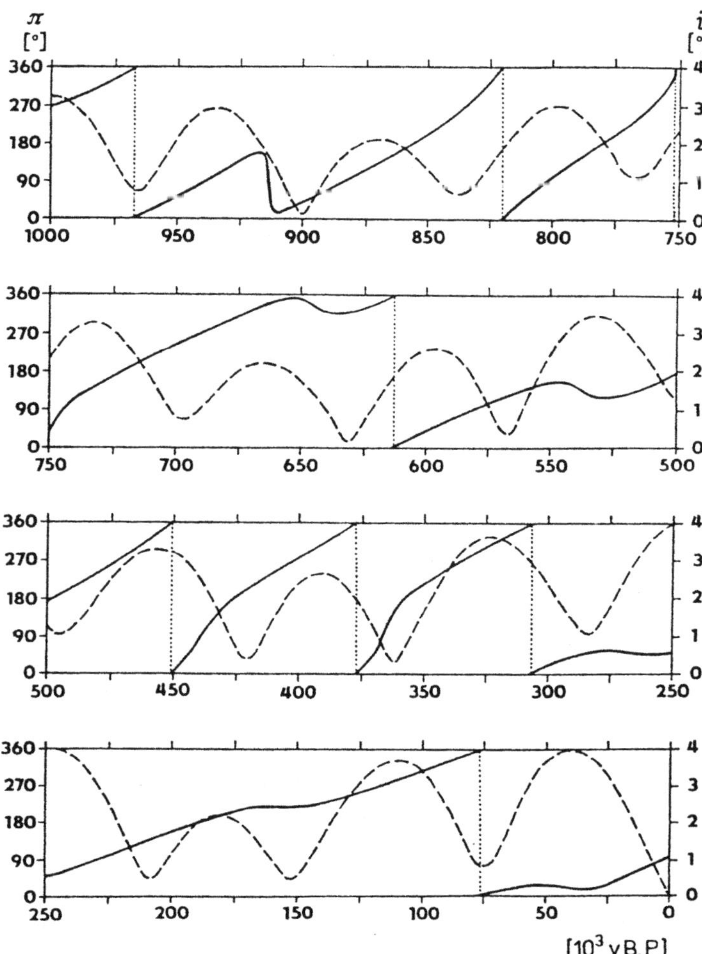

Fig. 6.2. Long-term variations of the Earth's orbital elements over the last 10^6 years and up to epoch 1950.0 (after Berger 1977). *Solid curve* Longitude of the perihelion relative to the equinox of the epoch (π); *dashed line* inclination i of the orbital plane to the plane of the ecliptic, epoch 1950.0

However, the actual conditions in the Solar System, the spherically non-symmetrical internal structure of the Earth and its state, differing considerably from perfect rigidity, and, finally, the existence of the atmosphere are responsible for its space trajectory being a very complicated curve and for its rotation being irregular.

Chapter 3 was devoted to rotation dynamics. We shall now present a brief overview of the changes in the Earth's orbital motion and, in particular, of the long-term changes which may be of interest to a number of Earth and Universe sciences. These changes are described by mean orbital elements which define the Earth's mean trajectory. The latter represents a certain average over a span of several centuries, and the mean elements thus differ considerably from the

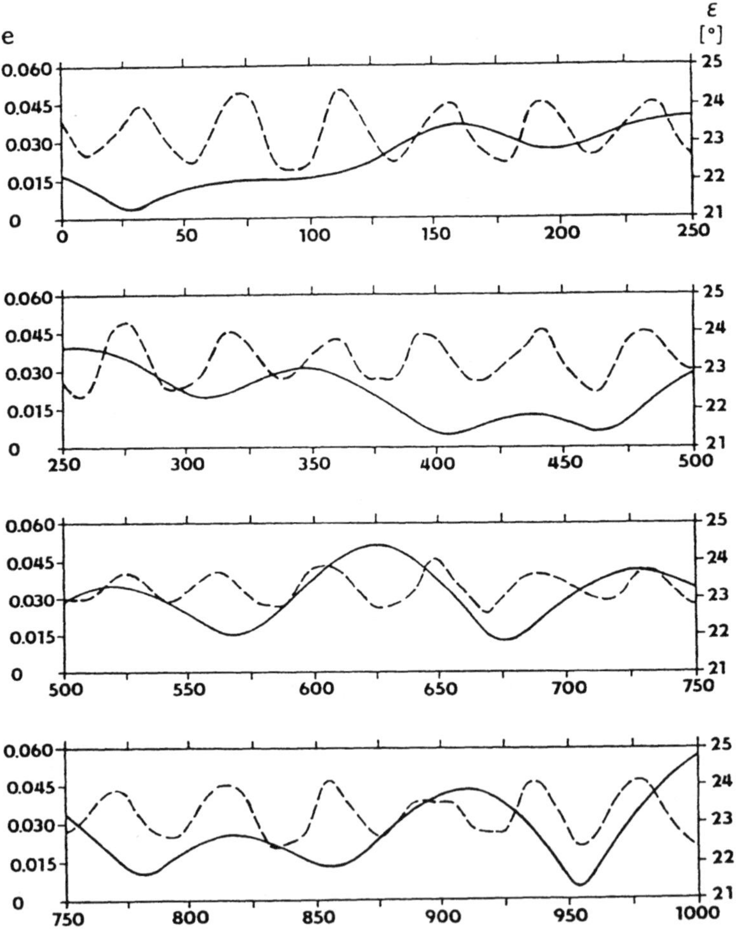

Fig. 6.3. Prediction of long-term variations of the Earth's orbital elements *e* (*solid curve*) and ε (*dashed curve*) over the next 10^6 years from epoch 1950.0 (after Berger 1977)

osculating elements which provide the position and components of the body's velocity for a particular instant.

In a number of fields, e.g. in the theory of astro-climate (palaeoclimatology) and in geology, the Earth's orbital elements are required for very long intervals of time in the past. Long-term variations of the elements for intervals of 1 million years into the past and future can be found, e.g. in Berger (1977), where some of the earlier solutions are also criticized.

Figures 6.1 and 6.2 show schematic diagrams of the long-term variations of the Earth's orbital elements over an interval of 1 million years from 1950.0 into the past, and Figs. 6.3 and 6.4 show analogous quantities for an interval of 1 million years from 1950.0 into the future. The diagrams indicate, for example,

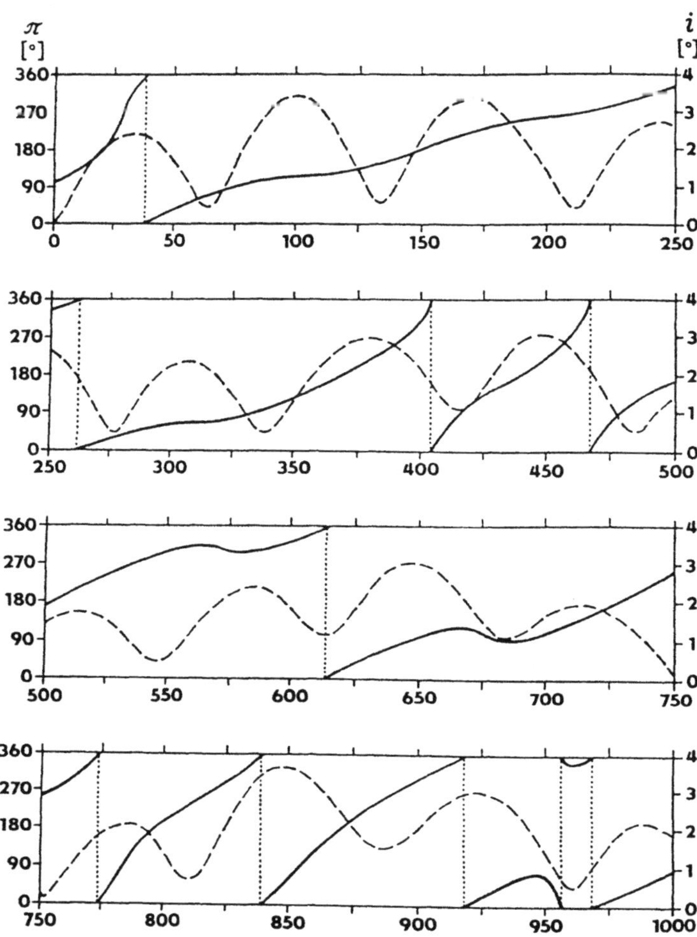

Fig. 6.4. Prediction of long-term variations of the Earth's orbital elements π (*solid curve*) and i (*dashed line*) over the next 10^6 years from epoch 1950.0. (After Berger 1977)

that in the interval of 1 million years from 1950.0 into the past the eccentricity of the Earth's orbit varied within the range $0.000\,567\,3 < e < 0.053\,510\,9$. We are now at the beginning of a period of about 101 000 years in which the eccentricity will be relatively small. During the next million years the eccentricity will vary from $0.002\,687\,5$ to $0.056\,559\,8$. The eccentricity of the Earth's orbit will apparently never exceed the value $e = 0.075$ (Berger 1977).

An important quantity for palaeoclimatology is the orientation of the Earth's axis of rotation relative to the mobile ecliptic. If the plane of the ecliptic were constant in space, this axis (the vector of the angular momentum) would describe a cone with a period of 25 703 Julian years, i.e. the precession due to the angular momenta of the Moon and Sun acting on the equatorial tide of the Earth. In this ideal case, angle ε between the plane of the ecliptic and the plane of the Earth's equator would not change. However, the plane of the ecliptic moves as a result of planet perturbations and this is the cause of the changes in precession and in inclination ε of the plane of the Earth's equator to the plane of the ecliptic. In the 1 million years prior to 1950.0 quantity ε ranged from $22.029°$ to $24.434°$ with a mean period of 41 083 years; in the 1 million years after 1950.0 it will vary between $22.200°$ and $24.326°$ with a mean period of 39 375 years. The precise value for epoch 2000.0 is $23°26'21.448''$. According to Berger (1977) the length of the tropical year in the interval in question is between 365 days 05 h 49 min 33.6 s and 365 days 05 h 47 min 39.2 s. The variations of the Earth's orbital elements for 30 million years into the past can be found, e.g. in Sharaf and Budnikova (1969).

Appendix A: Current Representative Values of the Parameters of Common Relevance to Astronomy, Geodesy and Geodynamics (Rep. IAG SSG 5-100 1991)

A.1 Defining the Constant

A.1.1 Velocity of light in a vacuum

$$c = 299\,792\,458 \text{ m s}^{-1}.$$

This is the value defining the length of 1 m in terms of the SI second. It has been adopted by definition and that is why no uncertainty is given. SI units are to be used throughout and, consequently, by definition of c above.

A.2 Primary Parameters

A.2.1 Newtonian gravitational constant

$$G = (6.672\,59 \pm 0.000\,30) \times 10^{-11} \text{ m}^3 \text{ s}^{-2} \text{ kg}^{-1}$$

(IERS Standards 1989).

A.2.2 Primary parameters defining the geodetic reference system (GRS)

A.2.2.1 Geocentric gravitational constant (the mass of the Earth's atmosphere included)

$$GM = (398\,600.441 \pm 0.001) \times 10^9 \text{ m}^3 \text{ s}^{-2}$$

(Ries et al. 1989).

A.2.2.2 Mean angular velocity of the Earth's rotation

$$\omega = 7\,292\,115 \times 10^{-11} \text{ rad s}^{-1}.$$

This is the actual mean value (Table A.1; BIH, IERS). Because of the variability of ω the next digit depends on the averaging time. It is a convenient truncation of a precise value which defines the UT1 time scale. It happens to agree with the mean rotation rate of the Earth at the present time.

A.2.2.3 Second-degree zonal geopotential (Stokes') parameter (tide-free, conventional, not normalized, scale $a = 6\,378\,136$ m)

$$J_2 = (1082.6269 \pm 0.0006) \times 10^{-6}. \tag{A.1}$$

Table A.1. Mean values of the Earth's angular velocity of rotation in the period 1978–1990

Year	ω $(10^{-11}\ \mathrm{rad\ s^{-1}})$
1978	7 292 114.903
1979	7 292 114.925
1980	7 292 114.952
1981	7 292 114.964
1982	7 292 114.964
1983	7 292 114.954
1984	7 292 115.019
1985	7 292 115.025
1986	7 292 115.043
1987	7 292 115.032
1988	7 292 115.036
1989	7 292 115.018
1990	7 292 114.983

To be consistent with Resolution 16 of the XVIIIth General Assembly of the International Association of Geodesy (Tscherning 1984), the indirect tidal effect on J_2 should be included; hence (if the Love number $k_2 = 0.3$),

$$J_2 = (1082.6362 \pm 0.0006) \times 10^{-6} \tag{A.2}$$

(Marsh et al. 1989, 1990).

A.2.2.4 Parameter defining the Earth's dimension

In principle, there are three geodetic parameters, determined independently from observations, which can be taken to define the dimension of the Earth:

a — the equatorial radius of the reference ellipsoid (mean equatorial radius of the Earth);

W_0 — the gravity potential on the geoid [for definition refer to (2.18)] or at the origin of heights (major vertical datum) or

$R_0 = GM/W_0$ – the geopotential scale factor;

g_e — mean equatorial gravity.

$$a = (6\,378\,136.3 \pm 0.5)\ \mathrm{m} \tag{A.3}$$

(Rapp 1987; Rapp et al. 1991).
Geopotential scale factor

$$R_0 = GM/W_0 . \tag{A.4}$$

Since it depends on the definition of the geoid, $W = W_0$, the definition should read, 'The geoid is the equipotential surface fitting the topographical (quasi-

Table A.2. The geopotential scale factor derived from various geopotential models

Model	R_0 (m)	Root-mean-square error (m)
GEM-T2 (1990)	6 363 672.526	± 0.098
GEM-T1 (1988)	6 363 672.819	± 0.086
GRIM 4S1 (1991)	6 363 672.096	± 0.063

stationary) sea surface best'. This definition is promising with regard to the future use of satellite altimetry to determine parameters R_0, W_0, g_e and a, and it enables uncertainties in determining the geoid under continents to be avoided. The value computed on the basis of test satellite altimetry arcs is

$$R_0 = (6\,363\,672.5 \pm 0.5)\,\text{m}\,.$$

For various geopotential models refer to Table A.2.

The mean equatorial gravity

$$g_e = (978\,032.78 \pm 0.2) \times 10^{-5}\,\text{m s}^{-2}\ (\text{mgal})\,. \tag{A.5}$$

This value was obtained from terrestrial gravity data and satellite-implied data (Rapp 1987).

With regard to the problem of selecting four primary geodetic parameters, the set

$$GM, \omega, J_2, R_0$$

is the proper set to use in determining the four primary geodetic parameters which define the GRS in full. The 'hybrid set',

$$GM, \omega, J_2, a$$

was chosen for practical reasons but, logically, it is somewhat unsatisfactory. If R_0 is adopted as primary, W_0, a and g_e can then be computed (Table A.3).

A.3 Derived Parameters

The geopotential value on the geoid (2.18)

A.3.1

$$W_0 = GM/R_0\,,$$

$$W_0 = (6\,263\,6856 \pm 5)\,\text{m}^2\,\text{s}^{-2}\,.$$

Table A.3. Terms W_0, a and g_e as functions of R_0 $[GM = 398\,600.441 \times 10^9\ \text{m}^3\ \text{s}^{-2}]$

R_0 (m)	W_0 $(\text{m}^2\,\text{s}^{-2})$	a (m)	g_e $(10^{-5}\ \text{m}\,\text{s}^{-2})$
6 363 674.17	62 636 840	6 378 138.15	978 032.17
6 363 673.66	62 636 845	6 378 137.64	978 032.33
6 363 673.15	62 636 850	6 378 137.14	978 032.48
6 363 672.64	62 636 855	6 378 136.63	978 032.64
6 363 672.13	62 636 860	6 378 136.12	978 032.80
6 363 671.63	62 636 865	6 378 135.61	978 032.95
6 363 671.12	62 636 870	6 378 135.10	978 033.10
6 363 670.61	62 636 875	6 378 134.59	978 033.27

A.3.2 The Earth's polar flattening

1. Tide-free

$$\alpha = \frac{1}{298.258 \pm 0.001}\,.$$

2. Zero-frequency indirect tide included ($k_2 = 0.3$)

$$\alpha = \frac{1}{298.257 \pm 0.001}\,.$$

3. Zero-frequency direct plus indirect tides included ($k_2 = 0.3$)

$$\alpha = \frac{1}{298.253 \pm 0.001}\,.$$

A.3.3 Potential factor of centrifugal force

$$q = \frac{\omega^2 a^3}{GM} = (3461.390 \pm 0.002) \times 10^{-6}\,.$$

A.3.4 Triaxiality parameters

A.3.4.1 Equatorial flattening

$$\alpha_1 = \frac{1}{91\,500 \pm 100}\,.$$

For different models refer to Table 2.6.

A.3.4.2 Longitude of major axis of equatorial ellipse

$$\Lambda_a = -(14.95° \pm 0.05°)\,.$$

A.4 Long-Term Varying Parameters

A.4.1 Long-term decrease in the second-degree zonal geopotential coefficient

$$dJ_2/dt = -(2.8 \pm 0.3) \times 10^{-9}\,\text{cy}^{-1}$$

(IERS Standards 1989; Wakker 1990).

A.4.2 Tidal acceleration of the Moon

(Long-term variation in the mean motion n)

$$dn/dt = -(24.9 \pm 1.0)\quad \text{arcsec cy}^{-2}\,;$$

this value is based on LLR data (Newhall et al. 1990).

A.4.3 Long-term variations in the angular velocity of the Earth's rotation

A.4.3.1 Positive relative long-term acceleration due to dJ_2/dt [see (A.4.1)]

$$(d\omega/dt)_{\text{rel}} = +(1.29 \pm 0.28) \times 10^{-22}\,\text{rad s}^{-2}$$

(Newhall et al. 1990; Marsh et al. 1990).

A.4.3.2 Resulting long-term deceleration (total; rounded value)

$$d\omega/dt = -(5.0 \pm 0.3) \times 10^{-22}\,\text{rad s}^{-2}$$

(Marsh et al. 1990).

References

Aksenov EP (1977) Theory of motion of Earth's artificial satellites. Nauka, Moscow (in Russian)

Alessandriny B (1989) The hydrostatic equilibrium figure of the Earth: an iterative approach. Phys Earth Planet Int 54: 180–192

Allan RR (1971) Resonant effect on inclination for close satellites. RAE Tech Rep 71245, Ministry of Defence, Farnborough

Anderson DL (1991) Chemical boundaries in the mantle. In: Sabadini R, Lambeck K, Boschi E (eds) Glacial isostasy, sea-level and mantle rheology. Kluwer, Dordrecht, pp 379–402

Baumgardner JK (1988) Application of supercomputers to 3-D mantle convection. In: Runcorn SK (ed) The physics of the planets. Wiley, London, pp 199–231

Berger A (1977) Long-term variations of the earth's orbital elements. Celest Mech 15: 53–74

Birch F (1964) Density and composition of the mantle and core. J Geophys Res 64: 4377–4388

Birch F (1969) Density and composition of upper mantle: First approximation as an olivine layer. In: Hart PJ (ed) The Earth's crust and upper mantle. AGU Monogr (Washington) 13: 18–36

Bondi H, Gold T (1956) On the damping of the free nutation of the Earth. Mon Not R Astr Soc 115: 41–46

Brosche P, Sündermann J (eds) (1982) Tidal friction and the Earth's rotation II. Springer, Berlin Heidelberg New York

Brouwer D, Clemence GM (1961) Methods of celestial mechanics. Academic Press, New York

Buchar E (1958) Motion on the nodal line of the second Russian Earth satellite (1957β) and flattening of the Earth. Nature 182: 198

Bullen KE (1975) The Earth's denisty. Chapman and Hall, London

Burša M (1970) Foundations of space geodesy. Part II. Dynamic space geodesy. MNO, Prague (in Czech)

Burša M (1974) Tidal potential due to a non-spherical lunar body. Stud Geoph Geod 18: 1–7

Burša M (1979a) The force function of the Earth–Moon system. Publ Finn Geod Inst 89: 27–33

Burša M (1979b) Moments of external disturbing forces in the rotational motion of celestial bodies. Stud Geoph Geod 23: 103–113

Burša M (1983) Disturbances of the Earth's inertia tensor due to tidal and centrifugal forces. Bull Astr Inst Czech 34: 321–323

Burša M, Šidlichovský M (1985) Influence of time variation in the second zonal harmonic on polar motion. Bull Astr Inst Czech 36: 24–27

Čadek O (1989) Spherical tensor approach to the solution of the mantle stress problem. Stud Geoph Geod 33: 177–197

Čadek O, Matyska C (1991) Mass heterogeneities and convection in the Earth's mantle inferred from gravity and core-mantle irregularities. PAGEOPH 135: 107–123

Čadek O, Martinec Z (1991) Spherical harmonic expansion of the Earth's crustal thickness up to degree and order 30. Stud Geophys Geod 35: 151–165

Chandler SC (1892) On the variation of latitude. Astr J 12: 17–22, 65–72, 97–102

Currie RG (1974) Period and Q of the Chandler wobble. Geoph R Astr Soc 38:179–185

Darwin GH (1879) On the precession of a viscous spheroid and on the remote history of the Earth. Philos Trans R Soc 170: 447–538

Darwin GH (1880) On the secular change in the elements of the orbit of a satellite revolving about a tidally distorted planet. Philos Trans R Soc Lond 171: 713–891

Davies ME, Abalakin VK, Burša M, Hunt GE, Lieske JH, Morando B, Rapp RH, Seidelmann PK, Sinclair AT, Tyuflin YS (1989) Report of the IAU/IAG/COSPAR working group on cartographic coordinates and rotational elements of the planets and satellites: 1988. Celest Mech Dyn Astr 46: 187–204

Delaunay CE (1860) Théorie du mouvement de la Lune. Mémoires de l'Académie des Sciences de L'Institut Impérial de France, Paris, vol XXVIII.

Dicke RH, Goldenberg HM (1967) Solar oblateness and general relativity. Phys Rev Lett 18: 313

Dickey JO (1984) Atmospheric excitation of the Earth's rotation Bull 1, SSG 5-98. JPL California Inst of Technology, Pasadena

Dickey JO, Williams JG, Newhall XX (1991) Geophysical results from lunar laser ranging. Presented at XXth IAG Gen Ass Union Symp, Application of gravimetry and space techniques to geodynamics and ocean dynamics, Vienna (unpublished)

Dunthorne R (1750) No. 492, for the months of April, May, June 1749 Philos Trans R Soc 46: 162–172

Dziewonski AM (1984) Mapping the lower mantle: Determination of lateral heterogeneity in P velocity up to degree and order 6. J Geophys Res 89: 5929–5952

Dziewonski AM, Anderson AL (1981) Preliminary reference Earth model. Phys Earth Planet Int 25: 297–356

Dziewonski AM, Hales AL, Lapwood ER (1975) Parametrically simple Earth models consistent with geophysical data. Phys Earth Planet Int 10: 12–48

Euler L (1765) Theoria motus solidorum seu rigidorum ex ... Rostochii et Gryphiswaldiae. Roese, Greifswald

Fedorov EP, Rasulov RM (1981) Can it be proved that the Earth's secular motion exists? Pis'ma v AZh 7: 247–250 (in Russian)

Gaposchkin EM, Lambeck K (1970) 1969 Smithsonian standard Earth (II). SAO Spec Rep 315, Smithsonian Institution, Astrophysical Observatory, Cambridge, MA

Groten E (1970) On tidal effects in satellite gravity data. Observatoire Royal de Belgique, Strasbourg, Série Géophysique 96

Guinot B (1970) Short-period terms in universal time. Astr Astrophys 8: 26–28

Guinot B (1982) The Chandlerian nutation from 1900 to 1980. Geophys J R Astr Soc 71: 295–301

Halley E (1695) Some account of the ancient state of the city of Palmyra; with short remarks on the inscriptions found there. Philos Trans R Soc Lond 19: 160–175

Honkasalo T (1964) On the tidal gravity correction. Boll Geofis Teor Appl VI(21): 34–36

Idel'son N (1936) Potential theory with applications to the theory of the figure of the Earth and geophysics. ONTI, Leningrad (in Russian)

IERS Standards (1989) McCarthy DD (ed) IERS Tech Note 3, Observatoire de Paris

James R, Kopal Z (1963) The equilibrium figures of the Earth and major planets. Icarus 1: 442–454

Jeffreys H (1928) Possible tidal effects on accurate time-keeping. Mon Not R Astr Soc Geophys Suppl 2: 56–58

Jeffreys H (1939) The times of P, S and SKS and velocities of P and S. Mon Not R Astr Soc Geophys Suppl 4: 498–533

Jeffreys H (1949) Dynamic effects of the liquid core. Mon Not R Astr Soc 109: 670–687

Jeffreys H (1959) The earth. Its origin, history and physical constitution, 4th edn. Cambridge Univ Press Cambridge

Jeffreys H (1972) The variation of latitude. IAU Symp 48. Reidel, Dordrecht, pp 39–45

Jeffreys H, Bullen KE (1967) Seismological tables. Brit Assoc for the Advancement of Science, Gray-Milne Trust, London

Kant I (1754) Wöchentliche Frag und Anzeigungs-Nachrichten 23 and 24 June 1754. Kants Werke, Bd I, Vorkritische Schriften I, 1747–1756, p 185. Also (1755) Allgemeine Naturgeschichte und Theorie des Himmels. Johann Friedrich Peterson, Königsberg und Leipzig, pp 18 ff, 163 ff

Kaula WM (1964) Tidal dissipation by solid friction and the resulting orbital evolution. Rev Geophys Space Phys 2: 661–685

Kaula WM (1966) Introduction to satellite geodesy. Waltham, Blaisdell

Kinoshita H, Souchay J (1990) The theory of the nutation for the rigid Earth model at the second order. Celest Mech 48: 187–265

Kovalevsky J (1982) Hipparcos and the dynamics of the Solar System. Celest Mech 26: 213–220

Kozai Y (1959) The Earth's gravitational potential derived from the motion of satellite 1958 Beta Two. SAO Spec Rep 22, Cambridge

Lambeck K (1976) Lateral density anomalies in the upper mantle. J Geophys Res 81: 6333–6340

Lambeck K (1977) Tidal dissipation in the oceans: Astronomical, geophysical and oceanographic consequences. Philos Trans R Soc A287: 545–594

Lambeck K (1980) The variable Earth's rotation. Geophysical causes and consequences. Cambridge Univ Press, Cambridge

Lambert WD (1943) Notes on Earth tides. Geophysics 8: 51–56

Landau LD, Lifshifts EM (1965) Theoretical physics, vol 1. Mechanics. Nauka, Moscow (in Russian)

Laplace PS (1805) Traité de mécanique céleste, Tome second. L'Imprimerie de Crapelet, Paris

Larmor J (1909) The relation of the Earth's free precessional nutation to its resistance against tidal deformation. Proc R Soc A82: 89–96

Le Mouël JL, Courtillot V, Ducruix J (1979) Secular variation acceleration and minimums in the Earth's rotation rate: A correlation. Press XVIIth Gen Ass IUGG, Canberra

Lerch FJ, Klosko SM, Wagner CA, Bellot RP, Laubscher RE, Taylor WA (1978) Gravity model improvement using GEOS-3 altimetry (GEM 10A and 10B). Spring Annu Meet AGU, AGU, Miami

Liouville MJ (1856) Développements. Sur un chapitre de la mécanique de poisson. In: Connaissance des temps où des mouvements célestes, à l'usage des astronomers et des navigateurs, pour l'an 1859, publieé, par le Bureau des longitudes. Mallet-Bachelier, Imprimeur – Libraire du Bureau des longitudes, de l'École impériale polytechnique, Paris, also J Math Pures et Appl (1858) 3: 1–25

Listing JB (1873) Über unsere jetzige Kenntnis der Gestalt und Grösse der Erde. Nachr Kgl Ges Wiss 3, Verlag J Dieterischen Buchhandlung, Göttingen

Love AEH (1909) The yielding of the Earth to disturbing forces. Proc R Soc Lond A82: 73–88

MacDonald GJF (1964) Tidal friction. Rev Geoph 2: 467–541

Markowitz W (1960) Latitude and longitude, and the secular motion of the pole. In: Runcorn SK (ed) Methods and techniques in geophysics, Interscience Publishers, New York, pp 325–359

Marsh JG, Martin TV (1982) The SEASAT altimeter mean surface model. J Geophys Res 87: 3269–3280

Marsh JG, Lerch FJ, Putney BH, Felsentreger TL, Snachez BV, Klosko SM, Patel GB, Robbins JW, Williamson RG, Engelis TE, Eddy WF, Chandler NL, Chinn DS, Kapoor S, Rachlin KE, Braatz LE, Pavlis EC (1989) The GEM-T2 gravitational model. NASA Tech Memorandum 100746 Goddard Space Flight Center, Greenbelt, MD

Marsh JG, Lerch FJ, Putney BH, Felsentreger TL, Sanchez BV, Klosko SM, Patel GB, Robbins JW, Williamson RG, Engelis TL, Eddy WF, Chandler NL, Chinn DS, Kapoor S, Rachlin KE, Braatz LE, Pavlis EC (1990) The Gem-T2 gravitational model. J Geophys Res 95: 22043–22071

Martinec Z (1989a) The influence of the core-mantle boundary irregularities on the mass density distribution inside the Earth. Proc 7th Int Seminar Model optimization in exploration geophysics, 4, Freie Univ Berlin, Berlin, 233–256

Martinec Z (1989b) Program to calculate the spherical harmonic expansion coefficients of the two scalar field products. Comp Phys Comm 54: 177–182

Martinec Z (1991) On the accuracy of the method of condensation of the Earth's topography. Manusc Geod 16: 228–294

Martinec Z, Pěč K (1986) Normal density Earth models. Stud Geoph Geod 30: 124–147

Matyska C (1989) The Earth's gravity field and constraints to its density distribution. Stud Geoph Geod 33: 1–10

Melchior P (1983) The tides of the planet Earth. Pergamon Press, Oxford

Melchior P, Georis B (1968) Earth tides, precession-nutation and the secular retardation of Earth's rotation. Phys Earth Planet Int 1: 267–287

Migal' NK, Markovich MN (1977) On the determination of the secular motion of the Earth's poles from the results of duplicate levelling. Izvestiya VUZ, Geodeziya i aerofotos"emka 6: 59–66 (in Russian)

Mikhailov AA (1970) On the secular motion of the Earth's poles. Astr Zh 47: 1296–1299 (in Russian)

Mikhailov AA (1971) On the motion of the Earth's poles. Astr Zh 48: 1301–1304 (in Russian)

Molodenskii CM (1980) Tides in a spherically non-symmetric Earth. Theory of Earth tides. Nauka. Moscow

Molodenskii MS (1953) Elastic tides, free nutation and some problems of the Earth's structure. Tr Geofiz Inst AN SSSR 19: 3–52 (in Russian)

Moritz H (1973) Ellipsoidal mass distribution. Ohio State Univ, Dept Geod Sci Rep 206, The Ohio State University Research Foundation, Columbus, OH

Morrison LV, Stephenson FR (1981) Determination of "decade" fluctuations in the Earth's rotation 1620–1978. In: Gaposchkin EM, Kołaczek B (eds) Reference coordinate systems for Earth dynamics. Reidel, Dordrecht, pp 181–185

Munk WH, MacDonald GTF (1960) The rotation of the Earth. Cambridge Univ Press, Cambridge

Nakiboglu SM (1982) Hydrostatic theory of the Earth and its mechanical applications. Phys Earth Planet Int 28: 302–311

Newcomb S (1891) On the periodic variation of latitude, and the observations with the Washington prime-vertical transit. Astr J 11: 81–82

Newhall XX, Williams JG, Dickey JO (1990) Tidal acceleration of the Moon. In: Brosche P, Sündermann J (eds) Earth's rotation from eons to days. Springer, Berlin Heidelberg New York, p 51

Newton I (1687) Philosophiae naturalis principia mathematica, 1st edn. 1687, 2nd edn. 1714, 3rd edn. 1727. Transl R Soc London 1966, Univ of California Press

O'Keefe JA, Eckels A, Squires RK (1959) Vanguard measurements give pear-shaped component of Earth's figure. Science 129: 565–566

Pariiskii NN (1978) A study of Earth tides. Fizika Zemli 9: 43–54 (in Russian)

Pariiskii NN, Kuznetsov MV, Kuznetsova LV (1972) The effect of oceanic tides on the secular deceleration of the Earth's rotation. Izv Acad Sci USSR, Fizika Zemli 2: 3–12 (in Russian)

Pěč K, Martinec Z (1984) Constraints to three-dimensional non-hydrostatic density distribution in the Earth. Stud Geoph Geod 28: 364–380

Peters CAF (1845) Von den kleinen Ablenkungen der Lothlinic und des Niveaus, welche durch die Anziehungen der Sonne, des Mondes und einiger terrestrischen Gegenstände hervorgebracht werden. Astronom Nachr 22: 33–42

Pick M, Pícha J, Vyskočil V (1973) Theory of the Earth's gravity field. Academia, Prague

Poinsot L (1851) Théorie nouvelle de la rotation des corps. J Math Pure Appl 16: 9–130, 289–336

Poma A, Proverbio W (1981) Fluctuations in the seasonal variations of the length of the day and the Earth's wobble. Veröff ZIPE 63, I: 148–160

Press F (1968) Earth models obtained from Monte Carlo inversion. J Geophys Res 73: 5223–5234

Press F (1970) Earth models consistent with geophysical data. Phys Earth Planet Int 3: 3–22

Radau R (1885) Remarques sur la théorie de la figure de la Terre. Bull Astr 2: 157–161, C R Acad Sci Paris 100: 972–974

Rapp RH (1987) An estimate of equatorial gravity from terrestrial and satellite data. Geoph Res Lett 14: 730–732

Rapp RH, Nerem RS, Shum CK, Klosko SM, Williamson RG (1991) Consideration of permanent tidal deformation in the orbit determination and data analysis for the Topex/Poseidon mission. NASA Memorandum, NASA, Greenbelt, MD

Rapport, Bureau International de l'Heure, 1967–1974, Annu Rep Published for ICSU, Paris

Report of IAG Special Study Group 5-100 (1991) Parameters of common relevance of astronomy, geodesy, and geodynamics 1987–1991. Presented at the XXth Gen Assoc IUGG/IAG, Vienna. Bull Geod 66(2), 1992: 193–197

Resal MH (1884) Traité élémentaire de mécanique céleste. Gauthier-Villars, Paris

Ricard Y, Vigny C (1989) Mantle dynamics with induced plate tectonics. J Geophys Res 94: 17543–17559

Ricard Y, Vigny C, Froidevaux C (1989) Mantle heterogeneities, geoid and plate motions: A Monte-Carlo inversion. J Geophys Res 94: 13739–13754

Ries JC, Eanes RJ, Schutz BE, Shum CK, Tapley BD, Watkins MM, Yuan DN (1989) Determination of the gravitational coefficient of the Earth from near-Earth satellites. Geophys Res Lett 16: 271–274

Ringwood AE (1969) Phase transformations in the mantle. Earth Planet Sci Lett 5: 401–412

Roche E (1854) Note sur la loi densité a l'intérieur de la Terre. Mém Acad Sci Montpelier, Paris, C R 39

Rochester MG (1973) The Earth's rotation. EOS 54: 769–780

Rochester MG, Smylie DE (1965) Geomagnetic core-mantle coupling and the Chandler wobble. Geophys J 10: 289–311

Sabadini R, Lambeck K, Boschi E (eds) (1991) Glacial isostasy, sea-level and mantle rheology. Kluwer, Dordrecht

Scheibe A, Adelsberger U (1936) Schwankungen der astronomischen Tageslänge und der astronomischen Zeitbestimmung nach den Quarzuhren der Phys.-Techn. Reichsanstalt. Phys Z 37: 185–191

Seidelmann PK, Doggett LE, DeLuccia MR (1974) Mean elements of the principal planets. Astr J 79: 57–60

Sharaf Sh G, Budnikova NA (1969) Secular changes in the elements of the Earth's orbit and the astronomical theory of climate fluctuations. Tr Inst Teor Astr 14: 48–84

Shida T, Matsuyama M (1912) On the elasticity of the Earth and the Earth's crust. Mem Coll Sci Engng Kyoto Univ 4: 277–284

Šidlichovský M (1978) The force function of two general bodies. Bull Astr Inst Czech 29: 90–97

Sidorenkov NS (1975) Non-uniform rotation of the Earth as derived from astronomical observations. Astr Zh 52: 1108–1112 (in Russian)

Sitter W de (1927a) Further note on the causes of the discontinuous changes of the Earth's rotation and their possible effect on the intensity of gravity. Bull Astr Inst Neth 130: 69–70

Sitter W de (1927b) On the secular accelerations and the fluctuations of the longitudes of the Moon, the Sun, Mercury and Venus. Bull Astr Inst Neth 124: 21–38

Smith DE, Kolenkiewicz R, Dunn PJ, Robbins JW, Torrence MH, Klosko SM, Williamson RG, Pavlis EC, Douglas NB, Fricke SK (1990) Tectonic motion and deformation from satellite laser ranging to LAGEOS. J Geophys Res 95: 22013–33041

Sokolnikoff IS (1971) Tensor analysis. Nauka, Moscow (in Russian)

Stoyko A (1970) La variation séculaire de la rotation de la Terre et des problèmes connexes. Ann Guebhard 46: 293–316

Subbotin MF (1941) A course of celestial mechanics, vol 1. OGIZ, Leningrad (1937) vol 2. ONTI NKTP, Leningrad (1949) vol 3. Gostekhizdat, Moscow (in Russian)

Thomas PC (1991) Planetary geodesy. Rev Geophys Suppl, US Natl Rep to IUGG 1987–1990: 182–187

Thomson W, Tait PG (1879) Treatise on natural philosophy. Cambridge Univ Press, Cambridge

Tisserand F (1891) Traité de mécanique céleste, t. 2. Théorie de la figure des corps céleste et de leur mouvement de rotation. Gauthier-Villars, Paris

Tscherning CC (ed) (1984) The geodesist's handbook 1984. Bureau Central De l'Association Internationale de Geodesie, Paris. Bull Géod 58, 3

Varshalovich DA, Moskalev AN, Khersonskii VK (1975) The quantum theory of the angular momentum. Nauka, Moscow (in Russian)

Vondrák J (1982) On the direct influence of the planets on the precession and nutation of the Earth's axis of rotation. Bull Astr Inst Czech 33: 26–32

Vondrák J (1984) On the motion of the principal axes of inertia within the elastic tri-axial Earth. Bull Astr Inst Czech 35: 92–104

Vondrák J (1985) Long-period behaviour of polar motion between 1900.0 and 1984.0. Ann Geophys 3: 351–356

Wakker KF (1990) Report by the subcommittee on intercomparison and merging of geodetic data of the Topex/Poseidon. Science Working Team, Delft University of Technology, Delft

Wang CY (1972) A simple Earth model. J Geophys Res 77: 4318–4329

Wilson JT (1965) A new class of faults and their bearing on continental drift. Nature 207: 343–347

Wittmann AD (1982) Deceleration of the Earth's rotation from old solar observations. In: Brosche P, Sündermann Y (eds) Tidal friction and the Earth's rotation 2nd edn. Springer, Berlin Heidelberg New York, pp 51–66

Woodhouse JH, Dziewonski AM (1984) Mapping the upper mantle: Three-dimensional modeling of Earth structure by inversion of seismic waveforms. J Geophys Res 89: 5953–5986

Woolard EW (1953) Theory of the rotation of the Earth around its center of mass. US Government Printing Office, Washington

Yatskiv YaS (1974) Chandler wobble and free diurnal nutation derived from latitude observations. Veröff ZIPE 30, 1: 143–149

Yoder CF, Williams JG, Dickey JO, Schutz BE, Eanes RJ, Tapley BD (1983) Secular variation of Earth's gravitational harmonic J_2 coefficient from Lageos and nontidal acceleration of Earth rotation. Nature 303: 757–762

Yumi S (1969) Annual report of the international polar motion service for the year 1967. Published for ICSU, Mizusawa

Yurkina MI (1983) L. Euler's contribution to the development of the theory of the figure and rotation of the Earth. Geodeziya kartogr 9: 55–59

Zadro MB, Marussi A (1973) On the static effect of Moon and Sun on the shape of the Earth. In: Commissione Geodetica Italiana ATTI Proc V Simp sulla Geodesia Matematica, Firenze, Oct 1972. Istituto Geografico Militare, Firenze, pp 249–254

Zhongolovich ID (1971a) Determination of a geocentric system of coordinate axes fixed to the Earth. Systems of coordinates in astronomy. FAN Uzbekskoi SSR, Tashkent, pp 8–25

Zhongolovich ID (1971b) The prime geocentric meridian, longitudes and universal time. Astr Zh 48: 1308–1313

Zschau J (1978) Tidal friction in the solid Earth: Loading tides versus body tides. In: Brosche P, Sündermann J (eds) Tidal friction and the Earth's rotation. Springer Berlin Heidelberg New York, pp 62–94

List of the Most Important Symbols

A	Principal moment of inertia of the Earth's body relative to the largest axis of the ellipsoid of inertia
A_{jm}	Complex geopotential coefficient of degree j and order m
$A_n^{(k)}$	Coefficient in the expansion of the geoid radius-vector
a	Semimajor axis of the satellite's osculating orbit
$a = a_1$	The largest semi-axis of the triaxial Earth ellipsoid
a_0	Factor of length (e.g. the semimajor axis of an ellipsoid representing a celestial body)
$\bar{a}$	The semimajor axis of the Earth's ellipsoid of rotation
B	Principal moment of inertia of the Earth's body relative to the medium axis of the ellipsoid of inertia
C	Principal moment of inertia of the Earth's body relative to the smallest axis of the ellipsoid of inertia
C	Modulus of the vector integral of areas
$C_m^{(\lambda)}$	Ultraspherical (Gegenbauer's) polynomial
$C_{j_1 m_1 j_2 m_2}^{jm}$	Clebsch–Gordan coefficient
$\bar{C}_{jm} = \bar{J}_j^{(m)}$	Fully normalized geopotential coefficient
D	Product of inertia relative to axis x_1
$\hat{\mathbf{D}}(\alpha, \beta, \gamma)$	Rotation operator defined by Euler angles (α, β, γ)
$D_{mm}^j(\alpha, \beta, \gamma)$	D-matrix of rotation
dm_0	Element of mass of the central body
dm_j	Element of mass of the j-th perturbing body
E	Eccentric anomaly
E	Product of inertia relative to axis x_2
E_{jm}	Complex coefficient in the expansion of the surface radius-vector
e	Eccentricity of the osculating orbit of a satellite or planet
e	Eccentricity of the meridional section of the triaxial ellipsoid in which the largest semi-axis lies
e_i	General symbol for an osculating element $(i = 1, 2, \ldots, 6)$
e_1	Equatorial eccentricity of the triaxial ellipsoid
F	Product of inertia relative to axis x_3
$F_{jm}^{(k)}$	Coefficient in the expansion for density
G	Gravitational constant
$\mathbf{G}$	Moment of external gravitational forces

g	Gravity, acceleration of gravity
$g_n^{(k)}$	Coefficient in the expansion of gravity on the geoid
H	Hamiltonian, energy integral
H	Coefficient in the precession constant
H	Gauss' curvature
h	Love number
$\mathbf{I}$	Tensor of inertia
i	Inclination of satellite's orbital plane to the equatorial plane
i	Imaginary unit ($i = \sqrt{-1}$)
J	Mean surface curvature
$J_n^{(k)}$	Geopotential coefficient of degree n and order k
K	Integral of areas
k	Gauss' gravitational constant
k	Order of spherical harmonics
k, k_2	Love number
k_s	Secular Love number
L	Lagrangian
$\mathbf{L}$	Angular momentum
l	Shida–Lambert number
M	Mass of the Earth
M	Mean anomaly
M_S	Mass of artificial satellite
M_i	Mass of and symbol for i-th perturbing body ($i = 0, 1, 2, \ldots, n$)
M_0	Mean anomaly for osculation epoch
M_0	Central body and its mass
M_p	Mass of p-th planet
$N_n^{(k)}$	Norm of spherical harmonic of degree n and order k
n	Degree of Legendre polynomials
n	Mean motion of planet or satellite
$\bar{n}$	Maximum degree of Legendre polynomials in spherical harmonics series
$\mathbf{n}$	Normal to the osculating plane of a satellite's orbit
O	Centre of mass of the central body (Earth)
O'	Centre of mass of the Moon
O''	Centre of mass of the Sun
P	Perigee
P	Potential point
$\dot{P}$	Areal velocity
$P_n^{(k)}$	Legendre associated function of degree n and order k
$P_n^{(0)}$	Legendre polynomial of the n-th degree
$P_n(x)$	Legendre polynomial of the n-th degree
$\tilde{P}_n(x)$	Shifted Legendre polynomial of the n-th degree
p	Parameter of the ellipse

Q	Potential of centrifugal forces
Q	Dissipation function
q	Parameter in the expression for the potential of centrifugal forces
q_r	Generalized coordinates
R	Perturbing function
R	Mean radius of the Earth
$\mathbf{R}$	Radius-vector in system x_j
R_E	Radius of curvature of the ellipsoid
R_H	Radius of curvature of the geoid
$\bar{R}_{OS}$	Perturbing potential
R_c	Geopotential scale factor related to geopotential W_c
R_0	Geopotential scale factor
$\mathbf{r}$	Radius-vector
$\mathbf{r}_S$	Centric radius-vector of the satellite
r_{ik}	Distance between elements dm_i and dm_k
r_{Sj}	Distance of satellite from element dm_j of the j-th perturbing body M_j ($j = 1, 2, \ldots, n$)
S	Satellite
S	General symbol for geopotential Stokes constant
$S_n^{(k)}$	Geopotential coefficient of degree n and order k
T	Satellite's orbital period
T	Double kinetic energy
T	Astronomical epoch
T	Sidereal period of the Earth's rotation
T_E	Euler period
T_{CH}	Chandler period
T_S	Hour angle of satellite relative to prime meridian
U	Potential energy
u	Angle between satellite's radius-vector and line of nodes
V	Force function; gravitational potential
V_s	Tidal potential
v	True anomaly
v	Satellite's velocity
W	Gravity potential (geopotential)
W_0	Value of gravity potential on geoid surface
X_1, X_2, X_3	Geodetic reference coordinate system
$\left. \begin{array}{l} x_1, x_2, x_3, x_j \\ (x^1, x^2, x^3) \end{array} \right\}$	Coordinate system whose origin is in the mass centre of the central body
$\bar{x}_j$	Heliocentric coordinate system
x, y	Coordinates of the pole
α	Right ascension
α	Flattening of the meridional section of the triaxial ellipsoid in whose plane the large semi-axis lies
$\bar{\alpha}$	Flattening of the rotational ellipsoid

α_h	Hydrostatic flattening
α_1	Equatorial flattening of the triaxial ellipsoid
Ω	Right ascension of ascending node of satellite's orbit
Δ_{so}	Distance of satellite from mass centre of central body M_o
Δ_{S_j}	Distance of satellite from mass centre of the j-th perturbing body M_j
Δ_{jm}	Distance of mass centres of bodies M_j and M_m
Δ_{op}	Heliocentric distance of the p-th planet
δ	Angle between vector of instantaneous rotation ω and vector of resultant angular momentum
δ_s	Centric (geocentric) satellite declination
δ'_s	Selenocentric satellite declination
δ''_s	Heliocentric satellite declination
$\delta(0, k), \delta_{ik}$	Kronecker delta (Kronecker symbol)
ε_{ijk}	Levi–Civita tensor (Levi–Civita epsilon)
η	Component of deflection of the vertical in the plane of the prime vertical
Υ	Vernal equinox
Θ	Hour angle of vernal equinox relative to prime meridian; Greenwich sidereal time
Θ	Angle between vector of instantaneous rotation ω and geocentric radius-vector of a given point on the Earth's surface
ϑ	Angle of nutation
$\vartheta = 90° - \phi$	Geocentric polar distance
$\varkappa_{(1)}, \varkappa_{(2)}$	Normal curvature along principal directions of curvature
Λ	Centric (geocentric) longitude
$\Lambda_{n,k}$	Phase angle of harmonic term in geopotential expansion of degree n and order k
λ	Order of ultraspherical (Gegenbauer's) polynomials
$\bar{\lambda}$	Geographic (astronomical) longitude
$\lambda^\alpha_{(1)}, \lambda^\alpha_{(2)}$ $(\alpha = 1, 2)$	Principal directions of curvature
μ	Reduced mass
μ	Shear modulus
ξ	Component of deflection of the vertical in meridional plane
ϱ	Radius-vector
ϱ, ϱ_s	Geocentric radius-vector of satellite
ϱ'_s	Selenocentric radius-vector of satellite
ϱ''_s	Heliocentric radius-vector of satellite
ϱ_e	Radius-vector of ellipsoid surface
σ	Density
$\bar{\sigma}$	Mean density
σ_E	Euler frequency of free nutation
σ_{CH}	Chandler frequency
ϕ	Geocentric latitude

φ	Geographic (astronomical) latitude
φ	Angle of natural rotation
ψ	Precession angle
Ω	Symbol for angular spherical coordinates (ϑ, Λ)
ω	Argument of perigee
$\tilde{\omega}$	Longitude of perihelion
$\omega_{\oplus} = \omega$	Angular velocity of the Earth's rotation